Architectural Material & Detail Structure

建筑材料与细部结构

(西)费尔南多·佩雷斯 编 常文心 译

Metal 金属

Preface 前言

ARCHITECTURE IN METAL
Towards a dialogue between industry and crafts

金属建筑
工业与工艺之间的对话

Associated to concepts of lightness, speed and industrialisation, the connection between steel and architectural production has strengthened and intensified since the second half of nineteenth century, allowing for the manufacturing in large quantities when necessary. The nature of steel construction has the capacity to accelerate construction processes, providing rapid response to growth or reconstruction of cities, in addition to its structural relation of optimisation of spatial usage of consolidated urban areas, steel structures have been involved major architectural productions in the last 150 years. (Image 1)

From the use of steel as linear elements, to its ability to work in tension; from conventional applications, to complex building solutions; steel structures extend the range of formal and spatial possibilities. Whether it be visible, or hidden, steel structure is able to free floor-spaces, provide for large spans and cantilevers, hanging or suspending various elements, and above all allow for great heights; consecrating the verticality as a typology indissolubly linked to modernity.

Today, only the need for fire protection limits the expressive possibilities and the role of steel in formal and spatial definition of architecture. The steel alone does not provide the necessary guarantees of stability in case of fire, and this requires the need for cladding and protective solutions, hence conditioning the expressive and visual possibilities offered by metal structures.

Steel has not only provided solutions for structural requirements, but when combined with other metal elements, has also been present in the building envelopes contributing to define the image of the architecture. While copper, bronze or zinc had always been used in roof coverings or finishing, there had been few examples of steel castings present on building façades in the turn of century into the twentieth. It is not and third decades of the twentieth steel façade systems were elopment of prefabricated metal façades.

Richard Paulik, prototype wall panels clad with sheet steel. The early example of modular prefabricated system and façade panels, is expanded by Walter Gropius, firstly in the project of Stutgart Kleinhaus (1927) and later in residential designs for Hirsch Kupfer Copper Berlin. The prefabrication system of these houses were designed with façade panels, consisting of exterior-facing corrugated sheet copper cladding, and interior-facing corrugated sheet aluminium. It allowed for a spatially flexible house, expanding and contracting when needed, all produced in factories, delivered and moved by trucks. The truly innovative solution represents a turning point in the architectural innovation, so important for later developments. (Image 2)

Although were Buckminster Fuller with the Dymaixon houses in 1944, the Wichita House in 1947, and finally, Jean Prouvé with les Maisons Tropicales in 1949 who consolidate the use of aluminum in architectural, notably, Jean Prouvé was who developed more accurately the characteristics and requirements of the façade panel and profile elements that became the curtain wall. Example of this can be seen in the Maison du Peuple façade in Clinchy (1936-1939), or the aluminum extruded profiles and façade panels of the Fédération du applied Bâtiment (1953-1954) and in the housing of Mozart (1953-1954) square in Paris. (Image 3, Image 4)

Soon, the development and incorporation of different types of metals, such as titanium, bronze, copper, and zinc, gave way to numerous metallic coating systems and enclosure designs, which continue to be present today.

With challenges such as optimisation of manufacturing and application, corrosive related durability issues, as well as the sealing of building envelopes, many systems and patents have been developed. Supported by steel's advantages like malleability; its ability to be folded, punched, or textured with finishes, and finally along with the development of digital fabrication methodologies, steel had increased the scope of architectural possibilities. Furthermore, these systems have radically transformed the construction procedures, and with them, the philosophy, concept and organisation of the construction site itself.

人们总是将金属与轻质、速度和工业化联系起来。由于大批量生产的可行性，钢材与建筑之间的联系19世纪下半叶变得更加紧密。钢结构的性质能加速建造流程，快速地对城市成长或重建做出反应。除了与城市区域的空间优化相联系之外，钢结构在过去的150年中与主要的建筑建造密不可分。（图1）

从钢的直线特征到它的延展性，从传统应用到复杂的建筑方案，钢结构不断拓展自己的应用形式和空间潜能。无论是外露还是隐藏起来，钢结构都能解放楼面空间、实现大跨度和悬臂造型、悬挂各种元件；最重要的是，它能支承超高建筑，保证了现代化建筑典型特征——垂直性。

现在，只有防火需求能限制钢材在建筑形式和空间造型中的表现力。钢材本身并不能在火灾中提供必要的稳定性保证，需要添加一些覆层处理和保护措施，因此限制了金属结构的视觉表现力。

钢材不仅能为结构需求提供解决方案，当它与其他金属元素结合起来时，还能呈现在建筑表皮上，实现建筑的外观改造。铜、青铜和锌经常被应用在屋面覆盖层或立面上，然而，在19世纪末20世纪初，很少有钢铸件出现在建筑立面上。直到20世纪20、30年代，第一个钢立面系统才获得专利[1]，而一些预制住宅也开始采用金属立面构造。

1926年，在理查德·波利克的协助下，乔治·蒙克（格罗皮乌斯工作室）利用钢结构和模块化钢覆层墙板建造了一座住宅原型。沃尔特·格罗皮乌斯进一步拓展了早期模块化预制系统和立面板材的应用，首先在斯图加特小住宅（1927）小试牛刀，随后将其应用在库伯住宅（柏林）的设计中。这些住宅的预制系统采用立面板材，由外饰面铜覆层波纹板和内饰面波纹铝板构成。该系统可实现灵活的住宅空间，根据需要随时扩张或收缩。所有结构都在工厂制造，由卡车运到现场。这个极富创意的策略代表着建筑创新的转折点，对后面的发展至关重要。（图2）

虽然巴克敏斯特·富勒建造的黛米克森住宅（1944）和威奇托住宅（1947）以及让·普罗维打造的热带住宅（1949）都选择铝材作为建筑的外包围材料，但是让·普罗维的立面板材和型构件显然更加精细，并最终形成了幕墙。这一应用还体现在民众之家（克林奇，1936–1939）的外墙、应用建筑联合会（1953–1954）的挤制铝型材和外墙板以及莫扎特住宅（巴黎，1953–1954）。（图3、图4）

很快，钛、青铜、铜、锌等不同类型金属的开发和结合开始让步于数不胜数的金属涂层系统和外壳设计，后者一直沿用至今。

在制造和应用优化、防腐性能以及建筑外壳密封性等挑战下，许多系统和专利进行了进一步开发。由于钢材具

Image1 William LeBaron Jenney, Home Insurance Building. Chicago, 1895, the first building contructed with steel structure
Source: unknown

Image2 W. Gropius, Copper Houses. Berlin, 1931.
Source: Faber and Faber (http://www.faber.co.uk/)

Image3 Buckminster Fuller, Dymaixon House. 1944
Source: Bucky and the Dymaxion © Bettmann/Corbis via britannicacom.jpg

Image4 J. Prouvé, Maisons Tropicales. Brazzaville, 1951
Source: lesmaisonstropicales.blogspot.com LMT 10_twoo houses in brazzaville 1951

图1 威廉·勒巴伦·詹尼，房屋保险大楼（芝加哥，1895），第一座钢结构建筑
来源：未知

图2 W·格罗皮乌斯，库伯住宅（柏林，1931）
来源：Faber and Faber (http://www.faber.co.uk/)

图3 巴克敏斯特·富勒，黛米克森住宅（1944）
来源：Bucky and the Dymaxion © Bettmann/Corbis via britannicacom.jpg

图4 J·普罗维，热带住宅（布拉柴维尔，1951）
来源：lesmaisonstropicales.blogspot.com LMT 10_twoo houses in brazzaville 1951

The possibility of reuse or disassembly, reversibility and transformation of spaces contribute to revitalise the idea of the building as a monument, to understand it as an organism in continuous evolution. The reconfigurable architectural space generates questions associated with spatial authorship, placing architects and their tools in a radically more open and dynamic position.

Yet at the same time, the construction processes of steel and metal architectures increasingly reinforce the systems of predesigned and prefabricated objects. Structures, partitions, and cladding are incorporated but are not produced on site, introducing new conditions and possibilities in the temporal and spatial organisation of the architectural work, issues that inevitably impact and affect major project decisions.

Nowadays, digital calculation and numerical control software systems make possible the realisation of almost all forms and surfaces with metallic elements. These advances optimise fabrication processes, and have facilitated the emergence of an even greater formal and expressive repertoire. However, they have also brought the construction of arbitrary shapes, absurd structures and banal use of the expressive possibilities.

Paradoxically, the obsession and fascination in the construction of increasingly complex shapes that these systems let conceive and resolve, has been what has most contributed to "increase in the gap between the space which can be represented, and space which can be effectively built"[2]. Hence, the increase of formal complexity, the greater the technology required for both design and fabrication of its components, thus placing these before the techniques and procedures of traditional manual labour and craft.

It is precisely in this tension, between the different phases of the spatial production, where the architectural design plays a crucial role. Architecture has the ability to influence and consolidate the space between industry and traditional crafts. The project can define and propose a method of materialisation, able to zoom and enhance technological capabilities, yet utilising existing traditional industrial resources, in the construction of the built environment. Therefore, enlivening the sense of cultural value, and the knowledge and expertise of its places where architectures operate.

One of the most important and urgent challenges for industrialisation is the urgent need to reduce consumption of non-renewable energy and emission of greenhouse gas. In this respect, metallic materials already offer significant advantages: 50% of new steel comes from of scrap metal smelting; almost all their products allow reuse, disassembly and transportation with low waste production. Nevertheless, the processes for obtaining raw materials for producing metal components still consume a lot of non-renewable energy, while also generating contaminating residues.

Therefore, through responsible use of the metals, a production that ensures long-term durability of the products, as well as the reuse and recycling at the end-of-life of the components, will certainly ensure the rapid development of metallic materials for construction, offering many new advantages and opportunities for the future built environment.

References
注释
【1】In 1929, it was patented the Fillod system for steel façades consisting of external and internal face in sheet metal panels fixed to a double tubular steel supporting structure spaced each other and filled with insulating material.
【2】AVELLANEDA, Jaume. "Revestimientos metálicos en fachadas y cubiertas". TECTONICA Nº32 (Mayo, 2010) p.4-17
【1】1929年，费尔罗德系统（Fillod system）获得了专利，该立面系统由内外层金属板固定在双层钢管支撑结构上构成，钢管的间隙填充了隔热材料。
【2】阿韦亚内达，若姆。《立面与屋顶的金属涂层》. TECTONICA Nº32（2010.5）p.4-17

Text by Fernando Pérez Blanco & Marta Pelegrín Rodriguez (MEDIOMUNDO Arquitectos) Translate by Chapman Kan (architect)
原文：费尔南多·佩雷斯·布兰可·玛尔塔·佩莱格林·罗德里格斯（MEDIOMUNDO建筑事务所）；英文翻译：查普曼·肯（建筑师）

有延展性，可折叠、可穿孔、可制作纹理表面，随着数字制造的发展，钢材为建筑带来了更多的可能。此外，这些系统从根本上改变了建造流程以及建造法则、概念和施工场地的组织方式。

再利用性、可拆卸性、可逆性和空间改造为建筑概念注入了新的活力，使建筑变成了不断进化的有机体。可重组的建筑空间衍生了与空间原创权相关的问题，把建筑师和他们的设计工具放在了更开放、更活跃的位置。

同时，金属建筑的建造流程逐步强化了预设计和预制组件系统。结构、隔断和覆层被合为一体，但是并不在现场制造，在建筑工作的时间和空间组织中引入了新的条件和可能，不可避免地影响了重要的项目决定。

如今，数字计算和数值控制软件系统让金属元件几乎可以实现任何造型和表面设计。这些发展优化了制造流程，有助于打造更大、更具表现力的建筑。但是，它们也带来了随意的形状、荒谬的结构和表现力的滥用。

在通过这些系统所实现的日渐复杂的造型中，困扰与魅力并存，它们已经成为解决"空间表现力和建造效率之间的差距"[2]的最大障碍。因此，造型越复杂，对预制组件的设计和制造技术的要求越高，它们的地位就会超越传统手工劳动和工艺的技术和流程。

恰恰是在这种空间制造的不同阶段之间所形成的张力中，建筑设计扮演了决定性的角色。建筑有能力影响和联合工业和传统工艺之间的空间。项目能明确提出一种物化方案，从而提升技术能力，但是在建筑环境中仍然使用传统工业资源。在这一过程中，建筑所在地的文化价值和知识技术都能得到活化。

工业化进程所面临的最重要、最紧急的挑战之一就是减少不可再生能源的消耗和温室气体的排放。在这方面，金属材料已经呈现出极大的优势：50%的新钢材来自于废金属冶炼，几乎所有钢产品都能实现再利用，拆卸和转化，减少了废料的产生。虽然如此，金属原材料的获取过程仍然需要消耗大量不可再生能源，并且会产生污染残留物。

因此，负责的使用金属、生产经久耐用的产品以及回收利用旧组件都能保证金属材料在建造过程中的快速发展，为未来的建筑环境带来更多的优势与机遇。

Contents 目录

9	**Overview**
	概述
15	**Chapter 1 Basic Information of Steel**
	第一章 钢
22	Ferretería O´Higgins
	法雷特里亚·欧希金斯公司
26	Roman Villa La Olmeda
	拉欧尔米达罗马别墅遗址
30	Broadcasting Place, Leeds
	利兹广播大厦
34	Administrative Centre Jesenice
	耶塞奈斯行政中心
38	Olympic Energy Centres
	奥林匹克能源中心
42	Wyckoff Exchange
	威科夫交流商店
46	Looptecture Fukura
	福良海啸控制预防中心
50	Palmiry Museum Place of Memory
	帕尔米瑞博物馆
54	South West Institute of TAFE, Warrambool Campus – Stage 3
	西南职业技术学院瓦拉姆布尔校区三期工程
58	Stockholmsmässan AE-hallen
	斯德哥尔摩会展中心AE大厅
62	Terminal in Ven
	芬岛航运站
66	New City School, Frederikshavn
	腓特烈港新城市学校
68	Prince Housing Sales Centre
	太子馥接待中心
72	71 Council and Private Flats
	塞斯港71号住宅
76	Sant Miquel Special Education School
	圣米克尔特殊教育学校
78	Institute of Functional Biology and Genomics
	功能生物学与基因学研究院

82	School of Art and Design in Amposta	

82 School of Art and Design in Amposta
    安波斯塔艺术设计学院

86 Social Cybercentre Macarena Tres Huertas
    玛卡瑞纳特里斯胡尔塔斯社会数码中心

90 The "Coslada" Hybrid Complex
    科斯拉达综合体

94 151 Viviendas, Locales Comerciales y Garaje en Mieres
    米耶雷斯151住宅、商店和车库

98 Full of Triangles
    三角楼

102 KUKJE Gallery
    库卡画廊

104 Vocational Education Centre
    职业教育中心

108 Yapı Kredi ACCR
    亚比信贷银行ACCR大楼

112 Neiman Marcus at Natick Collection
    纳蒂克内曼·马库斯百货商店

116 FRIEM Headquarters
    弗利姆总部

120 Office Building, Ravezies
    莱夫西斯办公楼

124 Gnome Parking Garage
    小矮人停车楼

129 **Chapter 2 Basic Information of Aluminium**
    第二章 铝

134 Cluj Arena
    克鲁日体育场

138 Maison des Sciences De l'Homme de Dijon, Université de Bourgogne
    勃艮第大学人类科学楼

142 Public Library in Ceuta
    休达公共图书馆

146 TIZIANO 32
    蒂奇亚诺32

150 Centre for Manufacturing Innovation, Metalsa CIDeVeC
    曼特沙公司工业创新中心

154 Castle of Skywalkers
    天行者城堡

158 Parametric Fragment
    参数碎片住宅

162 Office Building in Barcelona
    巴塞罗那办公楼

166 Science Park
    科学园

170	Hotel Centar	
	辛塔尔酒店	
172	School Isabel Besora	
	伊莎贝尔贝索拉学校	
174	CASP 74	
	卡斯普74办公楼	
176	Jinan Vanke Marketing Centre	
	济南万科营销中心	
180	A Change of Skin	
	维特鲁夫广场项目表皮翻新	
182	IMED Elche Hospital	
	埃尔切IMED医院	
184	Policemen House	
	警察之家	
188	Supermarket in Athens	
	雅典超市	
192	Innsbruck Furniture Store	
	因斯布鲁克家具店	
196	Knox Innovation, Opportunity and Sustainability Centre (KIOSC)	
	诺克斯创新机遇可持续中心	
200	Precinct Energy Project	
	区域能源项目	
204	Brno Observatory and Planetarium	
	布尔诺天文馆	
208	Head Offices of the Telecommunications Market Commission, CMT	
	CMT总部办公楼	
212	Nursery in Zarautz	
	萨劳特斯托儿所	
216	LoMa Chapalita	
	洛马大厦	
218	Nursery in the Jardines De Malaga in Barcelona	
	巴塞罗那马拉加花园托儿所	
222	ZAP' ADOS	
	活力滑板场	
226	Student Housing in Delft	
	代尔夫特学生宿舍	
230	Library of South University of Science and Technology of China	
	南方科技大学图书馆	
234	Extension of Two Elementary Schools	
	两所小学的扩建	
238	Hotel Well	
	维尔酒店	

242	Reconversion Post Site
	邮局大楼改造
246	Parking Garage 'de Cope' at Papendorp
	帕潘多普德科普停车楼
250	Statoil Regional and International Offices
	挪威国家石油公司办公楼
254	YJP Administrative Centre
	于家堡工程指挥中心
258	Ice Rink of Liège
	列日滑冰场
263	**Chapter 3 Basic Information of Copper, Zinc and Titanium**
	第三章 铜、锌和钛
268	City library in Seinäjoki
	塞伊奈约基市图书馆
272	Sports Hall St. Martin
	圣马丁体育馆
276	Tower Euravenir
	欧拉维尼尔大厦
280	Hiiu
	希尤住宅
284	Louisiana State Museum and Sports Hall of Fame
	路易斯安那州博物馆和体育名人堂
288	YapıKredi Banking Academy
	雅皮科里迪金融学院
290	Bu Yeon Dang
	浮烟堂
294	Théâtre 95
	95剧场
298	Platform of Arts and Creativity
	艺术与创意平台
302	Tales Pavilion
	朝阳态思故事厅
304	Socio-Cultural Centre in Mulhouse
	牟罗兹社会文化中心
308	Municipal Market
	市政市场
312	Le Carré en Seine
	塞纳河畔卡雷住宅
316	Cultural Civic Centre
	市民文化中心
318	**Index**
	索引

Overview

概述

1. Definition

A metal (from Greek "μέταλλον" – métallon, "mine, quarry, metal"[1][2]) is a solid material (an element, compound, or alloy) that is typically hard, opaque, shiny, and features good electrical and thermal conductivity. Metals are generally malleable – that is, they can be hammered or pressed permanently out of shape without breaking or cracking – as well as fusible (able to be fused or melted) and ductile (able to be drawn out into a thin wire).[3] 91 of the 118 elements in the periodic table are metals.

Metals in general have high electrical conductivity, high thermal conductivity, and high density. Typically they are malleable and ductile, deforming under stress without cleaving.[4] In terms of optical properties, metals are shiny and lustrous. Sheets of metal beyond a few micrometres in thickness appear opaque, but gold leaf transmits green light.

1. 定义

金属（在希腊语中写作μέταλλον，意为"矿产、开采、金属"[1][2]）是一种具有光泽、坚硬、不透明、容易导电、导热的物质。金属一般具有可锻性，可锤炼压延成永久的造型，并且不会破裂、断裂。金属还具有可熔性（能够熔化、熔解）和延展性（能够被拉成细丝）。[3]在元素周期表的118种元素中，91种元素都是金属。（见表1）

金属通常都具有良好的导电性、导热性和较大的密度，可锻造、可延展，在压力下能够变形且不破裂。[4]从光学特性来讲，金属具有光泽。厚度在几微米以下的金属板仍是不透明的，但是金箔能传送绿光。

2. Categories

Base Metal

In chemistry, the term base metal is used informally to refer to a metal that oxidises or corrodes relatively easily, and reacts variably with dilute hydrochloric acid (HCl) to form hydrogen. Examples include iron, nickel, lead and zinc. Copper is considered a base metal as it oxidises relatively easily, although it does not react with HCl. It is commonly used in opposition to noble metal.

In alchemy, a base metal was a common and inexpensive metal, as opposed to precious metals, mainly gold and silver. A longtime goal of the alchemists was the transmutation of base metals into precious metals.

In numismatics, coins in the past derived their value primarily from the precious metal content. Most modern currencies are fiat currency, allowing the coins to be made of base metal.

Ferrous Metal

The term "ferrous" is derived from the Latin word meaning "containing iron". This can include pure iron, such as wrought iron, or an alloy such as steel. Ferrous metals are often magnetic, but not exclusively.

Precious Metal

A precious metal is a rare metallic chemical element of high economic value.

Chemically, the precious metals are less reactive than most elements, have high luster and high electrical conductivity. Gold, silver, platinum and palladium each have an ISO 4217 currency code. The best-known precious metals are gold and silver. While both have industrial uses, they are better known for their uses in art, jewelry, and coinage. Other precious metals include the platinum group metals: ruthenium, rhodium, palladium, osmium, iridium, and platinum, of which platinum is the most widely traded.

3. Metal's Application in Architectural Facade

Sheet metals have a wide application and various types, including aluminium sheet, aluminium-plastic sheet, colour plate, copper plate, stainless steel plate, etc. There are many advantages of sheet metal: metal is highly recyclable and environment-friendly; sheet metal is easy to be processed preciously, which shortens construction time and reduces labour cost; with various types, rich colours and good ductility, sheet metal can express complex geometric shapes, textures and patterns; sheet metal is weather resistance and can meet most buildings' requirements in durability. Different metals bring different visual effects: smooth aluminium sheet and stainless steel sheet show the beauty of modern technology; copper plate achieves the combination between the present and history; corrugated plate brings rich details; raw steel plate is easily marked by nature and shows influences of time; titanium plate expresses a magic feeling…(See Figure 1 to Figure 4)

2.分类

基本金属

在化学中,基本金属指相对容易氧化或腐蚀、与稀盐酸能发生反应产生氢的金属,例如铁、镍、铅、锌。尽管铜不与稀盐酸发生反应,但由于它相对容易氧化,也属于基本金属。基本金属的概念主要与贵重金属相对。

在炼金术中,基本金属指普通且价格较低的金属,与金、银等贵重金属相对。炼金术士的终生目标就是把基本金属变成贵重金属。

在钱币学中,过去硬币的价值主要取决于其中贵重金属的含量。现代货币基本为法定货币,硬币也大多由基本金属制成。

含铁金属

"含铁"的英文"ferrous"来自于拉丁文。含铁金属包括纯铁(如锻铁)和铁合金(如钢)。含铁金属通常都具有磁性,但是也有例外。

贵重金属

贵重金属是稀有而具有很高经济价值的金属化学元素。

在化学中,贵重金属比大多数元素更具惰性、更有光泽、更易导电。金、银、铂、钯都具有ISO 4217(国际标准化组织)货币代码。最著名的贵重金属是金和银。虽然二者在工业中也有应用,它们更以其在艺术品、珠宝和钱币中的应用而闻名。其他贵重金属包括铂族金属:钌、铑、钯、锇、铱、铂,其中铂是最常见的一种。

3.金属在建筑饰面中的应用

金属板材的应用广泛,种类很多,包括铝板、铝塑板、彩钢板、铜板、不锈钢板等。金属板材有很多优点:回收率高,是名副其实的环保材料;金属板材易于加工,精度高,可以缩短工期,降低人工成本;金属板材形式众多,色彩丰富,易于延展成型,能够表现出各种复杂的立体造型、纹理及质感;金属板材还具有良好的耐候性、耐久性,能够适应大部分建筑耐久性的要求。各种金属板材的使用会带来不同的视觉效果:平滑的铝板、不锈钢板能体现现代技术及工艺美;铜板能达到现代感与历史感的结合;波纹板则给建筑带来更丰富的细部;自然未处理的钢板更容易留下自然的印记,显示出时间对建筑的影响;钛板带给人一种亦幻亦真的绝妙体验……(见图1~图4)

Figure 1 A market in Greece enveloped by Aluminium
Figure 2 Galvanised steel encloses an apartment in France
Figure 3 Copper covers a sports hall
Figure 4 Zinc is used to envelop a cultural Centre

图1 希腊一座有铝制外壳的市场
图2 法国一座由镀锌钢作为外壳的公寓
图3 由铜覆盖的体育馆
图4 由锌覆盖的文化中心

Table1: Physical Property Comparison between Metal and Other Building Materials 常用金属建筑材料与其他建筑材料物理性质对照

	Cu 铜	**Al 铝**	**Concrete 普通混凝土**	**Glass 普通玻璃**
密度（g/cm^3） Density (g/cm^3)	8.92	2.7	1.9~2.5	2.5~3
熔点(℃) Melting point (℃)	1083	660		无固定熔点，大约在 1,400~1,600℃ No definite melting point, approximate 1,400 to 1,600℃
导热系数（W/m.K） Thermal conductivity coefficient	293~364（20℃）	110	1.28~1.74	0.76
线性膨胀系数(mm/m.K) Linear expansion coefficient	0.022	0.024	0.10~01.015	0.005−0.009

(Note: This table is only a reference. The results may vary according to different institutions' test.)　　（注：本表仅供参考，不同机构测试结果允许略有差异）

Metal Corrosion Prevention Measures

Corrosion is one of the main causes of metal products' damages. As metal is key material of many machinery components, special measures should be taken to prevent metal from chemical corrosion. The followings are some measures to prevent metal from corrosion.

Structural Changes

One of the common corrosion prevention measures is to change the metal's structure. Some heavy metals is not chemically active and difficult to corrode; when combined with other metals, some metals' corrosion resistance will improve effectively. For example, combined with chromium or nickel, steel or iron becomes stainless steel and gets great corrosion prevention.

Protective Layer

Protective layer is the most common measure for corrosion prevention. This anti-corrosive measure creates various protective layers on metals' surfaces to separate metal products from corrosive medium, thus achieving corrosion prevention. The protective layers can be achieved through painting, spraying, electroplating, hot-dip coating etc. There are various materials can be used as protective layer, including coatings such as bio-oil or oil paint and non-metallic materials such as ceramics or plastics. Electroplating or hot-dip coating uses metals difficult to corrode, such as zinc, tin, chromium or nickel, to form a

金属防腐常用措施

腐蚀是金属制品损坏的主要原因之一，而金属又是很多机械备件的主要构成材料，因此针对金属制品易被腐蚀的特点，要采取专门的防腐措施，适当有效的防止金属的化工腐蚀。金属的腐蚀防护主要方法有以下几种：

金属防腐的结构改变法

金属防腐的常见办法之一是改变金属的结构。金属的种类很多，一些重金属的化学活性低，不易受到其他物质的腐蚀，也有部分金属与其他金属配合使用能有效提高防腐能力，例如在普通钢铁中加入铬、镍等材料制成不锈钢，就能获得较好的防腐效果。

金属防腐的保护层法

金属防腐的保护层法使用范围最为广泛。这种防腐方法是在金属的表面制造各种材质的保护层，将金属产品与外界的腐蚀介质隔离开来，从而达到防止腐蚀的效果。金属防腐的保护层可以通过涂抹、喷涂、电镀、热镀、喷镀等方法形成。金属防腐的防护层材料

protective layer on the surface. Another measure is to from a ferroferric oxide film on the surface of steel to protect it from corrosion.

Electrochemistry Protection

Electrochemistry protection is based on primary cell theory to prevent metal from corrosion. According to primary cell theory, once primary cell reaction which causes chemical corrosion is removed, metal corrosion prevention is achieved. Electrochemistry protection can be divided into anodic protection and cathodic protection and the latter is more commonly used.

Corrosive Medium Treatment

Metal corrosion prevention can be achieved through corrosive medium treatment. This measure focuses on eliminating corrosive medium, which means keep the metal dry. For example, wiping water on the metal's surface, placing desiccant, adding corrosion inhibitor in corrosive medium all belong to this category.

Typically, metal used in building façade is treated to prevent corrosion. Arranged by electric potential from high to low, metals commonly used in buildings are copper, lead, tin, iron, zinc and aluminium. To prevent metals with different activities from electrochemical corrosion, they should be installed from exterior to interior according to their electric potentials (from high to low).

When metal is contacted with other materials such as wood, asphalt paint can be used as a spacer-layer to prevent corrosion. In addition, architects can prevent direct contacts between corrosive materials, such as rainwater and dust, and metal through structure design.

Reference/参考文献
1. μέταλλον Henry George Liddell, Robert Scott, A Greek-English Lexicon, on Perseus Digital Library.
2. metal, on Oxford Dictionaries.
3. metal. Encyclopædia Britannica.
4. b c d e f Mortimer, Charles E. (1975). Chemistry: A Conceptual Approach (3rd ed.). New York:: D. Van Nostrad Company.

很多，常见的有机油、油漆等涂料和陶瓷、塑料等耐腐蚀的非金属材料。金属防腐的电镀和热镀一般是使用不易腐蚀的金属，如锌、锡、铬和镍等。金属防腐的另一种材料是钢铁在表面形成的氧化膜，也就是黑色四氧化三铁薄膜，同样能起到防腐作用。

金属防腐的电化学保护法

金属防腐的电化学保护法是以原电池理论为原理对金属进行防腐保护的方法。根据原电池理论，只要能够消除引起化学腐蚀的原电池的反应，就可以实现金属的防腐。金属防腐的电化学保护法分为阳极保护和阴极保护两个类别，其中阴极保护应用较多。

金属防腐的腐蚀介质处理法

金属防腐可以通过对腐蚀介质的处理来完成。这种方法着重消除腐蚀介质的存在，也就是保持金属的干燥，例如经常擦干金属上的水分、放置干燥剂、在腐蚀介质中添加缓蚀剂等都是属于这种防腐方法。

建筑立面中常用的金属材质通常是经过不同的防腐方式进行处理的。建筑中常用金属的正常电势由高到低的顺序是：铜、铅、锡、铁、锌、铝。为防止不同活动性的金属贴在一起产生电化腐蚀，应从外向内按活动性从强到弱的顺序设置安装。

金属材料与其他材质接触时，如木质构件，可以使用沥青涂料等作为中间隔层，防止其中的酸碱物质对其进行腐蚀。此外，还可以通过构造设计避免雨水、灰尘等腐蚀性物质与金属直接接触。

Chapter 1
Basic Information of Steel

第一章 钢

Steel is an alloy of iron, with carbon being the primary alloying element, up to 2.1% by weight. Carbon, other elements, and inclusions within iron act as hardening agents that prevent the movement of dislocations that naturally exist in the iron atom crystal lattices. Varying the amount of alloying elements, their form in the steel either as solute elements, or a precipitated phases, retards the movement of those dislocations that make iron so ductile and so weak, and so it controls qualities such as the hardness, ductility, and tensile strength of the resulting steel. Steel can be made stronger than pure iron, but only by trading away ductility, of which iron has an excess.

Steel is used widely in the construction of roads, railways, other infrastructure, appliances, and buildings. Most large modern building structures, such as stadiums and skyscrapers, bridges, and airports, are supported by a steel skeleton. Even those with a concrete structure employ steel for reinforcing.

The following types of steel are commonly seen in the modern building skins, weathering steel (Cor-ten), galvanised steel, lacquer coated steel, and stainless steel.

钢或称钢铁、钢材，是一种由铁与其他元素结合而成的合金，当中最普遍的是碳。碳的含量最高可达钢总重量的2.1%。碳与其他元素有硬化剂的作用，能够防止铁原子的晶格因原子滑移过其他原子而出现位错。调整合金元素的量及其存在与钢中的形式（溶质元素及参与相），就能够控制钢成品的特性，例如硬度、延展性及强度。加了碳的钢比纯铁更硬更强，但是这种钢的延展性会比铁差。

钢在公路、铁路、其他基础设施、电器、建筑中都有广泛的应用。体育场、摩天楼、桥梁、机场等大多数大型建筑结构都采用钢骨架进行支撑。混凝土结构建筑业经常采用钢筋进行加固。

以下是几种现代建筑表皮中常见的钢材：耐候钢（科尔坦钢）、镀锌钢、涂层钢、不锈钢。

Table 1.1 Chemical Composition of Cor-ten Grades　　表1.1 不同等级的科尔坦钢的化学成分

Grade 等级	C 碳	Si 硅	Mn 锰	P 磷	S 硫	Cr 铬	Cu 铜	V 钒	Ni 镍
Cor-ten A 科尔坦A	0.12	0.25~0.75	0.20~0.50	0.01~0.20	0.030	0.50~1.25	0.25~0.55		0.65
Cor-ten B 科尔坦B	0.16	0.30~0.50	0.80~1.25	0.030	0.030	0.40~0.65	0.25~0.40	0.02~0.10	0.40

1.1 Weathering Steel (Cor-ten)

Definition

Weathering steel, best-known under the trademark COR-TEN steel and sometimes written without the hyphen as "Corten steel", is a group of steel alloys which were developed to eliminate the need for painting, and form a stable rust-like appearance if exposed to the weather for several years. United States Steel Corporation (USS) holds the registered trademark on the name COR-TEN.[1] Although USS sold its discrete plate business to International Steel Group (now Arcelor-Mittal) in 2003,[2] it still sells COR-TEN branded material in strip-mill plate and sheet forms.

The original COR-TEN received the standard designation A 242 ("COR-TEN A") from the ASTM International standards group. Newer ASTM grades are A 588 ("COR-TEN B") and A 606 for thin sheet. All alloys are in common production and use. (See Table 2.1)

Properties

"Weathering" means that due to their chemical compositions, these steels exhibit increased resistance to atmospheric corrosion compared to other steels. This is because the steel forms a protective layer on its surface under the influence of the weather.

The corrosion-retarding effect of the protective layer is produced by the particular distribution and concentration of alloying elements in it. The layer protecting the surface develops and regenerates continuously when subjected to the influence of the weather. In other words, the steel is allowed to rust in order to form the "protective" coating.

Strength

Combined with microelements such as phosphorus, copper, chromium and nickel, the surface of weathering steel generates a compact and adhesive protective film, preventing the surface from rusting deeply and retarding the corrosion rate. This compact oxide film prevent oxygen and water in atmosphere from permeating into the substrate of steel, retarding the rust's development and enhancing steel's atmosphere corrosion resistance significantly. Weathering steel can be used thinning, exposed or with simple coating. As a weather-resistant and economic

1.1 耐候钢（科尔坦钢）

定义

耐候钢中最著名的品牌就是"科尔坦钢"，是一种无需涂装的钢合金，在室外曝露几年后能在表面形成一层稳定的锈层。美国钢铁公司持有"科尔坦"注册商标。[1] 尽管美国钢铁公司于2003年将离散板材业务出售给了国际钢铁集团（现名为阿赛洛集团）[2]，它仍然出售科尔坦系列品牌的条板和薄板型钢。

最初的科尔坦钢由美国材料试验协会标准命名为A 242（科尔坦A）。新的美国材料试验协会命名系统将其命名为A 588（科尔坦B）和A 606（薄板）。所有耐候钢合金材料都采用普通生产和应用。（见表1.1）

属性

"耐候"意味着由于特殊的化学成分，耐候钢比其他钢材更耐大气腐蚀。这是由于在气候作用下，耐候钢的表面形成了一层保护层。

保护层的缓蚀效果来自于合金元素的颗粒分布和浓度。保护层在气候的影响下不断发展和再生。换而言之，耐候钢通过生锈来形成保护涂层。

优点

钢中加入磷、铜、铬、镍等微量元素后，钢材表面形成致密和附着性很强的保护膜，阻碍锈蚀往里扩散和发展，保护锈层下面的基体，以减缓其腐蚀速度。由于这层致密氧化物膜的存在，阻止了大气中氧和水向钢铁基体渗入，减缓了锈蚀向钢铁材料纵深发展，大大提高了钢铁材料的耐大气腐蚀能力。耐候钢是可以减薄使用、裸露使用或简化涂装，而使制品抗蚀延寿、省工降耗、升级换代的钢系，也是一个可融入现代冶金新机制、新技术、新工艺而使其持续发展和创新的钢系。

steel type, weathering steel can also be combined with modern metallurgy system, new technologies and new processes.

Application in Architectural Field

The first use of COR-TEN for architectural applications was the John Deere World Headquarters in Moline, Illinois. The building was designed by architect Eero Saarinen, and completed in 1964. The main buildings of Odense University, designed by Knud Holscher and Jørgen Vesterholt and built 1971-1976, are clad in COR-TEN steel, earning them the nickname Rustenborg. COR-TEN was used to build the exterior of Barclays Centre, made up of 12,000 pre-weathered steel panels engineered by Dissimilar Metal Design. The New York Times says of the material, "While it can look suspiciously unfinished to the casual observer, it has many fans in the world of art and architecture".[3] (See Figure 1.1 to Figure 1.5)

耐候钢在建筑中的应用

位于美国伊利诺伊州莫林市的约翰迪尔国际总部是第一座使用耐候钢（科尔坦钢）的建筑。该建筑由埃罗·萨里宁设计，完成于1964年。由克努德·霍尔斯彻和乔尔根·韦斯特霍尔特于1971-1976年间设计的欧登塞大学主楼全部采用耐候钢覆盖，因此被称作"锈堡"。巴克莱中心的外立面由12,000块预制耐候钢板（由异种金属设计公司设计）组成。《纽约时报》这样评论耐候钢："对普通人来说，它看起来像是没有完工，但是它在艺术和建筑领域有很多拥护者。"[3]（见图1.1~图1.5）

Figure 1.1 to Figure 1.5 Buildings with Corten steel exterior

图1.1~图1.5 采用耐候钢外墙的建筑

1.2 Galvanised Steel

Definition

Galvanised steel is steel that has gone through a chemical process to keep it from corroding. The steel gets coated in layers of zinc oxide because this protective metal does not get rusty as easily. The coating also gives the steel a more durable, hard to scratch finish that many people find attractive. For countless outdoor, marine, or industrial applications, galvanised steel is an essential fabrication component.

How It is Made

One of the most common ways of making steel resist rust is by combining (alloying) it with a metal that is less likely to corrode: zinc. When steel is submerged in melted zinc, a chemical reaction permanently bonds the zinc to the steel. Therefore, the zinc is not exactly a sealer, like paint, because it does not just coat the other metal; it actually permanently becomes a part of it. The most external layer is all zinc, but successive layers are a mixture of zinc and iron, with an interior of pure steel.

This process, known as hot-dipped galvanisation, is one of the most common methods of making galvanised steel, but it's not the only one. Continuous sheet galvanising runs a steel sheet or wire through molten zinc; it's also a form of hot-dipping, but leaves a thinner layer of the protective metal. Steel can also be painted or sprayed with zinc, although neither method creates the strong bond that hot-dipping does. Electro-galvanising or electro-plating uses electricity to create the bond between the two metals, which results in relatively thinner layer of zinc.

The degree of galvanising is usually described as the zinc's weight per surface area rather than its thickness, because this gives a better representation of how much metal has been applied. Steel often gets galvanised after individual parts have been formed, such as braces, nails, screws, beams, or studs. Continuous sheet galvanising is often done before the metal is used to create products, however, and can withstand some bending and forming without flaking.

Strength

Zinc protects steel in two ways. First, it is highly resistant to rust; iron, a major component of steel, reacts very easily with oxygen and moisture and will eventually disintegrate. The layer of zinc on the surface prevents those elements from reaching the steel so quickly. It also develops a patina a layer of zinc oxides, salts, and other compounds – that offers further protection. Zinc is also extremely durable and scratch resistant, and has a satiny appearance that many find attractive.

The outer layer also protects the steel by acting as a "sacrificial layer". If, for some reason, rust does take hold on the surface of galvanised steel, the zinc will get corroded first. Even in areas where the surface is scratched or damaged, the surrounding zinc will still corrode before the steel does. The lifespan of galvanised steel varies, but industrial steel with a 3.9 mil (0.0039 inches or 0.09906 mm) thick coating can last more than 70 years

1.2 镀锌钢

定义

镀锌钢是一种通过化工工艺来防止腐蚀的钢材。钢材表面上镀有一层氧化锌，保护金属不易生锈。涂层还能让钢材更持久耐用，表面更为坚硬。对户外工程、航海业、工业应用来说，镀锌钢都是必不可少的制造元件。

镀锌钢的制作流程

让钢不易生锈的常用方法就是使其与不易腐蚀的金属——锌结合起来。钢材被浸入熔化的锌液中，产生化学反应，将锌永久地镀在钢材上。因此，确切地说，锌并不是像涂料一样的密封层，因为它不仅仅是附着在其他金属表面，而是永久地成为了它的一部分。镀锌钢的最外层是纯锌，下层则是锌铁混合物，最里层是纯钢。

这一流程即热镀锌流程，是最常见的镀锌钢的制作方式，但是并不是唯一的方式。连续薄板镀锌将钢板或钢线放入融化的锌液中，也是一种热浸镀方式，但是所形成的保护金属层更薄。还可以在钢材上涂刷或喷涂锌液，但是这两种方法都没有热浸镀结合得更紧密。电镀锌利用电流来实现两种金属的结合，形成相对较薄的锌层。

通常以单位表面积内锌的重量（而不是厚度）来衡量镀锌的等级，因为这种方式更能体现所应用的金属量。钢材通常在独立零件成型后进行镀锌，例如支架、钉子、螺丝、梁、双头螺栓等。连续薄板镀锌则通常在金属制成产品前进行，镀锌层可以承受一定程度的弯折或成型而不会脱落。

优点

锌以两种方式来保护钢。首先，它非常耐锈；作为钢的主要组成元素，铁很容易与氧气和水分发生作用，然后逐渐碎裂。钢表面的锌层让这些元素不能快速地接近钢材。它还会形成由氧化锌、盐和其他成分组成的锈层，实现进一步的保护。同时，锌也是一种十分耐磨、耐划的元素，具有光滑诱人的外观。

外层镀锌还能以"牺牲层"的形式保护钢材。如果镀锌钢表面开始生锈，那么锌是最先腐蚀的。如果某个区域被划损或损坏，周围的锌也比钢先腐蚀。镀锌钢的寿命各有不同，但是带有3.9密耳（0.0039英寸或0.09906毫米）工业钢在普通未经保养的条件下可以维持70多年。

without maintenance under average conditions.

Application in Architectural Field

Galvanised steel can be found almost everywhere. Many individuals live in steel frame houses, and buildings are often made with galvanised steel roofs and skins. Besides being inexpensive, durable and effective, this metal is also popular because it can be recycled and reused multiple times. (See Figure 1.6, Figure 1.7)

1.3 Lacquer Coated Steel
Definition

To enhance steel sheet's weather resistance, cold-rolled galvanised steel sheet is treated through surface degreasing, phosphorisation, chromating, organic coating and dry-out procedure to become weather-resistant lacquer coated steel. The coating can vary through different conditions. PE, SMP, reinforced polyester resin, PVDF are commonly used and the colours are determined by additives.

Strength

Lacquer coated steel are usually coated in factory, so its quality is stable and the surface is more evened compared to painted or sprayed formed metal. The coating can last 15 to 20 years, while the overall durability of steel can last 20 to 30 years depending to different coating processes.

Application in Architectural Field

Types of lacquer coated steel include single-layer panel, composite panel, flat panel and corrugated panel. In China, it is limitedly used in industrial buildings, large-scale markets and temporary buildings. The lifespan of lacquer coated panel is about 20 to 30 years and can meet

镀锌钢在建筑中的应用

镀锌钢随处可见。许多人都居住在钢框住宅中，建筑也常使用镀锌钢屋顶和表皮。除了价格低廉、经久耐用、高效节能之外，镀锌钢的可回收性和再利用性也是一大优势。（见图1.6、图1.7）

1.3 彩色涂层钢
定义

为提升钢板耐候性，以冷轧热镀锌钢板（或镀铝锌合金钢板）为基板，经过表面脱脂、磷化、铬酸盐等处理，涂覆有机涂料、烘干处理形成耐候性更佳的彩色涂层钢板。彩色涂层可以根据不同的使用环境选择合适的树脂，如一般聚酯树脂（PE）、硅化聚酯树脂（SMP）、强化聚酯树脂、聚偏二氟乙烯（PVDF，通常称作氟碳涂层）等，色彩根据添加剂不同可有多种选择。

优势

彩色涂层钢板多采用工厂化方式涂装，其质量远比对成型金属表面进行单体喷涂或涂装稳定、均匀，一般可保持15~20年不褪色，整体耐久性根据不同的涂层工艺可达20~30年以上。

彩色涂层钢在建筑中的应用

彩色涂层钢板从单层板到复合板、从平板到压型板在建筑中一般局限于工业厂房、大型市场、临时建筑等

Figure 1.6, Figure 1.7 Buildings with galvanised steel exterior
图1.6、图1.7 采用镀锌钢外墙的建筑

most buildings' exterior finish or roof requirements. Designed with configuration of zinc panel or copper panel, the visual effect is also excellent. As an economic material, it goes especially well with buildings with low-budgets.

In Europe and America, lacquer coated steel is extensively used in exterior walls and roofs. The profile types and constructions even surpass common zinc panel and copper panel to meet the building requirements. In China, the use of lacquer coated steel still needs to be developed. (See Figure 1.8, Figure 1.9)

1.4 Stainless Steel

In metallurgy, stainless steel, also known as inox steel or inox from French "inoxydable", is a steel alloy with a minimum of 10.5% [4] chromium content by mass.

How It is Made

Most stainless steels are melted in an electric furnace or basic oxygen furnace and then refined in another steel furnace to reduce carbon content. In argon-oxygen decarburization, the mixture of argon and oxygen is sprayed in molten steel. With changing proportions of argon and oxygen, carbon is oxidised into CO and expensive chromium is maintained, thus carbon content is reduced to controlled level. Therefore, in the original fusion process, inexpensive material such as high-carbon ferrochromium can be used. Chromium and low carbon content work together to enhance steel's corrosion resistance and heat resistance. Other elements such as nickel, molybdenum, titanium, aluminium, copper, nitrogen, sulfur, phosphorus and selenium can also be added to generate a rustproof oxidation film to enhance steel's corrosion resistance and oxidation resistance. These elements can give steel special performance and the film can protect steel from corrosions from oxygen, water, certain acid and alkali.

Strength

Stainless steel does not readily corrode, rust or stain with water as ordinary steel does, but despite the name it is not fully stain-proof, most notably under low-oxygen, high-salinity, or poor-circulation environments.[5] There are different grades and surface finishes of stainless steel to suit the environment the alloy must endure. Stainless steel is used where both the properties of steel and resistance to corrosion are required.

Stainless steel differs from carbon steel by the amount of chromium present. Unprotected carbon steel rusts readily when exposed to air and moisture. This iron oxide film (the rust) is active and accelerates corrosion by forming more iron oxide, and due to the greater volume of the iron oxide this tends to flake and fall away. Stainless steels contain sufficient chromium to form a passive film of chromium oxide, which prevents further surface corrosion by blocking oxygen diffusion to the steel surface and blocks corrosion from spreading into the metal's internal structure, and due to the similar size of the steel and oxide ions they bond very strongly and remain attached to the surface. [6]

Application in Architectural Field

档次稍低建筑或构筑物中。所以，对这一材料的运用还有待于建筑师对此进行充分的了解。涂层钢板一般至少有20~30年的使用周期，可以满足大部分的建筑外墙饰面及屋面的要求。其如果按照锌板、铜板的构造形式设计，同样可以取得不俗的视觉效果，还可以节省大量的投资，特别适合投资相对较低的建筑。

国外建筑墙面及屋面用彩色涂层钢板类型较为丰富，其断面的种类、构造方式甚至超越了常见的锌板、铜板，极大地满足了建筑要求。而国内，在这一领域则稍有欠缺，仍有待提升。（见图1.8、图1.9）

1.4 不锈钢
在冶金学中，不锈钢是一种含铬量至少达到10.5%[4]的钢合金。

不锈钢的制作流程
大多数不锈钢先在电炉或氧气顶吹转炉中熔化，然后在另一炼钢炉中精炼，主要为了降低碳含量。在氩－氧脱碳法中，将氧和氩的气体混合物喷入钢水中。改变氧和氩的比例，通过将碳氧化为一氧化碳而不使昂贵的铬氧化和损失，来将碳含量降低到控制的水平。因此，在初始的熔化操作中可使用较便宜的原料，如高碳的铬铁。

铬与低碳含量相配合，可显示出明显的耐腐蚀性和耐热性，还可以加入镍、钼、钛、铝、铜、氮、硫、磷和硒，使其表面会产生防锈的氧化膜，以提高对特殊环境的耐腐蚀性和抗氧化性，并赋予特殊性能，从而保护钢材本身受到外界环境中的空气（尤指氧气）、水、某些酸、碱的氧化腐蚀。

优点
与普通钢相比，不锈钢不易腐蚀、生锈或被污染。但是不锈钢并不是完全不会生锈，尤其是在低氧、高盐或通风不良的环境中。[5] 不锈钢有不同的等级和表面处理方式，以应对不同的环境。不锈钢可应用于同时要求钢的属性和抗腐蚀性的环境中。

不锈钢与碳钢的区别在于铬的含量。无保护的碳钢在接触空气和水分时很容易生锈。氧化铁薄膜（锈）十分活跃，能通过形成更多的氧化铁来加速腐蚀，随着氧化铁的增加，铁锈会剥落。不锈钢中含有足够的铬，形成了一层氧化铬钝化膜，通过组织氧气扩散到钢表面来防止进一步的表面腐蚀，阻止腐蚀扩散到金属内部结构中。由于钢和氧化物的离子大小十分相

Stainless steel is used for buildings for both practical and aesthetic reasons. Stainless steel was in vogue during the art deco period. The most famous example of this is the upper portion of the Chrysler Building. Type 316 stainless steel is used on the exterior of both the Petronas Twin Towers and the Jin Mao Building, two of the world's tallest skyscrapers.[7] Because of the durability of the material, many of the buildings using stainless steel retain their original appearance.

Today, stainless steel is mainly used in the exterior of high-rise buildings. Through surface treatment, painting, and cladding, stainless steel avoids problems such as fingerprint and its application range is broadened. Japan has developed many stainless steel roofing materials, such as medium- or high-chromium ferritic stainless steel and various cladding panels. Super-clean stainless steel developed in Sweden is one of the metal materials of highest strength in architectural field. Chromium stainless steel featuring high mechanical strength and high ductility is easy for processing and manufacturing and can meet various requirements by architects and structural designers. (See Figure 1.10, Figure 1.11)

Reference / 参考文献
1. "Trademarks and Ownership". USSS. Retrieved 24 September 2010.
2. Plate Products, 2003-10-31, archived from the original on 2007-12-28, retrieved 2010-01-13
3. Harris, Elizabeth (27 August 2012). "Constructing a Façade Both Rugged and Rusty". NY Times. Retrieved 27 September 2013.
4. "The Stainless Steel Family". Retrieved 8 December 2012.
5. "Why is Stainless Steel Stainless?". Retrieved 6 September 2013.
6. Stainless Steels and Alloys: Why They Resist Corrosion and How They Fail". Corrosionclinic.com. Retrieved on 29 June 2012.

近，它们结合得十分紧密，保持一致附着在不锈钢表面。[6]

不锈钢在建筑中的应用

不锈钢在建筑中兼具实用性和美观性。不锈钢在装饰艺术时期十分盛行，最著名的例子是克莱斯勒大厦的上半部分。作为世界顶级摩天大厦，马来西亚石油双峰塔和金茂大厦的外部设计都选用了316型不锈钢。由于不锈钢经久耐用，大多数采用不锈钢的建筑都能保持最初的外观。

目前，不锈钢主要用在高层建筑的外墙。经表面处理、着色、镀层的不锈钢板，解决了触摸后易出现手印等问题，使不锈钢的应用范围进一步扩大。日本开发了多种不锈钢屋面材料，如耐腐蚀性能更好的高、中铬铁素体和各种镀层板。瑞典研制的super-clean刷光表面不锈钢还是建筑用金属材料中强度最高的材料之一，其中含铬不锈钢还集机械强度和高延伸性于一身，易于部件的加工制造，可满足建筑师和结构设计人员的需要。（见图1.10、图1.11）

Figure 1.8, 1.9 Buildings with lacquer coated steel exterior
Figure 1.10, 1.11 Buildings with stainless steel exterior
图1.8、图1.9 采用彩色涂层钢外墙的建筑
图1.10、图1.11 采用不锈钢外墙的建筑

Ferretería O´Higgins
法雷特里亚·欧希金斯公司

Location/地点: Santiago, Chile./智利,圣地亚哥
Architect/建筑师: Guillermo Hevia H., Gh+A Arquitectos
Photos/摄影: Nicolás Saieh (www.Nicosaieh.Cl), Guillermo Hevia H. (www.Guillermohevia.Cl), Javier González (Jge.Arquitectura@Gmail.Com)
Gross floor area/总楼面面积: 7,170m²
Completion date/竣工时间: 2011

Key materials: Façade – perforated Corten steel sheet
主要材料：立面——穿孔耐候钢板

Overview
Architecture simple and absolute, expresses itself by means of two volumes that are the result of a subtle support between solid, closed and geometrically leaned bodies. The new corporate building for Ferretería O´Higgins contains offices, services and storage space for high precission tools used in large scale mining. Administrative offices, services and sales space are located in the west end of the building, storage and operations space in the back. In the front, the building is divided into two areas by the trhee storeys high lobby, where the staircase stands out as the main element of the space, expressing itself as a suspended, light and transparent object.

Detail and Materials
The weight supporting structure is surrounded by a Corten steel double skin, reinforcing the idea of two solid bodies, and at the same time protecting the inside of exterior temperature and direct solar radiation. An air chamber is created betwen the exterior Corten steel skin and the interior insulated panel façade, which helps creating a very efficient thermar barrier.

Because of its outside material, the building changes colour according to the sunlight and the passing of time. The building reacts as a living body, showing different and changing shades of orange, brown and ocher in its skin. In the front the main façade stands out within the building, defining a transluscent veil made out of perforated Corten steel sheets that allows people to look outside controlling solar radiation at the same time. These sheets move slightly and constantly with of the wind, combined with the effect of the multipurpose water mirror located in their base, give the building a dinamic dimension.

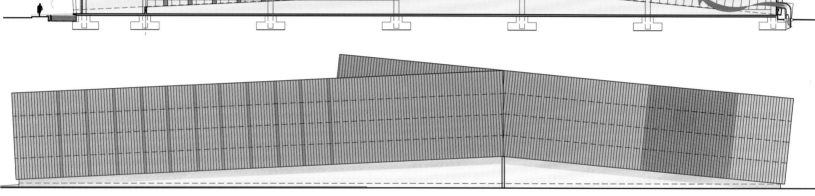

Façade detail
1. Insulated cover
2. Panel painting line 30 Corten steel H-D
3. Camera Panel ISOPOL 75mm
4. Corten steel plate
5. Air diffuser 18
6. Wall socket H.A.
7. Base H.A.
8. Perforated plate

立面节点
1. 隔热盖板
2. 涂漆板线30耐候钢H-D
3. 摄像头面板ISOPOL 75mm
4. 柯尔顿钢板
5. 空气扩散器18
6. 壁式插座H. A.
7. 底座H. A.
8. 穿孔板

项目概况
建筑简单而纯粹，由两个封闭的结构相互依靠而成。法雷特里亚·欧希金斯公司的新办公楼包括办公室、服务区以及用于大规模采矿活动的高精度设备储藏区。行政办公室、服务区和销售空间位于建筑的西侧，储藏区和运营区则位于建筑后部。建筑正面被三层高的大厅一分为二。楼梯是大厅内最显眼的元素，显得轻盈通透。

细部与材料
承重结构由双层耐候钢表皮包裹，突出了两个实体的概念，同时也保护建筑内部不受室外气温和阳光直射的影响。在耐候钢表皮和内层隔热板之间有一个气腔，帮助实现高效的隔热效果。
外部材料让建筑随着日光和时间的变化而变化。建筑像一个鲜活的生命，在表皮上呈现出橙色、棕色和赭色等变化的色调。建筑正面的穿孔耐候钢板像一层面纱，既让人能看到外部的景象，又能控制太阳辐射。这些钢板随风而动，与下方的多功能水镜相互作用，赋予了建筑动态的效果。

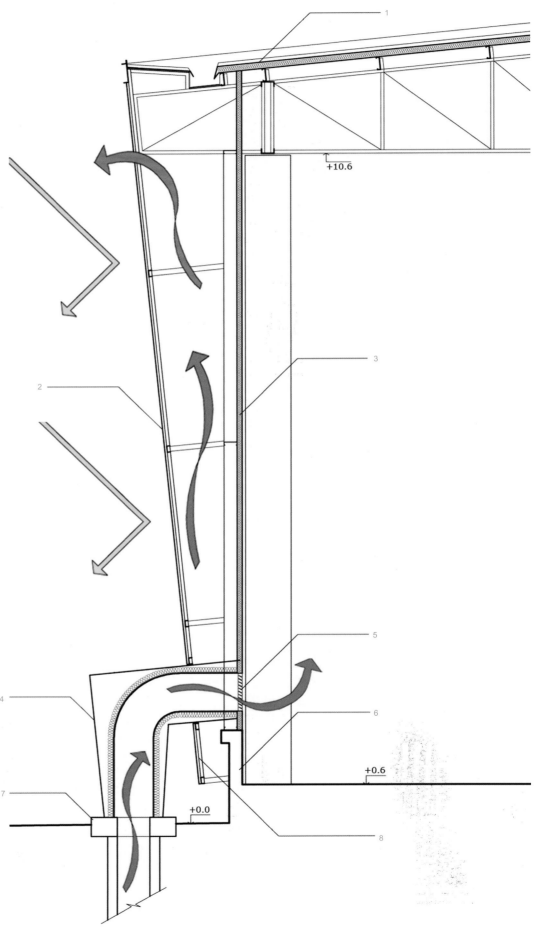

Roman Villa La Olmeda
拉欧尔米达罗马别墅遗址

Location/地点: Pedrosa de la Vega. Palencia. Spain/西班牙，佩德罗撒德拉维加
Architect/建筑师: Ángela García de Paredes, Ignacio Pedrosa/Paredes Pedrosa arquitectos
Project team/项目团队: Clemens Eichner, Álvaro Rábano, Eva Urquijo, Andrea Franconetti, Eva M. Neila
Photos/摄影: Luis Asín
Technical control/技术控制: Luis Calvo
Structure/结构: GOGAITE S.L.
Area/面积: 7,130m²

Key materials:
- Double façade. Exterior: perforated cor-ten steel plates. Interior: polycarbonate
- Roof structure rhomboidal steel-framework welded and painted (prefabricated modular system)
- Roof exterior clad with Aluminium clip sections
- Structure concrete plinth

主要材料:
– 双层立面：外层：穿孔耐候钢板；内层：聚碳酸酯
– 屋顶结构：涂漆长菱形钢架焊接安装（预制模块系统）
– 屋顶外部：铝夹型材覆盖
– 结构：混凝土基座

Façade material producer:
外墙立面材料生产商：
Perforated cor-ten steel plates by Talleres Margallo: www.talleresmargallo.com
Polycarbonate façade: Danpalon palplastic s.l: www.palplastic.es

Overview

The work develops a large case placed in a natural landscape that protects a 4th century AD roman villa. In this intervention archaeology is confronted to modernity, and the building is confronted to nature. The program includes the construction of a covering for excavations, the exhibition of the magnificent mosaics and an archaeology centre. A wide metallic roof covers the area and the space of roman rooms is built up with a translucent perimeter of metal curtains that frame the mosaics. These concepts wish to make visitors aware of the unitary character that these fragments of the past had.

The integration of new architecture in the landscape is an aim of this project. The relation between architecture and environment, between the large sized building and the hidden geometry of nature required special sensibility and the building is dimensioned regarding the scale of the nearby groups of poplars.

The archaeological area is organised in four vaults; gathered under a continuous roof that rests on four metal columns delimiting the central patio. The segmental domes of the roof are built with a rhomboidal structural steel framework, clad on the outside with aluminium sheet, which show a coffered finish in the interior. The extensive occupation of the site and the need to preserve archaeology during the construction work, decided the use of a prefabricated modular system. The dimensions of the rhomboidal elements ease their transportation and facilitate the on-site assembly of full arches that were fixed with screws.

Sustainability is achieved by the building concept. Cost was tight and both structure and construction are finishing's and conditioning. Conditioned areas are only 4% of the total volume. The inside of the villa is shaded from the intense summer sun by the double façade that filters light as tree leaves and branches do. Artificial light is reduced to the essential above the mosaics. The cor-ten steel freestanding panels bear the patina of time and permit a natural maintenance of façade and roof. Two ventilated layers that isolate the non air-conditioned interior, compose vaults. Finally water from rain and snow on the large roof, is conducted outside to the green slope that embraces the building and to the central patio, where the ancient roman pipes eighteen centuries later work again.

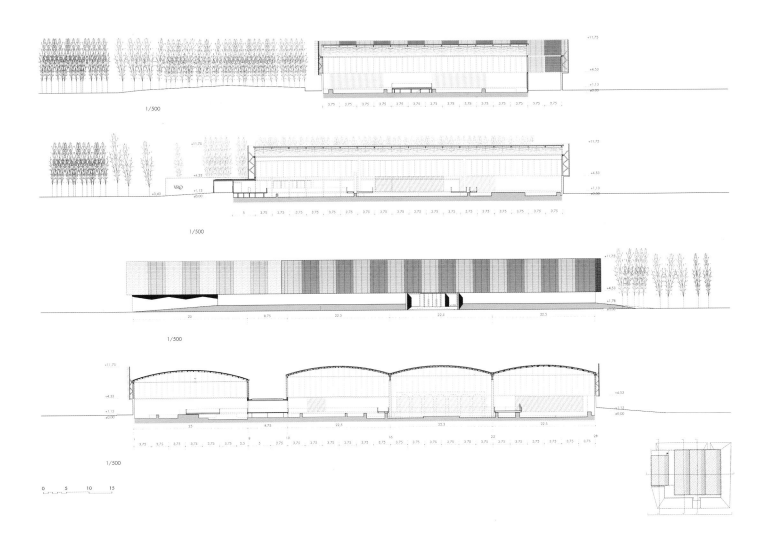

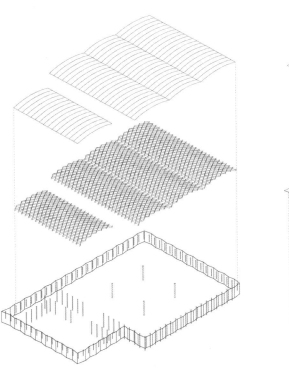

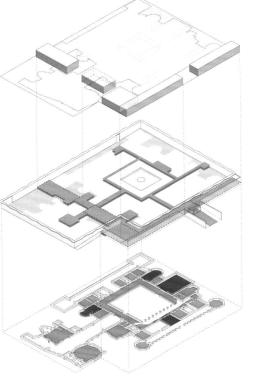

Detail and Materials

The double façade is cor-ten steel and polycarbonate. A raised wooden floor with transparent railings joins all the archaeological zones. From the roof hang a series of steel meshes that restore the position of the villa's original walls, favouring the views of the mosaics and showing its original layout.

On the outside, the upper part stands evenly over a concrete plinth surrounded by a folding perforated façade, separated from the structure, that attenuates solar irradiation and casts scattered shadows to the interior, responding in scale and appearance to the surrounding poplar tree cultivations. The metallic enclosure reduces the density of the pattern as it goes up from the plinth to the roof, blending with the natural surroundings and the natural ground slopes to embrace the building.

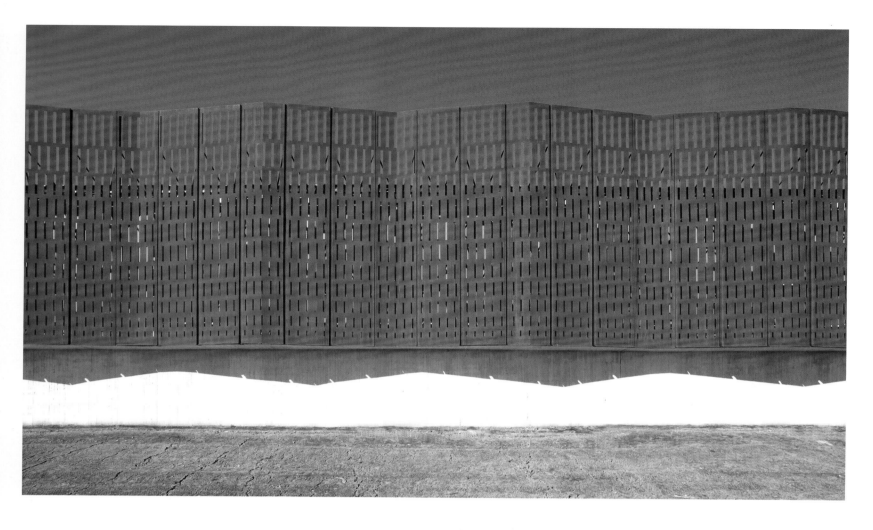

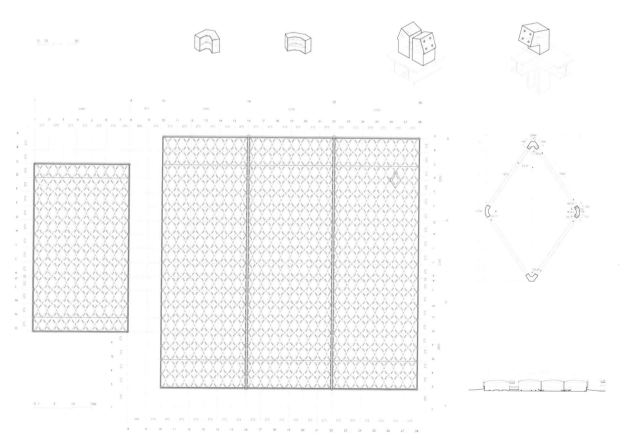

项目概况

项目为一个坐落在自然景观之中的4世纪罗马别墅遗址打造了展示空间。在项目中，考古与现代、建筑与自然都产生了强烈的冲突。项目包括为考古场地打造一个外部保护结构，一个展示华丽马赛克瓷砖的展览区和一个考古中心。宽大的金属屋顶覆盖了整个区域，罗马别墅外围的半透明金属幕帘将马赛克环绕起来。这些设计帮助游客将这些古老的碎片融合了起来。

让新建筑融入自然景观是项目的主要目标之一。建筑与环境、大规模建筑与隐藏的自然线条都要求设计具有独特的敏感性。建筑的尺寸和规模都考虑并参考了旁边的白杨树林。

考古区分为四个部分，通过连续的屋顶聚集起来。屋顶由四根金属柱支撑，划分出中央的天井空间。屋顶的分段部分由长菱形结构钢架建成，外层包有铝板，内部有平顶镶板装饰。场地的巨大面积和施工与考古工作的同时进行决定了预制模块系统的使用。长菱形元件的尺寸便于运输，同时也有助于现场安装。

Façade detail
1. 46/550mm Aluminium clip sections; 60MM glass-wool insulation; profile sheet steel
2. 200.80.8mm steel structure, welded and painted
3. Steel - flat angle connection, welded and painted
4. 80/80/4mm steel column
5. 45.4mm steel tube
6. 16mm polycarbonate sheeting
7. Steel structure, column and beam h section 200, welded and painted
8. 2mm perforated preoxidized - steel façade element
9. Steel structure welded and painted
10. 300mm white-pigmented reinforce concrete

立面节点
1. 46/550mm铝夹型材；60mm玻璃棉隔热层；成型钢板
2. 200×80×8mm结构，焊接+涂漆
3. 钢，平角连接，焊接+涂漆
4. 80/80/4mm钢柱
5. 45×4mm钢管
6. 16mm聚碳酸酯板
7. 钢结构，梁柱，H剖面200，焊接+涂漆
8. 2mm穿孔预氧化钢立面元件
9. 钢结构，焊接+涂漆
10. 300mm白色钢筋混凝土

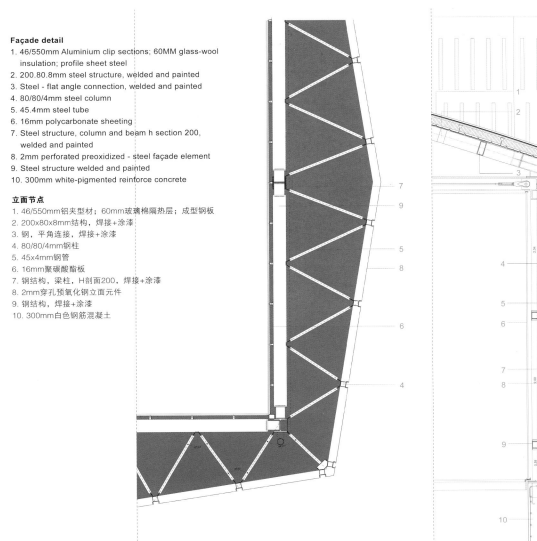
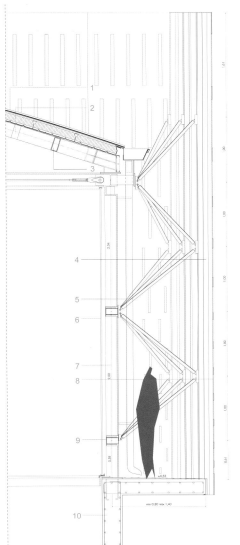
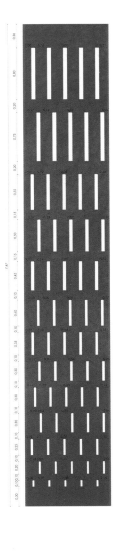

建筑实现了可持续设计。项目的预算有限，结构与施工都受到限制。建筑内使用空调的区域仅有4%。别墅内部通过双层立面实现了遮阳，立面的滤光效果与树枝树叶类似。马赛克上方的人工照明被缩减至最低。独立的耐候钢板会逐渐形成锈迹，保证了立面和屋顶的自然养护。两个通风层隔绝了无空调的室内空间，组成了拱顶。屋顶上的雨水和雪水被导流至环绕建筑的绿色坡地和中央天井，让18个世纪之前的罗马水管重新工作。

细部与材料

双层立面采用耐候钢和聚碳酸酯组成。抬高的木地板配有透明围栏，将所有考古区联系起来。屋顶上悬挂着一系列钢网，相当于原有别墅的墙面，既能展示马赛克，又能显示原始的布局。

在建筑外部，上半部分均匀地伫立在混凝土基座上，由折叠的穿孔立面环绕。立面保护建筑结构不受太阳辐射，为室内投下了分散的阴影，与四周杨树的比例和外观相互呼应。随着从基座向屋顶逐渐上升，金属外壳的图案密度逐渐减小，与周边的自然景观和天然缓坡相互融合，共同将建筑包围起来。

Broadcasting Place, Leeds
利兹广播大厦

Location/地点: Leeds, UK/英国，利兹
Architect/建筑师: Feilden Clegg Bradley Studios
Construction area/建筑面积: 13,562m²

Key materials: Façade – Corten steel
主要材料：立面——耐候钢

Overview
Broadcasting Place is a mixed use development close to Leeds city centre. Conceived as a public/private partnership for property group Downing and Leeds Metropolitan University, it provides approximately 110,000 square feet of new offices and teaching spaces together with 240 student residences in a landmark building rising to 23 storeys. A new Baptist Church completes the scheme on its northern edge. The building occupies a prominent location on the valley ridge running through the city centre. The site is bounded by roads to all sides and in particular to the south where the city's inner ring road is located in an open cutting. Two main structures wrap around the existing buildings and scale up towards the north eastern edge of the site. The tower element announces the building on Leeds' growing skyline.

Detail and Materials
The buildings are conceived as solid landscape forms which draw on Yorkshire's rich geological and sculptural heritage. Through this massive form, windows were conceived as the flow of water cascading through a rock formation. This design intent is reinforced by the selection of corten steel as a solid, sculptural and weathering material, constructed as a rain-screen façade. A key success in the scheme is the innovative approach to design of each elevation. The architects developed their own software programme to undertake a rigorous computational analysis of each small section of the building façades. The result is a varied appearance highly specific to this scheme, optimising daylight and reducing solar penetration. This is a key central Leeds location and a new public space linking key urban spaces forms a significant landscape element.

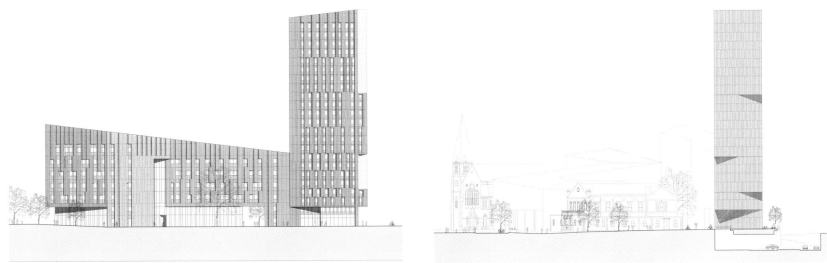

student room elevation 教室立面

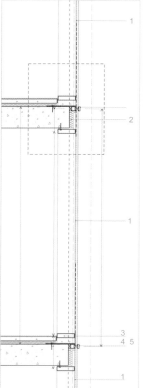
Section A-A' 剖面A-A'

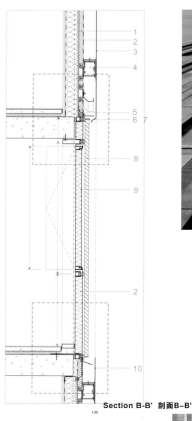
Section B-B' 剖面B-B'

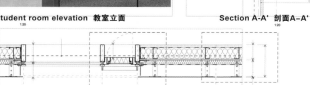
Section C-C' 剖面C-C'

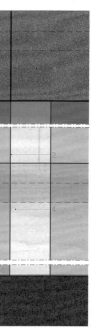
(Louvre – corten) 耐候钢百叶

Section A-A'
1. Glass vision panels
2. Fire and smoke stops
3. Aluminium frames
4/5. Pressure plates/pressure plate caps

剖面 A-A'
1. 玻璃可视板
2. 阻燃阻烟层
3. 铝框
4/5. 压板/压板盖

Section B-B'
1. Thermal insulation
2. Breather membranes
3. Weathering steel panels
4. Bracket
5. Perimetre seals and pressings
6/7. Pressure plates/pressure plate caps
8. Aluminium louvre panel
9. Opaque insulated aluminium framed vents
10. Aluminium panels

剖面 B-B'
1. 隔热层
2. 透气膜
3. 耐候钢板
4. 支架
5. 外围密封与冲压
6/7. 压板/压板盖
8. 铝百叶板
9. 铝绝缘框通风口
10. 铝板

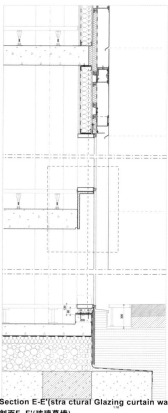

Ground floor elevation 一层立面
1. Alumnium frames   1. 铝框
2. Glass/glass corner details   2. 玻璃/玻璃角
3. Glass spandrel panels   3. 玻璃拱肩板
4. Aluminium panels   4. 铝板

Section E-E'(stra ctural Glazing curtain wall)
剖面E-E'(玻璃幕墙)

项目概况

广播大厦是一座紧邻利兹市中心的混合功能开发项目。项目归唐宁地产集团和利兹都会大学所有，在23层高的大厦中提供了约13,562平方米的办公、教学空间和240间学生宿舍。大厦北侧的浸礼会教堂也是项目的一部分。建筑在城市中心的山谷中占有重要的位置，四面都是街道，特别是南面还是城市内环的出口之一。两个主要结构将原有建筑包围起来，一直延伸至场地的东北边界。高大的塔楼使建筑在利兹的天际线上脱颖而出。

细部与材料

建筑被设计成稳健的景观形式，从约克郡丰富的地理环境和历史遗产中获得了许多参考。在宏伟的建筑上，窗户像水流一样奔涌而下。耐候钢的运用进一步突出了这种设计效果。这种厚重而具有雕塑感的材料被用作建筑的防雨立面。项目的主要成功之处在于各个立面的创新式设计。建筑师开发了专属的软件程序来对建筑立面的各个部分进行严谨的计算分析，最终形成了与建筑功能相呼应的特色外观，优化了日光照明并减少了日光渗透。项目是利兹市中心的地标，形成了引人注目的景观元素。

Detail section
1. Power pack
2. Diameter to cut in corten: 136mm
3. 140 camera
4. Soffit-mounted CCTV camera
   Honeywell HD4DX
   Powder-coated RAL 7024 black
5. Recessed light
6. Diameter to cut in corten: 175-180mm
7. Fitting

剖面节点
1. 电源组
2. 耐候钢上切口直径：136mm
3. 140摄像头
4. 底面安装闭路电视摄像头
   Honeywell HD4DX型
   粉末涂层RAL 7024黑色
5. 嵌灯
6. 耐候钢上切口直径：175~180mm
7. 固定零件

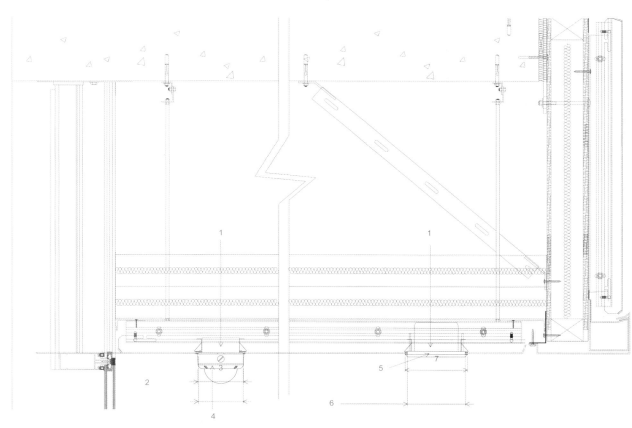

Administrative Centre Jesenice

耶塞奈斯行政中心

Location/地点: Jesenice, Slovenia/斯洛文尼亚, 耶塞奈斯
Architect/建筑师: Andrej Kalamar, Studio Kalamar
Area/面积: 2,980m²
Completion date/竣工时间: 2012

Key materials: Façade – corten steel
主要材料: 立面——耐候钢

Overview

Jesenice Administrative Centre is part of the Ministry of Public Administration project which aims to combine all state administrative functions (administrative unit, tax administration, surveying and mapping authority, inspectors and examination centre) in a single building. The Centre is situated at a central location of Jesenice, next to the municipality building, directly addressing a large roundabout with a steel worker memorial at its centre.

All offices are organised around an interior atrium with connecting bridges and natural illumination, the attractive shape and friendly atrium illumination works to remove a stigma too often associated with state offices. Atrium connects all the state functions into a single system, thus the visitor can perceive the state as a single function.

Building remains open into the environment while at the same time unobtrusively filtering everything that affects the wellbeing of its users. In its functioning, it puts users first: a satisfied user will be the best proof that including user comfort and providing a healthy working environment into sustainable investment behaviour is worth it.

Detail and Materials

The outer skin of the building establishes a dialogue with the steel-manufacturing industrial tradition of Jesenice. Even today, many of the houses in town are red-rusty, and the façade of the administrative centre is rusty indeed due to its Corten skin. A pulsing rhythm of façade openings creates a horizontal dynamics, establishing a dialogue with dynamics of the traffic. Natural light, natural ventilation, spontaneous transition of spaces are elements that create optimal working conditions for employees as well. By us-

ing efficient Corten panel shades, use of quality window systems and good thermal insulation the architect created a inert energy system which will ensure very low running costs. Façade system filters the aggressive contrast light into a softer diffuse light, thus encouraging use of natural light throughout the day.

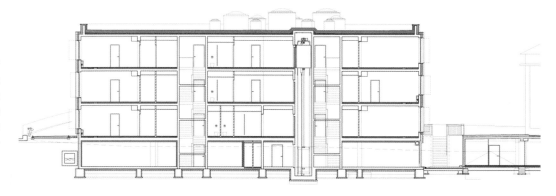

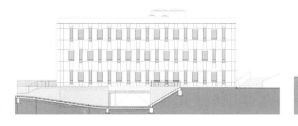
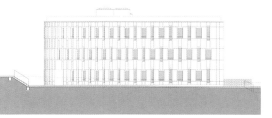

项目概况

耶塞奈斯行政中心是公共行政部项目的一部分，该项目目标是将各个州的行政功能（行政单位、税务管理、测绘机构、调研中心）结合在同一座大楼里。行政中心位于耶塞奈斯的市中心，紧邻市政大楼，临街呈现环岛造型，中心是一座钢铁工人纪念碑。

所有办公室都环绕室内中庭展开，以连接桥相连，配有自然采光。有趣的造型和友好而明亮的中庭让行政中心摆脱了传统的机关形象。中庭将所有政府功能连接成一个统一的系统，让访客将其看成一个完整的设施。

建筑既面向环境开放，又过滤了一切不利于楼内人员健康的因素。在功能上，它将使用者放在首位：满意的使用者才是健康舒适的工作环境的最好证明，项目的可持续投资物有所值。

细部与材料

建筑的表皮与耶塞奈斯的钢铁制造业传统形成了对话。直至今日，城中的许多住宅还有着锈迹斑斑的外观；耐候钢表皮让行政中心同样呈现了这种效果。建筑立面开口的韵律感让建筑活跃起来，与交通形成了联系。自然采光、自然通风、空间的自然过渡为员工提供着最佳的工作条件。高效的耐候钢板遮阳、高品质的窗户系统和良好的隔热层实现了被动式能源系统，大大降低了建筑的运营成本。立面系统过滤了强烈的光线，使其成为柔和的漫射光，从而推动了室内的日光运用。

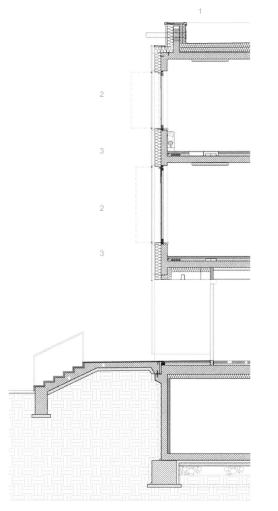

1. FLAT ROOF
- flexible poiolefin membrane
- thermal insulation - mineral wool
- vapor barrier
- construction

2. WINDOWS
- shutter- expanded metal, manual opening
- ventilated space
- outer screen
- double thermal glazing, Aluminium frame

3. FAÇADE
- expanded metal skin
- aluminium sub-construction grid
- ventilated space
- plaster
- thermal insulation-mineral wool
- construction

1. 平屋顶
- 弹性聚烯烃膜
- 隔热层，矿物棉
- 隔汽层
- 结构

2. 窗户
- 百叶窗，金属网+手动开窗
- 通风空间
- 外层遮阳屏
- 双层隔热玻璃，铝框

3. 立面
- 金属网表皮
- 铝制下部结构网格
- 通风空间
- 石膏板
- 隔热层，矿物棉
- 结构

Olympic Energy Centres
奥林匹克能源中心

Location/地点: London, UK/英国，伦敦
Architect/建筑师: John McAslan + Partners, Architect and Landscape Architect
Photos/摄影: Huffton + Crow Photography
Area/面积: 7,500m²
Completion date/竣工时间: 2012

Key materials: Façade – corten sheet Structure – SIPS substrate panels
主要材料：立面——耐候钢板
结构——结构保温板

Overview
The design of the Energy Centres at King's Yard and Stratford City responds strongly to a primary contextual issue. The Stratford City Energy Centre is clearly visible from Stratford town centre and housing estates east of the Olympic site, while the King's Yard Energy Centre dominates views from Victoria Park, Hackney Wick and parts of Bow.

The practice's response to site, and public perception of power generation, has been to develop an architecture for the Energy Centres that is highly distinct in both form and function – a strong 21st century industrial aesthetic featuring a 45m screened flue. Key features such as the tall ventilation flues and the flanking walls have been given special attention.

The Energy Centres deliver new operational benchmarks for power generation. They meet the client's plant, maintenance, durability and expansion requirements and have been delivered as turnkey facilities.

The fabric of the building utilises material with low embodied energy and high re-cycle content.

The SIPS substrate panels are constructed from OSB made from recycled timber.

Aluminium panelling is limited to only a plinth around the building for protection from vehicles. The EPDM waterproof base layer provides a low maintenance material. Factory bonding the EPDM to the SIPS panel reduces wastage and time on site

The plant in the building supplies hot water, chilled water and electrical power via a series of bio-mass and conventional gas fired boilers, gas turbines, absorption and electrical chillers.

Detail and Materials
The concept was to create a flexible sustainable skin using SIPS panels (made from compressed timber) with EPDM sheeting bonded to its surface. The building is then wrapped in a veil of corten sheet, which provide a unity to the surface and a scaling device which aligns with the adjacent substation building.

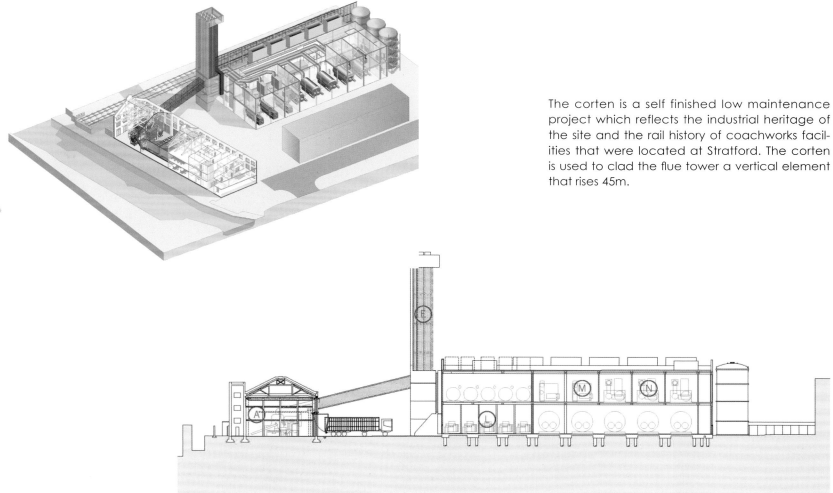

The corten is a self finished low maintenance project which reflects the industrial heritage of the site and the rail history of coachworks facilities that were located at Stratford. The corten is used to clad the flue tower a vertical element that rises 45m.

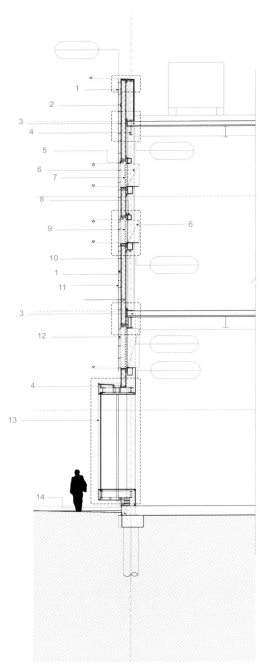

Section detail
1. Corten steel expanded metal mesh, fixed to SIPS panel using 75mm wide proprietary stainless steel brackets
2. SIPS panel fixed to primary steelwork. To be factory faced in EPDM membrane
3. Concrete floor slab
4. Line of primary steel frame
5. Lighting recess for fibre-optic lighting
6. Attenuation for ventilation as per process engineer's specification
7. Weather louvre recessed from external envelope, PPC finish, colour black, to run continuous outside of structural zone
8. Secondary steel for panel fixing to be advised by panel manufacturer
9. Weather louvre recessed from external envelope, PPC finish, colour black
10. Secondary steel to support louvre and attenuator to structural engineers details
11. Corten steel expanded metal panels to be approx 1200 mm (w) X 2400 mm (l) with horizontal orientation of long ways pitch. 6 no. LWP's per panel width. Horizontal and vertical spacing of brackets to be advised by mesh manufacturer
12. Anodized aluminium fixing supports for expanded metal, fixed through SIPS panel to secondary structure
13. Line of glazing
14. Concrete finish EC-01 to all vertical faces of ground floor slab

剖面节点
1. 耐候钢金属网、利用75mm宽专用不锈钢支架固定在结构保温板上
2. 固定在主钢结构上的结构保温板，上面覆有三元丙乙烯胶膜
3. 混凝土楼板
4. 主钢架线
5. 光导纤维照明嵌灯
6. 通风式（由工艺工程师说明）
7. 内嵌式百叶窗，氯化聚丙烯饰面，黑色，在结构区外连续排列
8. 辅助钢材，用于固定面板（根据面板制造商建议）
9. 内嵌式百叶窗，氯化聚丙烯饰面，黑色
10. 辅助钢材，用于固定面板支撑百叶窗（由结构工程师确定细节）
11. 1200mm（宽）x2400mm（长）耐候钢金属网，水平向倾斜。直接的水平和垂直间距根据金属网制造商的建议
12. 阳极氧化铝固定支架，用于安装金属网，穿过结构保温板安装到次级结构上
13. 玻璃线
14. 混凝土饰面EC-01，应用于所有一楼楼板的垂直面

项目概况

国王庭院和斯特拉特福德能源中心的设计都与场地有着密切的联系。从斯特拉特福德和奥林匹克场地东侧的居民区可以清晰地看到斯特拉特福德能源中心；而维多利亚公园、哈克尼威克和圣玛利亚教堂区的部分区域则都能看到国王庭院能源中心。

为了应对场地特征、打造发电厂的新形象，能源中心在造型和功能上都显得与众不同——呈现出强烈的21世纪工业美学，以45米高的丝网烟囱为特色。高大的通风烟囱和侧墙都独树一帜。

能源中心体现了发电行业的新标准。它们满足了客户的工厂、养护、耐久和扩建要求，是一座全包式设施。

建筑利用低能源含量的回收材料。结构保温板由回收木材制成的定向刨花板建成。铝板被使用在建筑四周的基座上，用来保护建筑不受机动车碰撞。三元乙丙胶防水基层是一种无需过多维护的材料。三元乙丙胶层与结构保温板在工厂组装，有效缩短了现场施工的时间，减少了废料的产生。

建筑内的工厂通过一系列生物量和常规天然气锅炉、燃气涡轮、电气制冷机组来提供热水、冷却水和电能。

Façade

1. Corten steel expanded metal mesh, fixed to SIPS panel using 75mm wide proprietary stainless steel brackets
2. Line of louvre behind mesh
3. Corten steel expanded metal panels to be approx 1200 mm (w) X 2400 mm (l) with horizontal orientation of long ways pitch. 6 no. LWP's per panel width. Horizontal and vertical spacing of brackets to be advised by mesh manufacturer
4. Line of primary steel frame
5. EPDM cladding
6. EPDM cladding on OSB fixed to top hat sections fixed to blockwork beyond
7. Glazing system

立面

1. 耐候钢金属网、利用75mm宽专用不锈钢支架固定在结构保温板上
2. 金属网后百叶窗线
3. 1200mm（宽）x2400mm（长）耐候钢金属网，水平向倾斜。直接的水平和垂直间距根据金属网制造商的建议
4. 主钢框线
5. 三元乙丙胶包层
6. 定向刨花板上的三元乙丙胶包层，固定在顶帽截面上，与后方的砌块墙固定
7. 玻璃幕墙系统

Façade detail

1. Line of expanded metal mesh
2. Line of secondary steelwork to form parapet and support external wall system and cladding to S.E. detail
3. Underside of steel beam
4. SIPS panel fixed to primary steelwork. To be factory faced in EPDM membrane
5. Weather louvre recessed from external envelope, PPC finish, colour black, to run continuous outside of structural zone
6. SIPS panel fixed to primary steelwork.
7. Line of 203 X 203 X 46 UC secondary steel to support louvre and attenuator, to structural engineer's detail
8. Line of primary steel frame
9. Line of concrete slab beyond
10. Anodized aluminium fixing supports for expanded metal, fixed through SIPS panel to secondary structure
11. EPDM cladding
12. EPDM cladding on OSB fixed to top hat sections fixed to blockwork beyond
13. Glazing system
14. Outer face of external cladding system/components to be flush

立面节点

1. 金属网线
2. 辅助钢结构线，形成护墙、支撑外墙系统和包层（由结构工程师确定细节）
3. 钢梁下面
4. 固定在主钢结构上的结构保温板，上面覆有三元丙乙烯胶膜
5. 内嵌式百叶窗，氧化聚丙烯饰面，黑色，在结构区外连续排列
6. 固定在主钢结构上的结构保温板
7. 203x203x46 UC辅助钢线，用于支撑百叶窗（由结构工程师确定细节）
8. 主钢框线
9. 混凝土泵板线
10. 阳极氧化铝固定支架，用于安装金属网，穿过结构保温板安装到次级结构上
11. 三元乙丙胶包层
12. 定向刨花板上的三元乙丙胶包层，固定在顶帽截面上，与后方的砌块墙固定
13. 玻璃幕墙系统
14. 外部包层结构外面/组件平齐

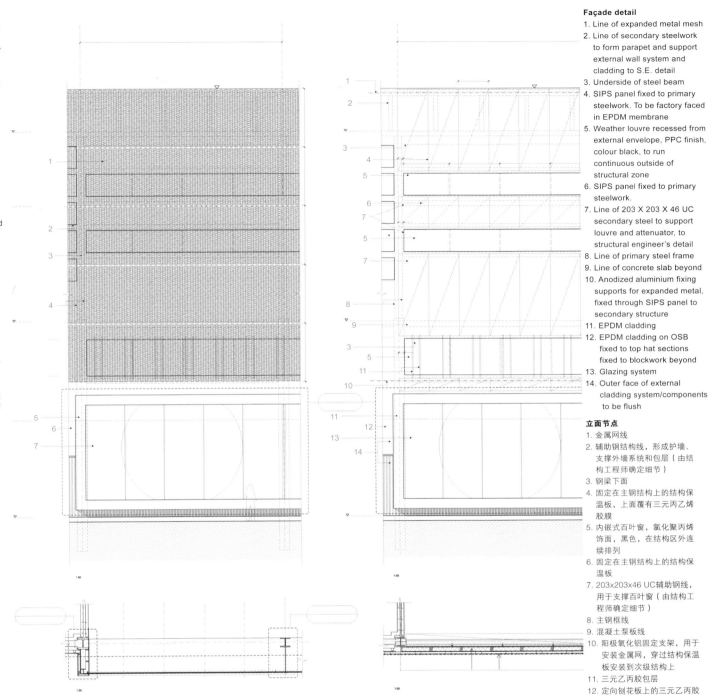

细部与材料

设计利用结构保温板（由压缩木板制成）和三元乙丙胶层打造了灵活的可持续表皮，然后将建筑包裹在一层耐候钢板内，既保证了表面的统一感，同时也使建筑与周边的变电大楼匹配起来。

耐候钢是这一种低维护材料，能反映场地的工业特征以及斯特拉特福德的制造传统。耐候钢被用在45米高的烟囱包层上。

Wyckoff Exchange
威科夫交流商店

Location/地点: Brooklyn, NY, USA/美国，纽约
Architect/建筑师: Andre Kikoski Architect
Photos/摄影: AKA, Renderings, Francis Dzikowski/ESTO
Construction area/建筑面积: 929m²
Completion date/竣工时间: 2011

Key materials: Façade – textured corten steel, light-gauge stainless steel
主要材料：立面——纹理耐候钢、轻不锈钢

Overview
The Wyckoff Exchange is an economical and adaptive re-use of two abandoned warehouses to create 10,000 square feet of retail and cultural space in Bushwick, Brooklyn. This place is marked by the strong traces of a gritty industrial past, and is rapidly transforming into a Centre of art and creativity.

The Wyckoff Exchange is dramatic and highly tactile. At once both simple and complex, the design uses a modest kit of parts of technology, material and light to create a sophisticated yet playful building, offering a fresh, bold, and different understanding of a retail venue.

Detail and Materials
The design solution offers an innovative response of what a modest retail building could be. They designed a trademark façade that responds to the place and purpose of this commission, paying careful attention to the resolution of formal and technical issues with extremely modest means. The choice of materials and technologies in this project is highly considered.

The design relies upon five pairs of motorised scissor doors/panels, whose technology is adapted from warehouses. The position of the façade panels creates a dynamic expression of purpose within: by day the panels fold up to create awnings for the stores and to shelter pedestrians; by night they fold down to secure the shops.

The panels consist of a steel frame that is clad in a double layered skin. The outer layer is textured corten steel. The inner layer is shimmering light-gauge stainless steel. Each layer is laser cut with a different gradient pattern. And each double-layer panel is internally illuminated by LED's.

While industrial in nature, the texture of the corten steel responds to the modulation of daylight. The sun set transforms the richly oxidised surface into a Rothko-like canvas. At night these simple materials and technologies create a contemporary glowing mural of light, 100 feet long, eighteen feet tall, and only two inches deep.

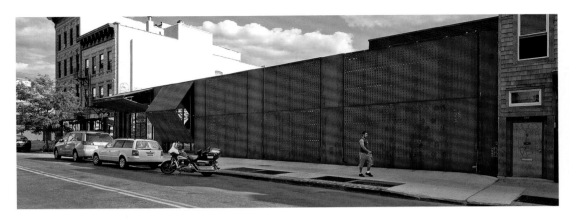

Closed façade
1. M-1 corten, TYP
2. Schweiss service door
3. M-1 corten security panel, varies per slope, VIF, TYP
4. Schweiss key access. Final location TBD. Refer specification

Open façade
1. 2'-0" allowance above storefront for signage band
2. M-1 corten steel without perforations
3. Signage band, TYP.
4. Schweiss bi-fold door w/metal mesh, M-1
5. Stone base, s-1 TYP.
6. Kawneer storefront TRIFAB VG 451-T framing system, M-5, TYP.
7. Schweiss bi-fold door key operated panel, TYP.

封闭的立面节点
1. M-1型耐候钢，TYP
2. Schweiss服务门
3. M-1型耐候钢安全板，斜率不同，VIF，TYP
4. Schweiss主入口（最终位置待定，参考详细说明）

打开的立面节点
1. 店面上方保留2'-0"净空用于安装招牌
2. 无孔M-1型耐候钢
3. 招牌带，TYP
4. Schweiss双折门w/金属网，M-1
5. 石底座，s-1型
6. Kawneer店面TRIFAB VG 451-T框架系统，M-5型
7. Schweiss双折门，带锁控制板，TYP

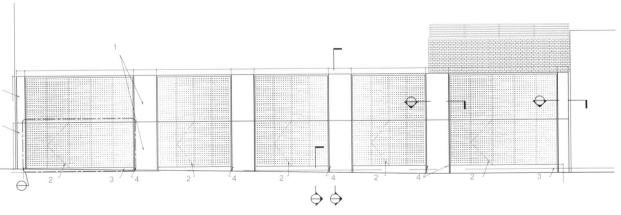
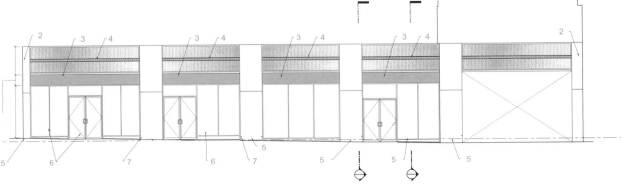

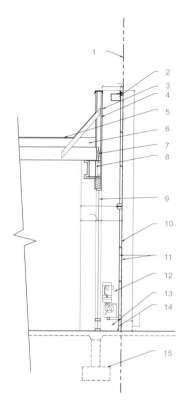 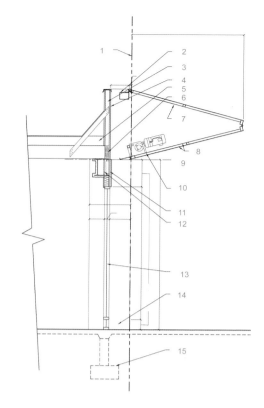 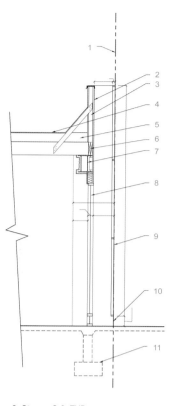

Closed façade
1. Property line
2. Structural header for schweiss bifold door, refer to structural engineer
3. Stucco, S-2, TYP.
4. New metal stud parapet
5. Sarnafil adhered roof system or approved equivalent
6. Existing metal docking and slab
7. Existing wood joist
8. Structural header salvaged from site, refer to structural engineer
9. Kawneer storefront trifab VG 451-T framing system, M-5, TYP.
10. Schweiss bi-fold door, SC-1, TYP.
11. Metal mesh, M-1
12. Schweiss bi-fold door motor, black painted w/ black straps
13. Stone base, S-1,TYP.
14. Metal kick, M-3, TYP.
15. Frost wall foundation footing, verify with structural engineer

Open façade
1. Property line
2. Structural header for schweiss bi-fold door, refer to structural engineer
3. New metal stud parapet
4. Stucco, S-2, TYP.
5. Existing metal docking and slab
6. Existing wood joist
7. Schweiss bi-fold door
8. Metal mesh, M-1
9. Adjust door angel to align bottom of schweiss bi-fold door with top of signage band
10. Schweiss bi-fold door motor, black painted w/ black straps
11. Signage band
12. Structural header salvaged from site, refer to structural engineer
13. Kawneer storefront trifab VG 451-T framing system, M-5, TYP.
14. Stone base, S-1,TYP.
15. Frost wall foundation footing, verify with structural engineer

Stucco section detail
1. Property line
2. Stucco, S-2, TYP.
3. New metal stud parapet
4. Sarnafil adhered roof system or approved equivalent
5. Existing metal docking and slab
6. Existing wood joist
7. Structural header salvaged from site, refer to structural engineer
8. Kawneer storefront trifab VG 451-T framing system, M-5, TYP.
9. Stucco, S-2, TYP.
10. Stone base, S-1,TYP.
11. Frost wall foundation footing, verify with structural engineer

封闭的立面
1. 界址线
2. Schweiss双折门结构顶（参考结构建筑师建议）
3. S-2型灰泥
4. 新金属立杆护墙
5. Sarnafil黏合屋顶系统或同等材料
6. 原有的金属平台和楼板
7. 原有的木龙骨
8. 原址上再利用的结构顶（参考结构建筑师建议）
9. Kawneer店面TRIFAB VG 451-T框架系统，M-5型
10. Schweiss双折门，SC-1型
11. 金属网，M-1型
12. Schweiss双折门电机，黑漆/黑糖糊色
13. 石底座，S-1型
14. 金属踢脚线，M-3型
15. 冻结壁基脚线（由结构工程师决定）

开放的立面
1. 界址线
2. Schweiss双折门结构顶（参考结构建筑师建议）
3. 新金属立杆护墙
4. S-2型灰泥
5. 原有的金属平台和楼板
6. 原有的木龙骨
7. Schweiss双折门
8. 金属网，M-1型
9. 调整门的角度，使双折门与顶部的招牌带相配合
10. Schweiss双折门电机，黑漆/黑糖糊色
11. 招牌带
12. 原址上再利用的结构顶（参考结构建筑师建议）
13. Kawneer店面TRIFAB VG 451-T框架系统，M-5型
14. 石底座，S-1型
15. 冻结壁基脚线（由结构工程师决定）

灰泥剖面节点
1. 界址线
2. S-2型灰泥
3. 新金属立杆护墙
4. Sarnafil黏合屋顶系统或同等材料
5. 原有的金属平台和楼板
6. 原有的木龙骨
7. 原址上再利用的结构顶（参考结构建筑师建议）
8. Kawneer店面TRIFAB VG 451-T框架系统，M-5型
9. S-2型灰泥
10. 石底座，S-1型
11. 冻结壁基脚线（由结构工程师决定）

项目概况

威科夫交流商店重新利用两座废弃的仓库在纽约布鲁克林打造了近930平方米的零售和文化空间。整个空间弥漫着坚忍不拔的工业气息，快速地成为了艺术和创意的中心。

威科夫交流商店十分引人注目。设计简约而不简单，利用零件组合技术、材料和光线打造了一座精致而充满趣味的建筑，形成了新鲜、大胆而与众不同的零售场所。

细部与材料

设计针对这座低调的零售建筑采用了创新的方案。他们设计了一个标志性立面，利用富有条理的技术方案和适度的方式实现了这一目标。项目所采用的材料和技术都经过了精挑细选。

设计以五对剪刀式门板为基础，沿用了仓库的技术设计。立面面板的位置帮助建筑实现了动态的外观：白天，面板向上折叠起来，成为商店的遮阳篷；夜晚，它们折下来保护商店。

面板由带有双层表皮的钢框组成，外层是耐候钢，内层是闪光的轻不锈钢。夜晚，这些简单的材料和技术会营造出发光的光影壁画，长30米，宽5.5米，而厚度只有5毫米。

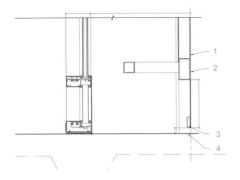

Section detail: bottom of Schweiss door
1. Corten, M-1, TYP.
2. Schweiss bi-folding door, TYP.
3. Additional strengthening metal behind corten, M-1; shape, size, spacing and attachment method per façade manuf.
4. Finished sidewalk

Details: door control panel & sidewalk spacing
1. G-1 insulated glazing refer schedule
2. Kawneer VG 451-T storefront system refer schedule
3. Sealant and backer rod, sealant to match storefront system, TYP.
4. Schweiss bi-fold door interior supporting truss
5. Kawneer break trim W/HP Flashing, M-2, TYP.
6. Schweiss bi-fold door, SC-1, TYP.
7. Metal mesh, M-1, TYP.
8. New 8'' CMU pier w/#5 verticals @ 8'' O.C in solid grouted cells, and DUR-0-wall every second course (TYP.)
9. #5 verticals (TYP.)
10. 1" Φ S.S thru bolt @ 24'' O.C (TYP.)
11. ~0'' Φ 'hilt' HY 20 ANCH, bolt w/4' min, embed @24'' O.C(TYP.)
12. Kawneer break trim W/HP Flashing, M-2, TYP.
13. ~P~8x17xcont.
14. Corten, M-1, pier
15. GC to anchor to CMU block, Centreed about hinged panel
16. Schweiss door control # PBTL-3 (7'Hx2.75Wx.35D)
17. Corten, M-1, hinged panel. W/Lock and concealed hinge

剖面节点：Schweiss门底部
1. M-1型耐候钢
2. Schweiss双折门
3. 耐候钢后面附加的加固金属件，M-1
（形状、尺寸、间距和安装方式由立面制造商确定）
4. 铺装人行道

节点：门控制面板与人行道间距
1. G-1隔热玻璃
2. Kawneer VG 451-T店面系统
3. 密封剂和泡沫棒，密封店面系统
4. Schweiss双折门内部支撑桁架
5. Kawneer切边/ HP防水板 M-2
6. Schweiss双折门，SC-1型
7. 金属网，M-1型
8. 新增8"CMU柱
9. #5垂直柱
10. 1" Φ S.S横穿带@ 24" O.C
11. 0" Φ 把手HY 20 ANCH, 螺栓w/4' min, 嵌入 @24" O.C (TYP.)
12. Kawneer切边/ HP防水板 M-2
13. ~P~8x17xxcont.
14. 耐候钢柱，M-1
15. GC固定在CMU砌块上，以可折面板为中心
16. Schweiss门控制# PBTL-3 (7'Hx2.75Wx.35D)
17. 可折耐候钢板，M-1/锁和暗式链条

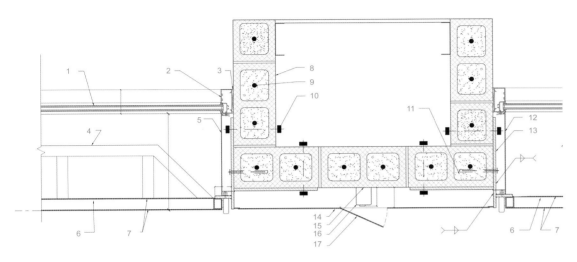

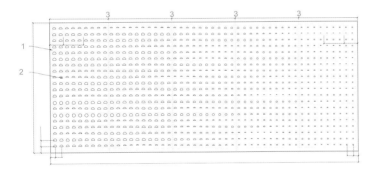

Front corten steel elevation detail
1. M-1, 18GA, stainless steel type 316 W/non-directional finish, mechanically fasten to back of frame, TYP., façade manuf to provide fastening specifications & tolerances
2. Schweiss bi-fold door frame behind
3. Of intermediate support, MTL façade seam and perforation spacing

Back stainless steel panel elevation detail
1. M-1, 18GA, stainless steel type 316 W/non-directional finish, mechanically fasten to back of frame, TYP., façade manuf to provide fastening specifications & tolerances
2. Schweiss bi-fold door frame in front
3. Of intermediate support, MTL façade seam and perforation spacing
4. Hinged panel 2'-9.25'' from CL of frame, VIF
5. Hinged panel for service of schweiss door equipment
6. Hardware & lockset to be coord, W/façade manuf.
7. 4.75'' O.C TYP.

正面耐候钢立面节点
1. M-1, 18GA不锈钢316W型/无直接饰面，机械安装在框架后（由立面制造商提供固定连接说明）
2. Schweiss双折门框后
3. 中间支承，MTL立面接缝和穿孔间隔

背面不锈钢板立面节点
1. M-1, 18GA不锈钢316W型/无直接饰面，机械安装在框架后（由立面制造商提供固定连接说明）
2. Schweiss双折门框前
3. 中间支承，MTL立面接缝和穿孔间隔
4. 可折面板2'-9.25"，VIF
5. Schweiss门设备的可折面板
6. 五金锁具安装位置
7. 4.75" O.C 类型

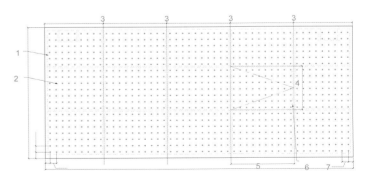

Looptecture Fukura
福良海啸控制预防中心

Location/地点: Minamiawaji, Japan/日本，福良
Architect/建筑师: Endo Shuhei
Built area/建筑面积: 310m²
Completion date/竣工时间: 2010

Key materials: Façade – weathering steel
Structure – steel
主要材料： 立面——耐候钢
结构——钢材

Overview
The function of this architecture is security and controlling all the floodgates located at port of Fukura, enlighten dangerous of the Tsunami for tourists, and use as a place of refuge in case of the Tsunami warning. For these reasons, ensure to keep the spaces of necessary and viewpoint for watching all over the port, also rational shape and structure to against of Tsunami and the drift came after disaster are necessary.

Main floor is placed higher level than assumed height of tsunami and opening the ground level floor allowed when waves passing through. Make the outside wall curved as efficient form to disperse stress. This form consists of 7.3m width belt (curved wall); continuous arcs are constructer and crossing by 6 different Centre of circle. Consequently, these arcs are closing at the same point both start and finish.

Detail and Materials
The steel that has a distinctive color is used to be the symbol of disaster prevention. As the characteristic of the steel is effectively utilized, the toughness of the steel is expressed as "restoring force" from disaster.

项目概况

建筑的功能是控制和预防福良港的防潮水闸，向游客们发布海啸预警以及在海啸预警中为游客提供避难所。因此，建筑必须保证必要的空间和视角来观察全港的景象，并且拥有可与海啸抗衡的造型与结构。

建筑的主楼面高于预期的海啸高度，底层是开放的，能让海浪穿过。建筑师将外墙设计成弧形，能有效地分散压力。整个建筑由7.3米宽的环形带（曲面墙）组成。6个同心圆形成了连续的弧面，完美地封闭了起来。

细部与材料

建筑采用了具有独特色彩的钢材来象征防灾预警。设计有效地利用了耐候钢的特色，用粗糙的表面象征着灾难的"恢复力"。

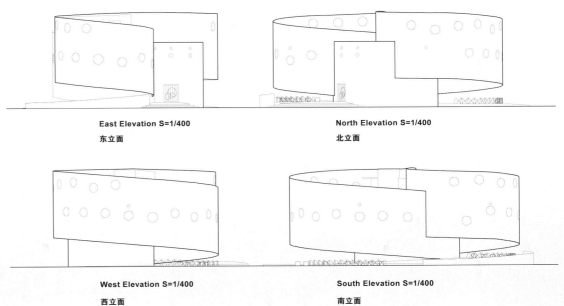

East Elevation S=1/400　东立面

North Elevation S=1/400　北立面

West Elevation S=1/400　西立面

South Elevation S=1/400　南立面

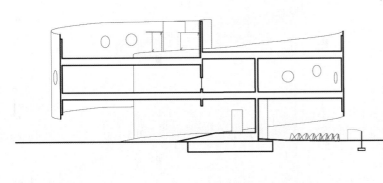

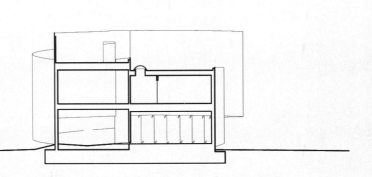

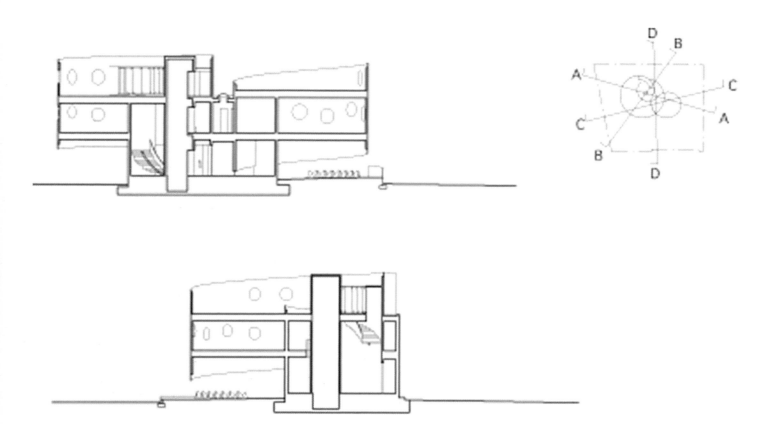

剖面节点
1. FB-125x9耐候钢板
2. T4.5耐候钢板
3. T50 Kawara
   T1.5防水膜
4. T9耐候钢板
   T35喷镀石棉
5. T15 50x50 Awajigawara砖
   T15砂浆
   T110承重板
   T25泡沫酸酯
6. H-350x175x7x11
   45x45@900
   T15 w105木板
7. T9耐候钢板
   T35喷镀石棉
8. 玻璃接缝
   玻璃混合处
9. 冲洗裸露骨料
10. T50 Kawara
11. H-250x125x6x9
    LGS
    T9.5石膏板，涂有水基漆
12. T9钢板
    钢油漆
13. T9 verding板
    T2.3镀锌网纹钢板
14. O-27.3x1.6钢油漆
15. T8曲面玻璃
    覆有防飞溅薄膜
16. 钢管D190xT12
    钢油漆
17. T15 50x50 Awajigawara砖
    T15砂浆
18. T4.5耐候钢
    T25泡沫酸酯
19. T9.5石膏板，涂有水基漆
20. T50冲洗裸露骨料
    T1.5防水膜
21. T15木板
22. 耐候金属网XS-33
23. Awaji砂岩块
24. T50 Kawara
    泥刀涂抹灰泥
25. T9耐候钢板
    T35石棉隔热材料
26. 多孔混凝土
27. 绿色屋顶
    T100土层
28. T9耐候钢板
    T25泡沫酸酯
29. 间接照明
30. Φ22耐候钢板
31. T12镀锌钢板

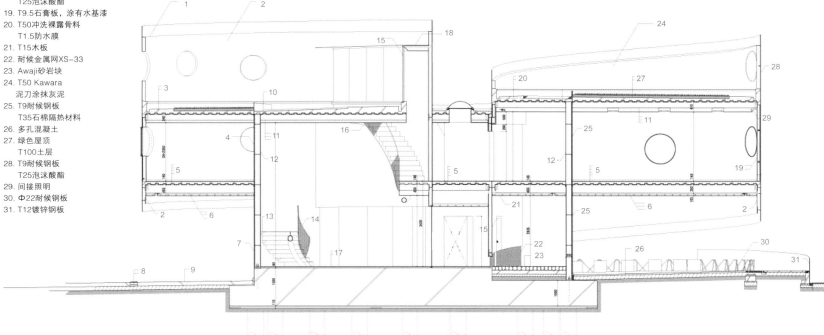

Section detail

1. FB-125x9 weathering steel plate
2. T4.5 weathering steel panel
3. T50 Kawara
   T1.5 water roofing sheet
4. T9 weathering steel panel
   T35 spraying rock wool
5. T15 50x50 Awajigawara tile
   T15 mortar
   T110 deck slab
   T25 foam urethane
6. H-350x175x7x11
   45x45@900
   T15 w105 wood panel
7. T9 weathering steel panel
   T35 spraying rock wool
8. Glass joint
   Glass mix
9. Exposing aggregate by washing
10. T50 Kawara
11. H-250x125x6x9
    LGS
    T9.5 plaster board water based paint
12. T9 steel plate
    Steel oil paint
13. T9 verding plate
    T2.3 galvanizing checker plate
14. O-27.3x1.6 steel oil paint
15. T8 curved glass
    Splash prevention film attach
16. Steel pipe D190xT12
    Steel oil paint
17. T15 50x50 Awajigawara tile
    T15 mortar
18. T4.5 weathering steel panel
    T25 foam urethane
19. T9.5 plaster board water based paint
20. T50 exposing aggregate by washing
    T1.5 water roofing sheet
21. T15 wood panel
22. Weathering expand metal XS-33
23. Awaji sandstone volume
24. T50 Kawara
    Mortar suppressed by Trowel
25. T9 weathering steel panel
    T35 rock wool insulating material
26. Porous concrete
27. Green roof
    T100 soil
28. T9 weathering steel panel
    T25 foam urethane
29. Indirect lighting
30. Ø22 weathering steel plate
31. T12 galvanising steel plate

Palmiry Museum Place of Memory
帕尔米瑞博物馆

Location/地点: Palmiry, Poland/波兰，帕尔米诺
Architect/建筑师: Zbigniew Wroński, Szczepan Wroński, Wojciech Conder/WXCA (www.wxca.pl)
Photos/摄影: Rafał Kłos (q-ph.blogspot.com)
Site area/占地面积: 8,738m²
Construction area/建筑面积: 1,145m²
Completion date/竣工时间: 2011.3

Key materials: Façade – corten(rusted)steel perforated panels fire-resistant glazing security glazing precast concrete panels reinforced with glass fiber
主要材料：立面——穿孔耐候钢板、耐火玻璃、安全玻璃、预制混凝土板（玻璃纤维加固）

Overview
The Palmiry Museum Place of Memory is a museum area complying with the pine-birch forest surrounding the cemetery. The museum building is a part of the Kampinos National Park, separated with glass and steel walls, covered with a green roof.

The exhibition space lies among trees – witnesses of past tragedies. The ascetic form of the building and the severity of the used materials form a background for the exhibition inside.

Form and Function
The interior and the exterior of the building form a whole. The used materials are a background for the exhibition. The architecture is not supposed to interfere with the emotional message of the personal items belonging to the victims buried at the cemetery. A clear simple division of functions has been suggested inside the building.

The entrance to the building is located at the ending of the main alley from the cemetery. The visitor enters the glazed space of the hall, with a reception and a cloakroom on the right, closed within a steel panel booth, so as not to restrict the inside cubature and the view from the outside. The end of the hall holds an education room and a corridor to the café, the restrooms, and the office area with a backstage.

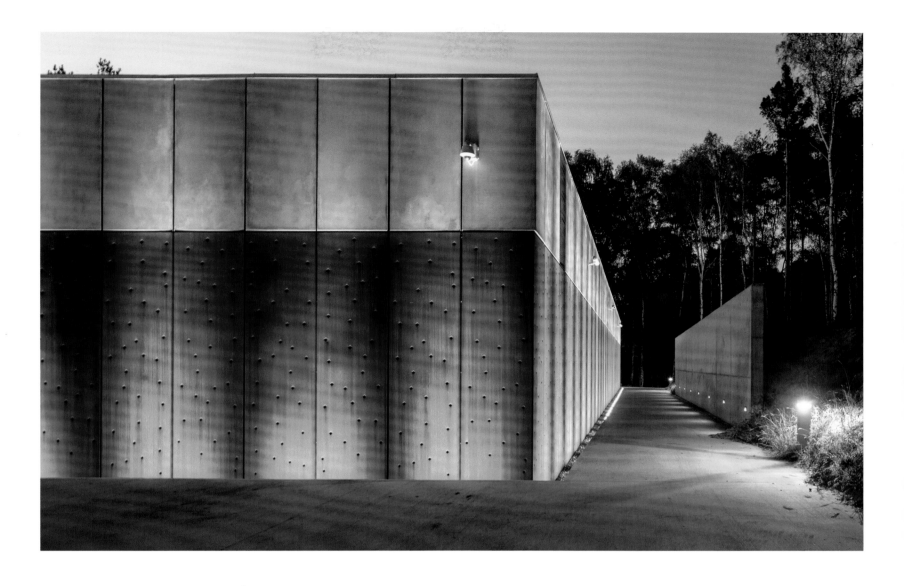

The rectangular exposition space is separated with a reinforced concrete wall. In the direction of the cemetery the exhibition space opens with a wide glassing, directing the eye of the visitor to the three white crosses at the end. The glass patios include pines. The patios enlight the exhibition area, separating particular parts of the exposition, and organizing the direction of the visit.

Detail and Materials
The applied materials match the importance of the place and its location. The severity of the materials such as concrete, glass, the corten steel, complies with the surroundings of the Kampinos National Park, and the concrete rough crosses of the cemetery-mausoleum.

The glass walls introduce the history of the place and nature, viewed from the outside they comply with the surroundings while the walls, fitted with the corten steel, with 2252 holes symbolizing bullets commemorates victims of the past.

The architects worked closely with the firm which produced the panels to determine the appropriate shape of the holes. There were several mockups done before they choose the proper ones. Corten is a material that changes with time. The grey steel surface covers with rust. The rust appears at first in places where water cumulates. First lines of rust were like tears of blood running from the bullet holes. Rain washes small parts of rust from the walls, that is why the building is surrounded by a gravel trough not to allow the concrete pavement to soil with rust.

项目概况

帕尔米瑞博物馆坐落在一片白桦林中,旁边是大片的纪念公墓。博物馆建筑是坎皮诺斯国家公园的一部分,拥有玻璃钢面墙和绿色屋顶。

展览空间坐落在树林中,见证着过去的惨剧(指二战时德军在此进行的屠杀)。建筑简单的造型和材料选择的沉重感为内部的展览奠定了基调。

形式与功能

建筑内外是一个整体,所选用的材料形成了展览的背景。建筑设计不会干涉遇难者遗物所传达出的信息,内部功能设计清晰简明。

建筑的入口位于公墓主通道的尽头。来访者进入玻璃大厅,左侧是前台和更衣室。更衣室由钢板隔间封闭,既不会破坏建筑的完整感,又保证了内部的隐私。大厅尽头是教室和通往咖啡厅、洗手间以及办公区的走廊。长方形的展览区通过混凝土墙隔开。在公墓方向,展览空间以玻璃幕墙向外开放,让参观者直接看到尽头的白十字架。玻璃天井内种有松树。天井为展览空间提供了自然采光,隔开了部分展览空间并且引导了参观的方向。

细部与材料

建筑所选的材料与其所在的地理位置和功能十分匹配。混凝土、玻璃、耐候钢等材料的沉重感与坎皮诺斯国家公园的环境和公墓的混凝土十字架有相互呼应的感觉。玻璃幕墙为建筑带来了空间的历史感和自然感。从外部看来,建筑完全融入了自然;耐候钢板上的2,250个孔洞代表着受害者所遭受的枪击。

建筑师与面板制造商紧密合作,确定了孔洞的合适造型。在最终确定之前,他们还做了一些模型。耐候钢会随着时间而变化,在灰色的钢表面形成红锈。锈迹最开始像是水滴堆积。一开始的锈迹看起来就像弹孔流下的血泪。雨水会将锈迹从墙面上冲刷下来,因此四周安装了碎石水槽,避免混凝土铺装被锈迹弄脏。

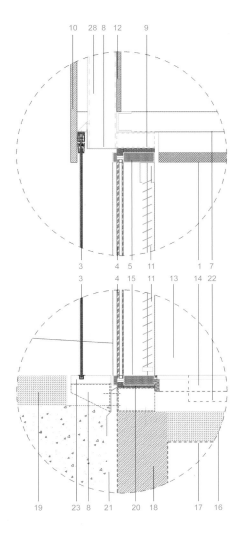
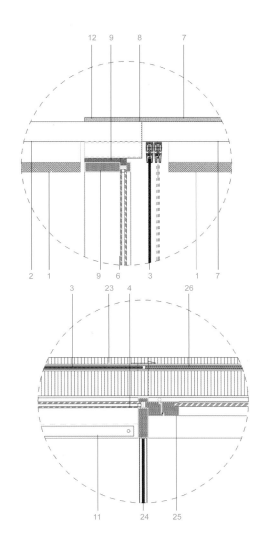
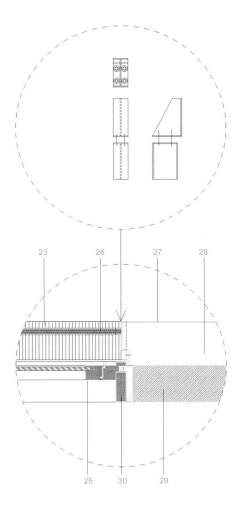

Detail
1. Expanded metal lath with steel substructure
2. Steel beam
3. Security glazing
4. Fire-resistant glazing
5. Aluminium façade upper frame
6. Flashing
7. Wind barrier membrane
8. Façade bracket
9. Mouting of the façade frame to the building steel construction
10. Concrete panel reinforced with glass fiber
11. Interior shutter
12. Fire-resistant oriented strand board
13. Durastone flooring, 2 cm
14. Screed with wire mesh reinforcement 6 cm
15. Aluminium façade lower frame
16. Foundation wall –watertight concrete 24 cm
17. Waterproofing
18. Concrete with fiber reinforcement
19. Concrete pavement 15 cm
20. Sealing fire-resistant layer
21. Styrofoam 5 cm
22. Styrofoam 10 cm
23. Stainless steel bridging grille
24. Inner glazing
25. Door frame in fire-resistant façade
26. Rubber seal-antydust protection
27. Korten (rust) steel perforated panel
28. Styrofoam 14 cm
29. Rainforced concrete wall 18 cm
30. Façade frame

节点
1. 金属网，配有下层钢结构
2. 钢梁
3. 安全玻璃
4. 耐火玻璃
5. 铝立面上层框架
6. 防水板
7. 防风膜
8. 立面支架
9. 立面框架安装在建筑钢结构上
10. 混凝土板，玻璃纤维加固
11. 内层百叶窗
12. 耐火定向刨花板
13. DURASTONE地板2cm
14. 钢丝网钢筋砂浆层6cm
15. 铝立面下层框架
16. 基础墙，不透水混凝土24cm
17. 防水层
18. 纤维强化混凝土
19. 混凝土铺面15cm
20. 密封耐火层
21. 泡沫聚苯乙烯5cm
22. 泡沫聚苯乙烯10cm
23. 不锈钢桥格栅
24. 内层玻璃
25. 耐火立面上的门框
26. 橡胶密封圈，防灰条
27. 穿孔耐候钢板
28. 泡沫聚苯乙烯14cm
29. 钢筋混凝土墙18cm
30. 立面框架

South West Institute of TAFE, Warrambool Campus – Stage 3
西南职业技术学院瓦拉姆布尔校区三期工程

Location/地点: Victoria, Australia/澳大利亚，维多利亚
Architect/建筑师: Lyons
Photos/摄影: Dianna Snape Photography
Site area/占地面积: 2,870m²

Key Materials: Façade – colorbond steel
主要材料：立面——镀铝锌钢板

Overview
This project is the third stage in the redevelopment of South West TAFE's Warrnambool Campus, and meditates between the civic space of the town, and the private interior space of the campus.

The primary circulation of the building is located on Timor Street, as an extension of the civic space – like the footpath repeated across three levels. This is combined with the concept of a self shading wall, which allows the street façade to be both shaded and transparent, making the circulation visible onto the street. The building is entered via a glazed wall formed underneath the major circulation stair, and leads to the student service centre, and to the rear campus courtyard space.

Detail and Materials
At an urban design scale, the rhythm of the façade geometry links up the street and the adjacent heritage buildings. Within the building, the internal circulation stairs are co-ordinated to the geometry of the façade, further amplifying the sensation of connecting the circulation with the street.

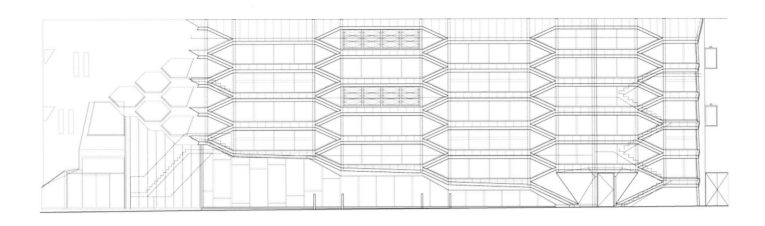

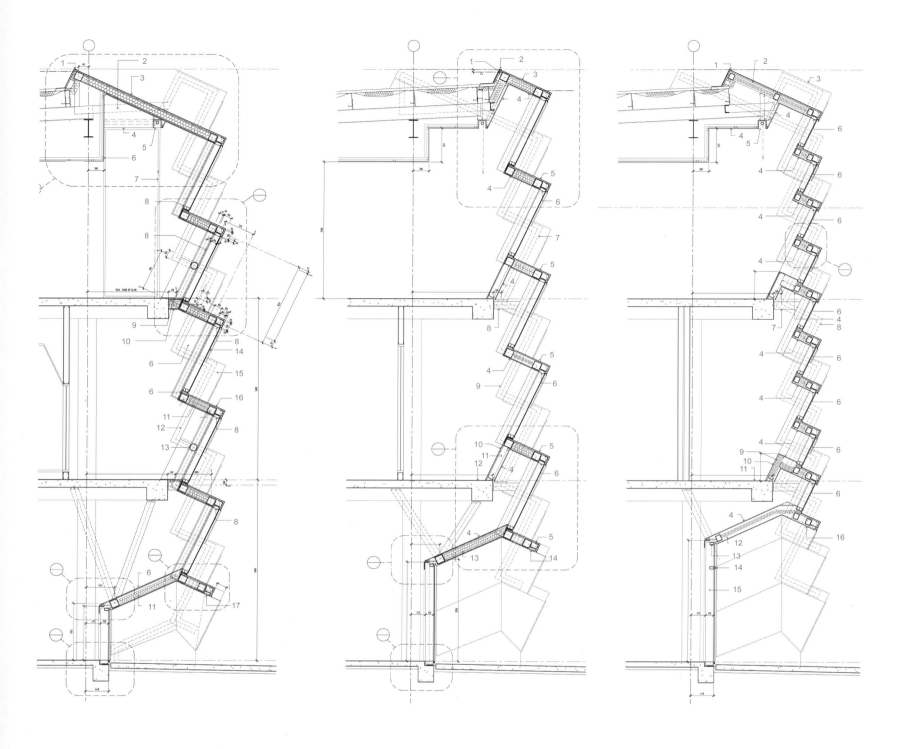

North façade 1
1. Colourbond capping
2. RJB beyond
3. Metal cladding finish as specified over 20mm PLY substrate
4. MDF pelmet continues around to bulkhead
5. Continuous MDF pelmet for roller blind:NOM, 150Hx130W
6. Plasterboard lining
7. Line of window
8. Window frame type A
9. Steel connection bracket refer structural DWG's.
10. Sound check plasterboard & insulation 32kg/m2
11. 150mm GIRT framing AT 450mm CTRS., refer to structural DWG's
12. Mullion 8/9 beyond shown dashed
13. 150 ø CHS handrail between mullions
14. 150mm SHS visible mullion
15. Metal cladding finish as specified beyond
16. Metal cladding finish as specified
17. Compressed sheet as specified soffit lining

North façade 2
1. 41 top span battens
2. Colourbond capping
3. Metal cladding finish as specified over 20mm PLY substrate
4. Plasterboard lining
5. Metal cladding finish as specified
6. Window frame type A
7. Mullion 6/7 beyond shown dashed
8. Steel connection bracket refer to structural DWG's.
9. Mullion 4/5 refer to structural DWG's
10. Vinyl wall lining over 20mm PLY substrate
11. 92mm steel stud framing
12. Vinyl floor finish
13. 150mm GIRT framing AT 450mm CTRS., refer to structural DWG's
14. Compressed sheet as specified soffit lining

Section through north façade junction B/W mullions
1. Colourbond capping
2. Metal cladding finish as specified over 20mm PLY substrate
3. Metal cladding finish as specified
4. Plasterboard lining
5. Continuous MDF pelmet for roller blind:NOM, 150Hx130W
6. Window frame type A
7. Steel connection bracket refer to structural DWG's.
8. Mullion 3 beyond shown dashed
9. 92mm steel stud framing
10. Vinyl wall lining over 20mm PLY substrate
11. Vinyl floor finish
12. 150mm GIRT framing AT 450mm CTRS., refer to structural DWG's
13. Window head
14. Window transom
15. Mullion
16. Compressed sheet as specified soffit lining

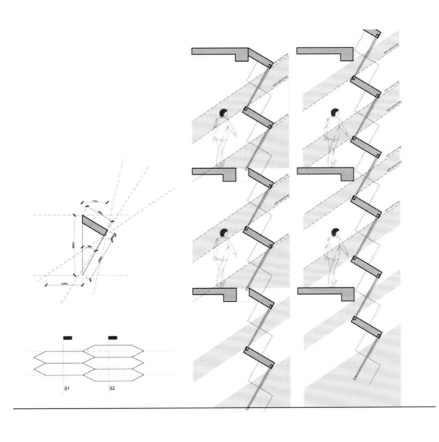

项目概况
项目是西南职业技术学院瓦拉姆布尔校区的三期工程，处在城市公共空间与校园内部空间之间，起到了过渡的作用。

建筑的主要内部交通流线位于帝汶街，作为城市空间的一部分，穿梭于三个楼层之中。交通流线与遮阳墙壁结合起来，让临街立面既通透又阴凉，使街上的行人也能看到建筑内部的交通。人们由位于主楼梯下方的玻璃幕墙进入，然后前往学生服务中心以及后方的校园庭院。

细部与材料
项目采用了城市设计的方式，富有韵律感的立面几何造型将街道与相邻的历史建筑联系起来。在建筑内部，交通流线和楼梯与立面造型相结合，进一步增强了建筑流线与街道之间的联系。

北立面1
1. 镀铝锌钢板顶盖
2. 柔性金属波纹膨胀节
3. 金属涂层，底层是20mm胶合板
4. 中密度纤维板窗帘盒
5. 连续的中密度纤维板百叶窗窗帘盒：NOM, 150H×130W
6. 石膏板衬线
7. 窗线
8. 窗框A型
9. 钢连接架（参考结构图纸）
10. 隔音石膏板和保温层32kg/m²
11. 150mm GIRT框安装在450mm CTRS上（参考结构图纸）
12. 窗框8/9在可见虚线后方
13. 150 Φ CHS扶手，位于窗框之间
14. 150mm SHS可见窗框
15. 金属涂层（详见设计说明）
16. 金属涂层（详见设计说明）
17. 压缩板，作为底衬

北立面2
1. 41顶跨板条
2. 镀铝锌钢板顶盖
3. 金属涂层，底层是20mm胶合板
4. 石膏板衬线
5. 金属涂层（详见设计说明）
6. 窗框A型
7. 窗框6/7在可见虚线后方
8. 钢连接架（参考结构图纸）
9. 窗框4/5（参考结构图纸）
10. 乙烯基墙衬，底层是20mm胶合板
11. 92mm钢立筋框架
12. 乙烯基地面铺装
13. 150mm GIRT框安装在450mm CTRS上（参考结构图纸）
14. 压缩板，作为底衬

透过北立面窗框的剖面
1. 镀铝锌钢板顶盖
2. 金属涂层，底层是20mm胶合板
3. 金属涂层（详见设计说明）
4. 石膏板衬线
5. 连续的中密度纤维板百叶窗窗帘盒：NOM, 150H×130W
6. 窗框A型
7. 钢连接架（参考结构图纸）
8. 窗框3在可见虚线后方
9. 92mm钢立筋框架
8. 钢连接架（参考结构图纸）
10. 乙烯基墙衬，底层是20mm胶合板
11. 乙烯基地面铺装
12. 150mm GIRT框安装在450mm CTRS上（参考结构图纸）

7. 钢连接架（参考结构图纸）
8. 窗框3在可见虚线后方
9. 92mm钢立筋框架
8. 钢连接架（参考结构图纸）
10. 乙烯基墙衬，底层是20mm胶合板
11. 乙烯基地面铺装
12. 150mm GIRT框安装在450mm CTRS上（参考结构图纸）
13. 窗头
14. 窗顶
15. 窗框
16. 压缩板，作为底衬

Stockholmsmässan AE-hallen
斯德哥尔摩会展中心AE大厅

Location/地点: Stockholm, Sweden/瑞典，斯德哥尔摩
Architect/建筑师: Alessandro Ripellino Arkitekter (formerly Rosenbergs Arkitekter)
Photos/摄影: Petter Karlberg
Site area/占地面积: 22,000m²
Construction area/建筑面积: 11,000m²
Completion date/竣工时间: 2010
Award name/获奖: Stålbyggnadspriset (Steel construction award) 2011 Honorary mention/ 2011 "钢结构建筑奖" 提名

Key materials: Façade – screen by perforated steel panel, glass
Structure – steel
主要材料: 立面——穿孔钢板幕墙、玻璃
结构——钢

Overview

The Stockholmsmässan International Fairs in Älvsjö is one of the world's leading organisers of meetings with 10,000 exhibitors and 1.5 million visitors annually. Rosenbergs architects have carried out a number of projects at the site since 1998. The latest addition to the premises is a new multifunctional space intended for conferences and large exhibitions, the AE-hall, which has now become one of the main venues of many fairs, such as the recent Stockholm Furniture Fair. The hall is connected to the existing complex by a gallery which has also been completely remediated, with new mirror-like ceilings and greenery walls.

From the glass façades of the AE-gallery, a pond with fountains is visible along almost the full 100 meters of the building. In the exhibition hall there are three atria where Himalayan birches are planted. All visible roofs are covered in a variety of sedum plants.

Detail and Materials

The whole building is wrapped in a façade screen; a giant metallic basket with an embossing effect which is intensified by the lighting fixtures integrated in the steel structure. This screen is made of about 1500 partly perforated steel panels. The walls of the pool structure are clad in expanded metal screens with integrated sliding gates, revealing entrances to the subterranean parking garage below.

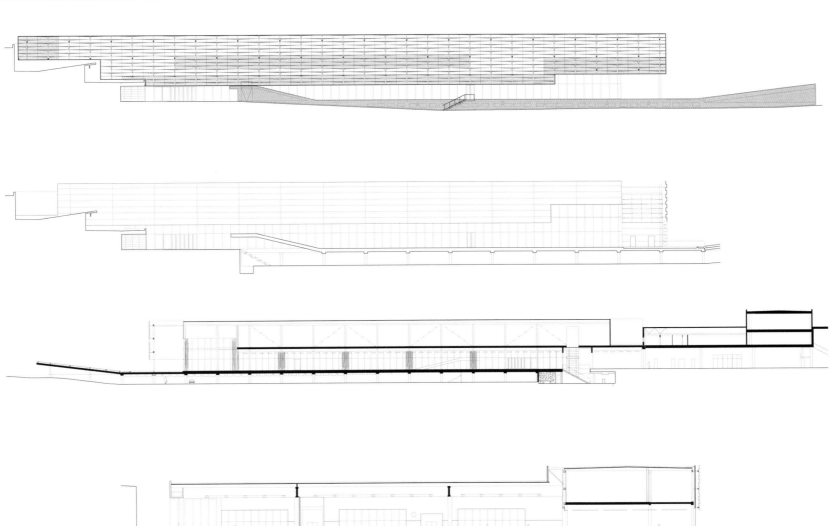

项目概况

斯德哥尔摩会展中心作为世界主要展会举办者之一，每年可接待10,000名参展商和150万名游客。自1998年开始，Rosenbergs事务所在这里打造了数个项目，最新的即为用于举办会议和大型展览的AE厅。这里已成为举办各种展览的主要场所，如斯德哥尔摩家具展等。AE厅与原有建筑通过走廊连结起来，走廊同样进行了翻新，配有镜面天花板和绿植墙壁。

透过走廊的玻璃外观可以看到沿建筑而打造的带有喷泉的泳池，泳池长近100米，几乎与建筑等长。大厅内的三个中庭内种植着喜马拉雅桦树。所有可见的屋顶上都种植着各种景天属植物。

细部与材料

整个建筑包裹在一个幕墙系统中：巨大的"金属篮子"在灯光的照射下呈现出浮雕般的效果。幕墙由1,500块穿孔钢板拼接而成。泳池空间外观墙壁采用金属网构造，带有滑动门，有道路直通地下停车场的入口。

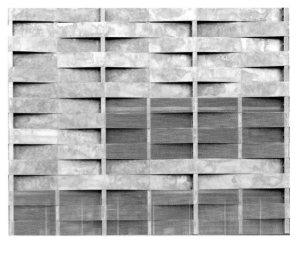

Vertical detail south façade and pond
1. Sandwich panel
2. 70mm insulation
3. 200x200x10mm galvanised steel console
4. Main beam
5. Convector
6. Thermal brake
7. Convector
8. Glass façade
9. Aluminium grating
10. Water jet
11. Bitumen water proofing
12. Steel sections 40x40x3mm
13. Water surface
14. Light fixture
15. Galvanised steel gutter
16. Steel tubes
17. Steel sections 40x40x3mm
18. Expanded metal cladding
19. Down spout
20. Outlet from pond

南立面与泳池的垂直节点
1. 夹层板
2. 7mm隔热层
3. 200x200x10mm镀锌钢架
4. 主梁
5. 对流散热器
6. 热闸
7. 换流器
8. 玻璃立面
9. 铝格栅
10. 喷水管
11. 沥青防水层
12. 型钢40x40x3mm
13. 水面
14. 灯具
15. 镀锌钢格栅
16. 钢管
17. 型钢40x40x30mm
18. 金属网包层
19. 下水管
20. 水池排放口

Section
1. Sandwich panel
2. Light fixture
3. Façade screen 3mm galvanised sheet metal
4. Space for steel grating service deck
5. 200x200x10 mm galvanised steel tube
6. Water surface
7. Light fixture norka bern
8. Expanded metal cladding

Outer elevation
1. Light fixtures
2. Perforated panels
3. Light fixtures along pond
4. Expanded metal cladding

Inner elevation
1. Sandwich panel
2. Recessed trim
3. Screen structure: 200x100x100mm steel tubes
   200x200x100mm steel tubes
   Connected to internal load bearing structure
4. Space for steel grating
   Service deck
5. Recessed flashing
6. Water surface
7. pond

剖面
1. 夹层板
2. 灯具
3. 立面幕墙，3mm镀锌金属板
4. 钢格栅服务平台空间
5. 200x200x10mm镀锌钢管
6. 水面
7. norka bern灯具
8. 金属网包层

外立面
1. 灯具
2. 穿孔板
3. 水池边灯具
4. 金属网包层

内立面
1. 夹层板
2. 嵌壁式切边
3. 幕墙结构：200x100x100mm钢管
   200x200x100mm钢管
   与内部承重结构相连
4. 钢格栅服务平台空间
5. 嵌入式防水板
6. 水面
7. 水池

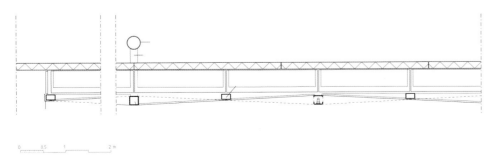

Terminal in Ven
芬岛航运站

Location/地点: Landskrona, Sweden/瑞典，兰斯克鲁纳
Architect/建筑师: FOJAB arkitekter
Photos/摄影: Ole Jais
Gross floor area/总楼面面积: 1,000m²
Completion date/竣工时间: 2012

Key materials: Façade – galvanised steel sheet, oiled oak panel
主要材料：立面——镀锌钢板、涂油橡木板

Overview
The Venterminal in Landskrona is designed and built for the people traveling to and from the Ven island, located on the Öresund channel between Sweden and Denmark. The design concept of the terminal is based on the idea of travelling, which is about expectations and waiting. The terminal consists of a simple core containing all serving functions. This core is surrounded with a layer of waiting areas, consisting of weather protected spaces on the quay for people to stay and pass the time, both indoors and outdoors.

The site requirements were defined and the new docks were already established on the quay. The buildings are carefully placed to complement to the dominating 17th century urban structure of Landskrona, yet set to operate as a freestanding pavilion at the vast openness of the Skeppsbron area. Given these site conditions, the buildings were designed not to have a backside; instead all façades were designed as frontages characterised by openness, provided with entrances.

Detail and Materials
The building volume is clad with sheets of galvanised steel that withstands the harsh conditions of the harbor. The exterior walls and ceilings of these spaces are covered with standing bars of oiled oak. Indoors, the wooden surfaces continue as grooved oak panels. On the outside waiting areas, the ground is covered with anthracite gray, precast concrete slabs. These choices of materials, together with floor heating and a coherent lighting strategy creates a continuous space, dissolving the boundary between indoor and outdoor.

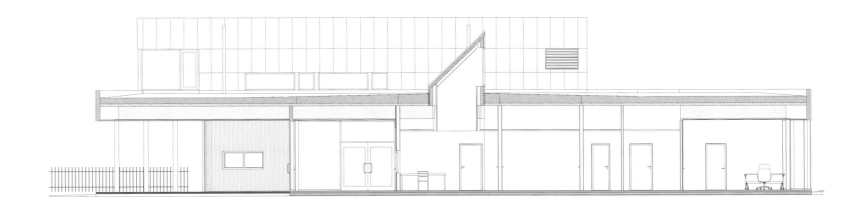

项目概况

兰斯克鲁纳的芬岛航运站专为来往于芬岛的人们设计，位于瑞典和丹麦之间的厄勒海峡。建筑的设计概念以旅行为基础，充满了期待与等候。建筑的简单内核内设置着所有服务功能。内核外环绕着一层候船区，外面的挡雨屋顶为排队、候船的人提供了保护。

场地要求十分明确，新的登船码头也建设完毕。建筑与兰斯克鲁纳17世纪的城市建筑风格十分相称，同时也以独立的形式进行运营。由于场地的要求，建筑并没有设计背面，所有里面都呈开放的姿态，均设有出入口。

细部与材料

建筑体块被镀锌钢板所包围，以应对海港严酷的环境。建筑的外墙和天花板上都覆盖着涂油橡木竖条。在室内，木制表面体现为沟槽式橡木板。露天候船区的地面上覆盖着煤灰色的预制混凝土板。这些材料的选择与地热和相干光源照明策略共同打造了整体环境，解决了室内外的界限问题。

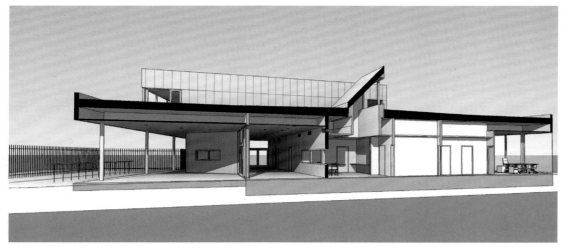

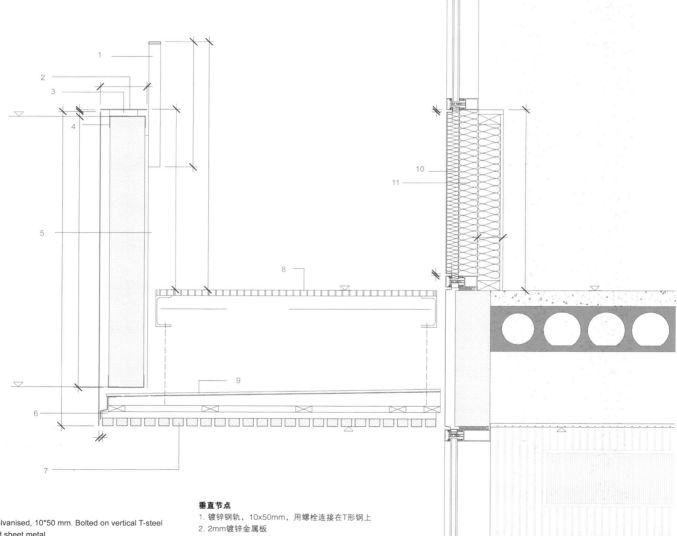

Vertical detail
1. Fat steel rail, galvanised, 10*50 mm. Bolted on vertical T-steel
2. 2mm galvanised sheet metal
3. 40 stainless hat channel c600 mm
4. Sheet metal mounting cased over sandwich element
5. 2 galvanised sheet metal
   40 stainless hat channel c600 mm
   150 sandwich element PIR
6. Sill flashing behind metal cladding
7. 34*45 oiled oak laths c60 mm
8. Galvanised flat bar grating
9. Sloping fiber cement siding
10. Tight filling, colored same as operable parts
11. Insulation between stud profiles

垂直节点
1. 镀锌钢轨，10x50mm，用螺栓连接在T形钢上
2. 2mm镀锌金属板
3. 40槽型不锈钢c600mm
4. 金属板，安装在夹层板外层
5. 2镀锌钢板
   40槽型不锈钢c600mm
   150夹层板PIR
6. 金属包层后的窗台挡雨板
7. 34x45涂油橡木条c60mm
8. 镀锌条钢格栅
9. 倾斜纤维水泥护壁板
10. 密封填充，色彩与其他部分相同
11. 螺柱之间的绝缘

New City School, Frederikshavn
腓特烈港新城市学校

Location/地点: Frederikshavn, Denmark/丹麦, 腓特烈港
Architect/建筑师: Arkitema Architects
Photos/摄影: Kontraframe and Arkitema Architects
Built area/建筑面积: 13,500m²
Completion date/竣工时间: 2012

Key materials: Façade – perforated galvanised steel
主要材料: 立面——穿孔镀锌钢板

Overview

With its star-like shape, all class rooms in the school can be placed at a façade with direct access to light and fresh air. At the same time the star shape gives an optimum background for natural ventilation and easy access to the playground. The school has two floors for the older students and one floor for the smallest children. On top of the lower level the architects have placed a roof terrace with 750 m2 of solar panels, making it possible for the school to reach low energy class 2020 according to the Danish building code (roughly equivalent to LEED gold). The central square of the school has naturally been placed in the centre of the star. This is the meeting point of all zones – a learning space that is enhanced by a big sculptural stairway. The square is the central shared space that can be accessed from all departments. It is the dynamo of the school where teaching, learning and social activities melt in an inspiring atmosphere. Each point of the star makes up a department for two grades - each with its own identity and furnishings, designed for exactly the age group it houses.

Detail and Materials

The New City School has galvanised steel on the entire façade. At the end of the building the steel is perforated. That way, different pictures of the landscape are created in the façade. At the end of each house end the façades has super graphic integrated in the panels. This graphic corresponds to the continent the house end points to – to the north the house end graphic has a theme reflecting the north pole and so on.

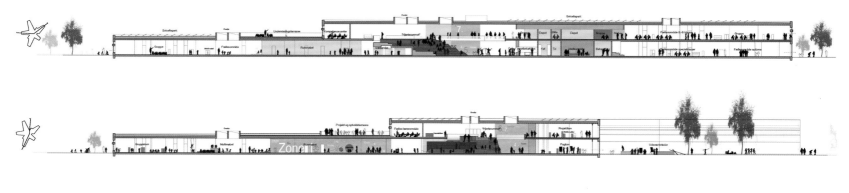

Façade elevation
1. Ribbon windows with partial automatic opening / night cooling
2. Screens integrated into the sheath of galvanised sheet
3. Galvanised steel plate mounted on steel rail
4. Glass / Aluminium party by common zone / graphic printing on glass
5. Relief of the façade cassette with integrated automatic shading
6. Wood / Aluminium lot of classes / graphic printing on glass
7. Exit from every room

立面节点
1. 带状窗，带有半自动开窗/夜间冷却效果
2. 与镀锌板外壳结合的幕墙
3. 安装在钢轨上的镀锌钢板
4. 公共区域的玻璃/铝元件/玻璃上印有图形
5. 立面盒结构上的浮雕，带有自主遮阳效果
6. 教室的木/铝区域/玻璃上印有图形
7. 房间出口

项目概况

星星造型让学校的所有教室都能享有光照和新鲜空气，同时也为自然通风和便捷地出入操场提供了条件。学校有两层楼给高年级学生，一层楼给低年级学生。建筑师在底层结构的顶部设计了一个屋顶平台，上面设有750平方米的太阳能电池板，让学校达到了丹麦建筑规定的低能源等级2020（相当于LEED绿色建筑金奖水平）。学校的中央广场被自然地设置在星星中央，是各个区域的集会中心，宏伟的楼梯提升了学习空间的感觉。人们可以从各个部门进入广场，广场相当于学校的发动机，将教学与社交活动融合在一起。星星的各个角都设有两个年级，各有特色，与学生的年龄相匹配。

细部与材料

新城市学校的立面全部采用镀锌钢板。在建筑的一端，钢板是穿孔的，从而形成了不同的风景画。在各个结构的尽头，钢板立面上都有巨大的平面图形。各个图形都与结构的朝向一一对应。例如，朝北结构上的图形就反映了北极的主题。

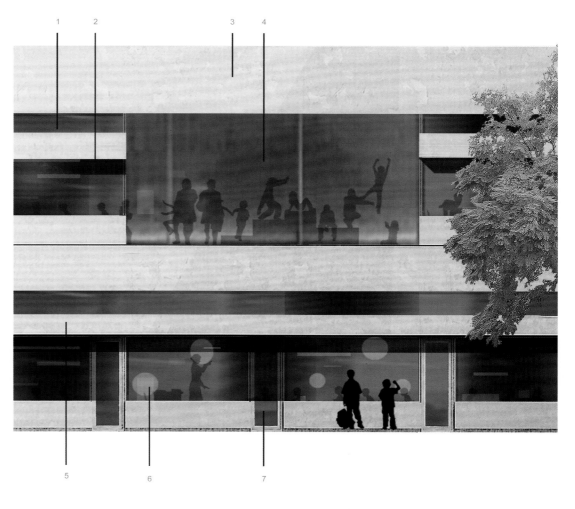

Prince Housing Sales Centre
太子馥接待中心

Location/地点: Taiwan, China/中国，台湾
Architect/建筑师: lab Modus/仲向国际设计顾问有限公司
Completion date/竣工时间: 2013
Photos/摄影: Wu, chih-ming
Area/面积: 1, 350m²

Key materials: Perforated metal louvers, granite, oak veneer, stainless steel
主要材料：冲孔镀锌钢板、南非灰花岗石、橡木皮、不锈钢发丝面

Overview
The site is located in a newly developed area near Taiwan High Speed Rail's Taoyuan Station. The scheme is composed of a two-storey rectangular volumes and a deformed double layered perforated metal louvers. Conceptualised of a "dancing box", the architects intend to give dynamic rhymes in this open field site. These dynamics present significant challenges for them working to shape a built environment that will meet commercial and perceptive needs both today and future. Giving maximum presence to the context, the massing of project applies advantage of the openness of the site to create a continuous translucency to all dimensions of the project. The irregular openings trimmed out from metal louver façade are to provide air and views to the corresponded interior spaces. Therefore, the interior can adapt the filtered sunlight and air into spaces.

Detail and Materials
Simplicity is beauty. The architects decide to abandon complicated decorations and start with simple geometry. Through the transformation and distortion of geometric elements, various spacial elements are created. The building volume expands from the optimal square volume and is softened by distortion and airy elements. Thus, perforated galvanised steel panel becomes the first choice in exterior design. It introduces light and air into the interior space. However, the angles and roughness of steel is a great challenge. The architects wish to integrate

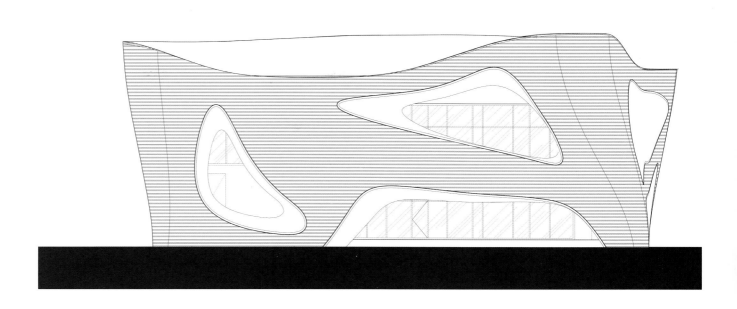

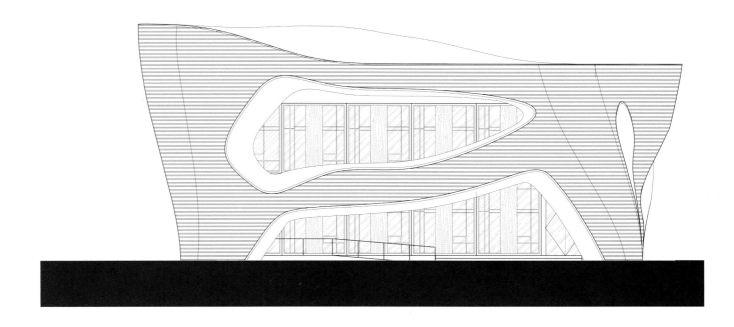

dynamic curves, so they combine woodwork and steel. With the help of digital parametrics, the computer calculates sizes and positions of each piece of materials, turning rough metal into soft waves. In daytime, the building looks like a shining dancing box. Sunlight cast on the white envelop and penetrate through punched holes into the multi-layer wood box. As a picture frame, the wood volume frames various living styles and presents them in an indirect way, symbolising the relationship between nature and human beings. At night, it is a completely different feeling. The shining metal turns to be darkened while the internal wood box becomes the leading role. It transcends the perforated envelope and presents an effect of hollow-out lantern, illuminating our lives.

From white dancing box, through air layer and into wood box, one steps into the interior space. The interior space launches a dialogue with time through its curves. Walking in the space is like walking in a time corridor. Light penetrates through windows to guide people to the brightness. In the discussion space, the architects select textured wood to match people's tactile sense. Details are carefully considered: the window opening sizes are maximised to express the additional value of multi-layered structure. The perforated steel panels keep off direct sunlight, leaving a pleasant visual experience for people. Sunlight, air and wood together convey a sophisticated touch through simple geometric elements.

项目概况

项目位于台湾高铁桃园站附近一块新开发的场地上，包含一个两层高的长方体结构和一个扭曲的双层冲孔金属框架。建筑师以"舞动的盒子"为设计理念，试图为开放的场地带来动感。设计对建筑师是巨大的挑战，他们必须打造一个既符合商业需求，又符合感官需求，既能体现现在，又能体现未来的建筑。未来最大限度地展现周边环境，项目在各个方向都采用了连续的通透设计。冲孔镀锌钢板上的不规则的开口为室内空间提供了气流和景观，使其获得了自然采光和新鲜空气。

细部与材料

生命因简单而喜悦，建筑师决定抛弃繁复的装饰，从几何的转化开始，象征万物的原型，将其付诸不同的空间元素，举凡扭转、透空、分割、拉伸，由最符合人使用的方型量体开始延展，使之柔软，于是借由的扭转设计，并套进空气感元素。冲孔镀锌钢板便成为外观材质的首选，让光影能够顺势引进室内，解读空气感，但钢材的坚实棱角成了这次设计中的一大挑战。建筑师们希望将圆润活力的曲线与之融合，于是结合了木作与钢料，利用计算机参数的运算技术，将每一片材质计算出属于它的位置和大小，将坚实的金属转变为柔软的粼粼波光，从日间来看，犹如随时在翩翩舞动的闪耀光盒。阳光照射在纯白色的建筑表面上，透过每一个冲孔穿透进复层的木盒子，木盒子量体像是画框一样，层层叠叠的堆栈出人与人生活的样貌，用另一种间接的方式呈现，象征自然与人的关系，我们与之共生，构筑想象未来生活的美好；而夜间的效果，又是另一种截然不同的感觉，金属从亮面转为建物的暗面，内层的木盒子成为外观量体的主角透过冲孔的外衣，画出宛如镂空灯笼的效果，照亮生活的点点滴滴。

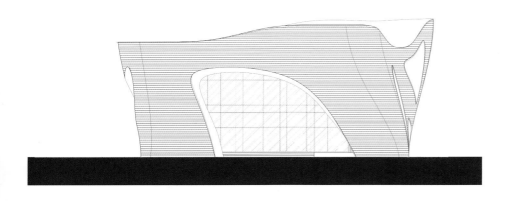
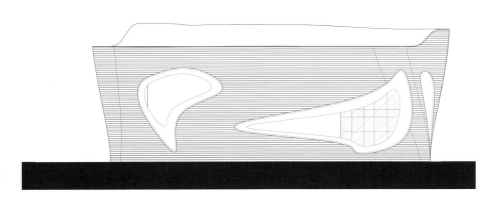

从舞动的白盒子，穿过空气层进入木盒子，开始走进室内属于"时空"的空间，建筑内部借由曲线与时空对话，走在空间中仿佛走进时空的河流，视觉的光影线顺流而下，汇集指向建筑外观的窗，象征光的引导，随之带领迎向光明。而停留时间较久的讨论空间，他们用最贴近人触感舒适的木材质，并注重每一个小空间的细节，将开窗放到最大，复层建筑的附加价值在此体现，半遮掩的冲孔钢板，挡住了强烈阳光的直射，留下的人与自然接近的视觉享受，阳光、空气、木，用简单的几何演绎每个元素间的极致触感。

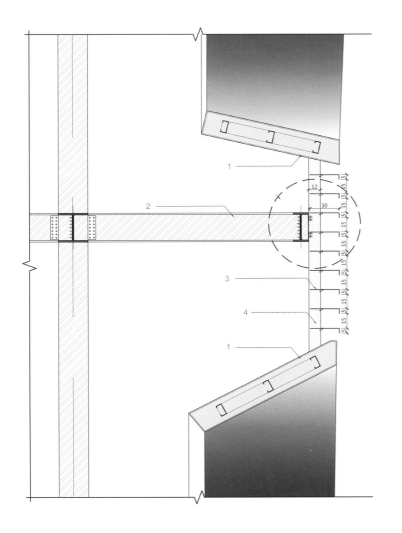

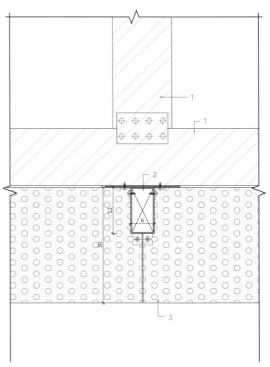

Section detail A
1. 30x15cm H beam, lacquered/white
2. 12x6cm square tube, baking finished/white
3. 30x5x0.3cm perforated louver, baking finished/white

剖面细部A
1. 30x15cm H梁，喷漆/白
2. 12x6cm方管，烤漆/白
3. 30x5x0.3cm冲孔板，烤漆/白

Façade section
1. Wood curved panel, lacquered/white
2. 30x15cm H beam, lacquered/white
3. 30x5x0.3cm perforated louver, baking finished/white
4. 12x6cm square tube, baking finished/white

立面剖面
1. 木作可弯板，喷漆/白
2. 30x15cm H梁，喷漆/白
3. 30x5x0.3cm冲孔板，烤漆/白
4. 12x6cm方管，喷漆/白

71 Council and Private Flats

塞斯港71号住宅

Location/地点: Port of Sète, France/法国，塞斯港
Architect/建筑师: Colboc Franzen & Associates
Photos/摄影: Cécile Septet
Site area/占地面积: 2,956m²
Completion date/竣工时间: 2011

Key materials: Façade – galvanised steel screen
主要材料：立面——镀锌钢幕墙

Overview

The building plot lies on the thin strip of land between the Étang de Thau and the Mediterranean Sea on the northern side of the old town, close to the commercial port and its huge industrial facilities. The project design is based on three blocks of flats set on a ground-floor base. The development comprises four distinct parts: 16 council flats in various configurations; 55 private two- and three-room flats; and shops and car parks to service all of the above.

These blocks also embody a principle of 'Mediterranean architecture' that allows for a lifestyle adapted to the local climate: outdoor living protected from intense heat. The arrangement of the blocks makes it possible to offer dual-aspect flats with particularly attractive corner living rooms. There are balconies running along the façade and these outdoor extensions allow occupants to walk around the outside of their flats. The corner of the living room is truncated, making the balcony wide enough for a table and chairs. A galvanised steel screen protects it during very hot weather and also provides a nice amount of privacy. It follows the curve created by the varying widths of the balconies. It lends harmony to the three blocks and makes them easier to interpret. They become gigantic steel cocoons whose materials remind us of the maritime world, while their shape is reminiscent of a ship's stem and the wind in the screen slats sounds like the jangling of masts in a port.

Detail and Materials

The galvanised steel screens spin a delicate architecture of shade and privacy that controls the sunlight by letting the oblique winter rays in while blocking the high-angle rays of summer. They also allow occupants to make appropriate their balconies without disturbing their neighbours, and to create a 'homely' feel while also enjoying the view and life in the town centre.

Enabling occupants to appropriate their flats also means offering them spaces that evolve with their needs. Hence there are no technical constrictions on the façade and as the building's weight-bearing core contains the shared amenities, most of the ducts are placed close to this core. The walls between the flats are also weight-bearing to ensure optimum sound-proofing. Ultimately, making sure that the project is flexible enough to adapt to future changes is a means of ensuring that the building lasts.

Each building's apparent structural complexity is outsmarted by optimising their geometry. Thus the curve of the metal screen, which changes as it adjusts to a variety of geometrical shapes, can be broken down into six distinct scenarios. This work is considerably simplified and speeded up by 3D computer modelling combined with the digital cutting technology developed by the metalwork company responsible for manufacturing the screen. The same method has been adopted for more traditional elements. There are only three different types of window, and the frames are the same, since the balconies all have a similar curve to them.

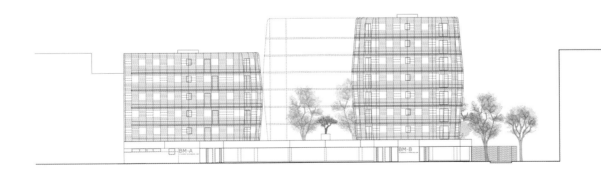

项目概况

建筑场地位于旧城区北侧的一个长条地带，紧邻商业港口和工业设施。项目由三座公寓楼组成，其中两座拥有一个共用的底座。开发分为四个独立部分：16所不同规格的公营公寓、55所2~3室的私人公寓以及服务于公寓的商店和停车场。

设计展现了地中海建筑特色，选择了与当地气候相符的生活方式：提供了有遮阳保护的露天生活区。建筑的布局让公寓拥有双向朝向和别具一格的转角客厅。建筑立面上布满了阳台，让住户可以走出自己的公寓享受户外空间。客厅转角被截平，让阳台的宽度足够摆放桌椅。镀锌钢幕墙保护阳台不受日光直射，同时也保证了住户的隐私。幕墙随着阳台所形成的曲线而动，赋予了公寓楼和谐而简单的诠释。幕墙像巨大的钢茧，让人们想起了海洋世界。此外，建筑的造型还让人想起了船首，而幕墙板条之间的风声则像是船桅的响声。

细部与材料

镀锌钢幕墙为沿着建筑盘旋上升，为建筑提供了遮阳和隐私保护：既能引入冬季倾斜的太阳辐射，又能阻挡夏季的高角辐射。幕墙还让住户在阳台上的活动不打扰邻居，营造出居家的感觉，同时还能享受市中心的风景和生活。

满足住户的需求还要求公寓空间能根据他们的需求而进化。因此，立面上没有技术限制，而大多数管道等公用设施都设置在建筑的承重核及其周围。公寓之间的墙壁也是承重墙，保证了最佳的隔音效果。保证项目根据未来需求调整的灵活性是保证建筑持久使用的方法之一。建筑的几何造型能掩盖自身结构的复杂性。金属幕墙的曲线随着建筑的几何造型而调整，可分为6种模式。三维计算机建模让这一设计过程变得快速简单，数字切割技术则由金属幕墙的制造商提供。同样的方法也被运用到更多的传统元素中。窗户只有3种类型，并且它们的窗框都是相同的，因为阳台的弧度是固定的。

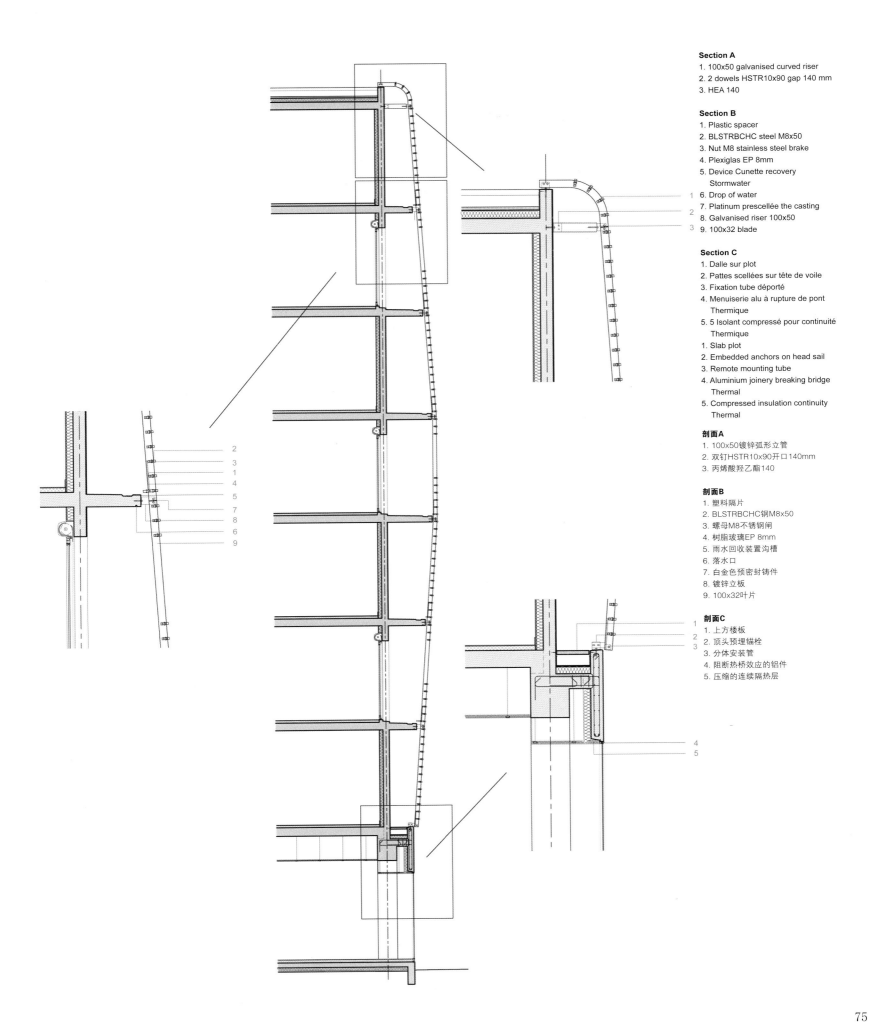

Section A
1. 100x50 galvanised curved riser
2. 2 dowels HSTR10x90 gap 140 mm
3. HEA 140

Section B
1. Plastic spacer
2. BLSTRBCHC steel M8x50
3. Nut M8 stainless steel brake
4. Plexiglas EP 8mm
5. Device Cunette recovery Stormwater
6. Drop of water
7. Platinum prescellée the casting
8. Galvanised riser 100x50
9. 100x32 blade

Section C
1. Dalle sur plot
2. Pattes scellées sur tête de voile
3. Fixation tube déporté
4. Menuiserie alu à rupture de pont Thermique
5. 5 Isolant compressé pour continuité Thermique
1. Slab plot
2. Embedded anchors on head sail
3. Remote mounting tube
4. Aluminium joinery breaking bridge Thermal
5. Compressed insulation continuity Thermal

剖面A
1. 100x50镀锌弧形立管
2. 双钉HSTR10x90开口140mm
3. 丙烯酸羟乙酯140

剖面B
1. 塑料隔片
2. BLSTRBCHC钢M8x50
3. 螺母M8不锈钢闸
4. 树脂玻璃EP 8mm
5. 雨水回收装置沟槽
6. 落水口
7. 白金色预密封铸件
8. 镀锌立板
9. 100x32叶片

剖面C
1. 上方楼板
2. 顶头预埋锚栓
3. 分体安装管
4. 阻断热桥效应的铝件
5. 压缩的连续隔热层

Sant Miquel Special Education School
圣米克尔特殊教育学校

Location/地点: Barcelona, Spain/西班牙，巴塞罗那
Architect/建筑师: Pepe Gascón/ Pepe Gascón Arquitectura
Photos/摄影: José Hevia
Area/面积: 14,9275m²
Completion date/竣工时间: 2011

Overview

The project employs volumetric force to resolve the complex, irregular geometry of the site. For instance, a triangular volume, with ground floor and basement, adjoins the north side (providing optimal lighting for study), whilst the rectangle of the existing building occupies the east corner. The triangular extension is low-rise in order to minimise the environmental impact of new works. Finally, a courtyard provides the rooms in the basement with natural light and ventilation.

All routes through the common zones in the school (stairs, landings, corridors and vestibules) are designed to become "places of social attraction" – spaces for encounters, meetings and exchanges, generously proportioned, enjoying natural lighting and, always, open to the city as backdrop. In order to frame views of the city itself, the perimeter structure is painted black, like a picture frame, whilst the interior finishes are given austere, light colours. It will be the future users, therefore, whose presence imbues the school with life and colour, leaving the architecture as a neutral, yet orderly, frame for the students and teachers.

In short, all the decisions taken to define the project seek to further these rich and subtle relations between the interior and the exterior. The solutions include, for example, the design of the steel net "non-fence", the patterned front, the light-filled common spaces, the new windows over the existing building and the longitudinal skylight that provides the overall architecture for communications. All these solutions are aimed particularly at generating relations between the inhabitants and the immediate landscape of the city, precisely so that this place can become a space for teaching and learning.

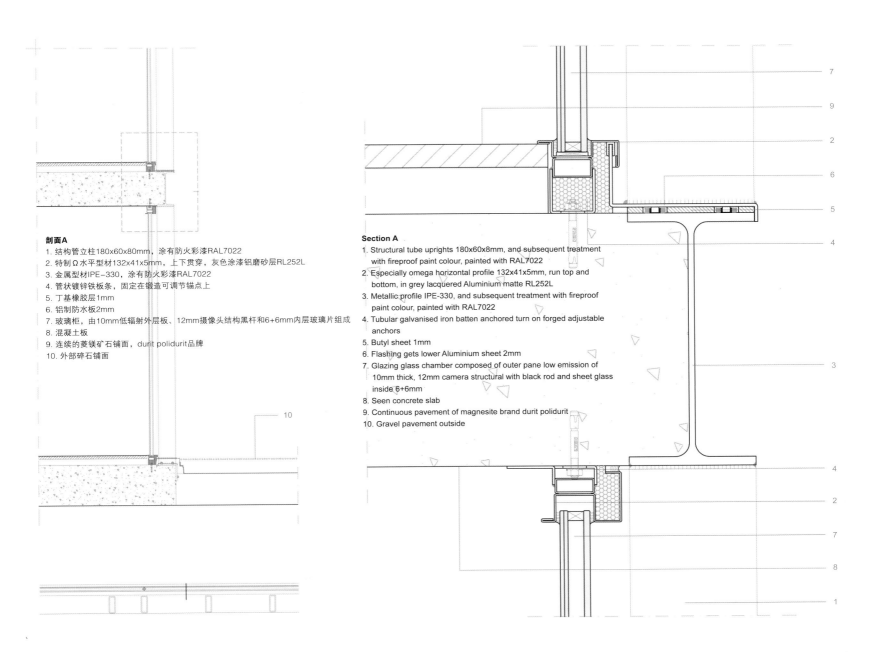

剖面A
1. 结构管立柱180x60x80mm，涂有防火彩漆RAL7022
2. 特制Ω水平型材132x41x5mm，上下贯穿，灰色涂漆铝磨砂层RL252L
3. 金属型材IPE-330，涂有防火彩漆RAL7022
4. 管状镀锌铁板条，固定在锻造可调节锚点上
5. 丁基橡胶层1mm
6. 铝制防水板2mm
7. 玻璃柜，由10mm低辐射外层板、12mm摄像头结构黑杆和6+6mm内层玻璃片组成
8. 混凝土板
9. 连续的菱镁矿石铺面，durit polidurit品牌
10. 外部碎石铺面

Section A
1. Structural tube uprights 180x60x8mm, and subsequent treatment with fireproof paint colour, painted with RAL7022
2. Especially omega horizontal profile 132x41x5mm, run top and bottom, in grey lacquered Aluminium matte RL252L
3. Metallic profile IPE-330, and subsequent treatment with fireproof paint colour, painted with RAL7022
4. Tubular galvanised iron batten anchored turn on forged adjustable anchors
5. Butyl sheet 1mm
6. Flashing gets lower Aluminium sheet 2mm
7. Glazing glass chamber composed of outer pane low emission of 10mm thick, 12mm camera structural with black rod and sheet glass inside 6+6mm
8. Seen concrete slab
9. Continuous pavement of magnesite brand durit polidurit
10. Gravel pavement outside

Detail and Materials

The horizontal structure is built out of reinforced concrete slabs measuring 30cm thick on the first and second floors, 35cm thick in the basement and 40cm thick on the ground floor, with 16-mm-gauge base reinforcement below every 20cm and 12-mm-gauge base reinforcement above every 20cm. In many areas, the slab is exposed and has no false ceilings, hence it is shuttered using phenolic panels.

The exterior façade structure consists of vertical uprights made of structural tubing measuring 160x60x8mm placed every 300/600/900 mm, welded at the bottom onto plate moulding of 4mm and 300mm maximum length and welded at the top to the beam, treated with fire-resistant paint and later painted with colour RAL 7022.

Exterior barrier made of connecting panels measuring 2250x1500mm, with 30x30mm mesh, screwed to galvanised steel vertical uprights measuring 80x8mm anchored to the concrete foundation.

项目概况

项目利用体积力来解决场地的复杂性和不规则性。例如，底层和地下室的三角形体量毗连场地北侧（提供了最佳的照明），而原有建筑的矩形造型则占据了场地东面。三角形结构相对较低，将新工程对环境的影响降到了最低。最后，庭院为地下室提供了自然采光和通风。

学校公共区域的路径（楼梯、梯台、走廊和门廊）都被设计成了具有"社会吸引力"的场所，人们可以进行会面、交流。空间设计比例得当，享有自然采光和城市风景。为了获取城市的景象，外围结构被漆成了黑色，像画框一样，而室内空间则采用简朴的浅色。学校未来的使用者将会为空间带来生气和色彩，建筑本身持有序的中立状态，为学生和教师的活动提供了背景。

总之，所有设计都帮助项目在室内外之间寻找丰富而微妙的联系。设计的解决方案包括钢网围栏、带有图案的建筑正面、充满日光的公共空间、原有建筑的新窗户和长条天窗。这些设计的目的是在建筑使用者和城市景观之间形成联系，让建筑成为独特的教学空间。

细部与材料

水平结构由钢筋混凝土板建成，二楼和三楼楼板为30厘米厚，地下室为35厘米厚，一楼为40厘米厚，每20厘米下方有16毫米底层加固，上方有12毫米底层加固。在许多区域，混凝土板直接暴露出来，没有安装假吊顶，利用酚醛树脂板遮挡起来。

外立面结构由间隔为300/600/900毫米，尺寸为160x60x8毫米的结构管材制成的垂直支柱组成，支柱焊接在4厘米厚的模板底部（最大长度300毫米）和横梁顶部，外层涂有耐火漆和RAL 7022色彩漆。

连接板制成的外层围栏尺寸为2250x1500毫米，带有30x30毫米网眼，以螺栓固定在80x8毫米的镀锌垂直支柱上，支柱则固定在混凝土地基上。

Institute of Functional Biology and Genomics
功能生物学与基因学研究院

Location/地点: Salamanca, Spain/西班牙，萨拉曼卡
Architect/建筑师: Salvador Mata Pérez (Mata y Asociados)
Design Team/设计团队: Myriam Vizacaíno Bassi, Javier Encinas Hernández, Eduardo García García, Jaime Pedruelo Sánchez, Stefania Augliera, David Emiliano Fernández Mateos
Photos/摄影: Juan K. Ayala
Completion date/竣工时间: 2010

Key Materials: Façade – steel
Structure – reinforced concrete
主要材料：立面——钢材
结构——钢筋混凝土

Overview

The project is located on a plot at the end of Salamanca University Campus Miguel de Unamuno. Taking the condition of the rugged topography, with a difference of up to 6m between the outermost points, the implantation proposal starts off from a topological approach with two different input levels, allowing access further segregation by types and uses (vehicle access and representative pedestrian access at the lower and upper level, respectively).

The solution adopted led to the construction of a unitary and compact building, with a continuous and semi-transparent mask inspired in genetic coding processes, after which the plan is organised around eight functional large vertical holes (four indoor and four attached to the façade), thereby obtaining a foamed interior provided with natural lighting. The compaction of the program also allowed us to liberate most of the plot's surface, and to project terraced/landscaped plates to allow the coexistence of the outdoor leisure area with controlled parking strips.

The auditorium of amoeboid form, located on the northern head of the building, is configured as a (formal and constructively) differential element that marks the main entrance to the building. At the other extreme, a large concrete bracket allows shelter the space for expeditions (the other major access area) and hold cryogenic gas tanks.

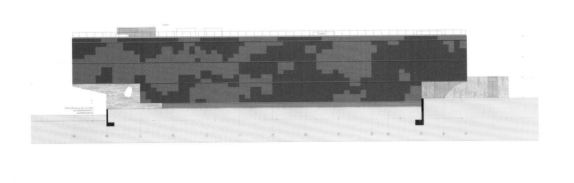
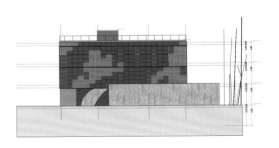
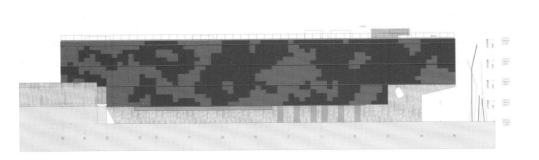
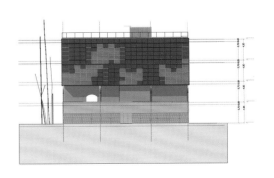

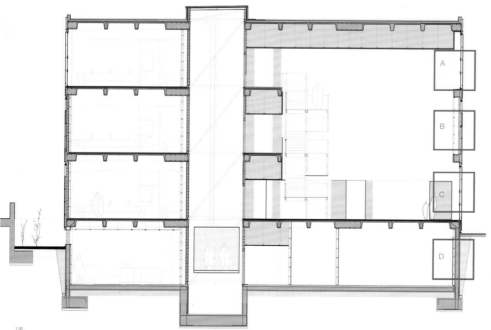

Detail and Materials
The façade consists of lacquered aluminium sandwich panels suspended from a metal structure. Ahead of windows and patios has been installed a lacquered micro-well plate in the same tones of red. The effect is a continuous and smooth "skin" glowing with sunlight, changing throughout the day. Only concrete looks out, occasionally, revealing the nature of the structure.

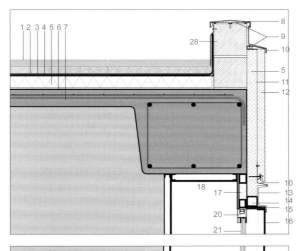

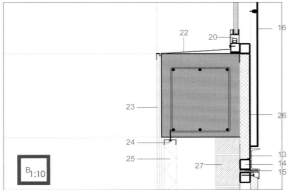

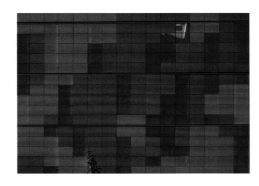

项目概况

项目位于萨拉曼卡大学乌纳穆诺校区的尽头。建筑场地地势高低不平，最远端高度差可达到6米，因此，设计选择将建筑分为两个结构，将车辆通道和步行通道分设在高低两层，实现分别出入。

设计最终实现了一座统一而紧凑的建筑，连续的半透明表皮从基因代码编程中获得了灵感。项目规划围绕着8个巨大的垂直孔洞（4个位于市内，4个附着在立面上）展开，形成了泡沫状的室内空间并实现了自然采光。精炼的项目规划释放了建筑场地的地面空间，从而建造了阶梯式景观平台，实现了户外休闲区和停车带的共存。

变形虫造型的礼堂位于建筑的北端，以与众不同的造型和构造标志出建筑的主入口。在另一端，巨大的混凝土支架标志出另一个入口并容纳了低温储气罐。

细部与材料

建筑立面由悬挂在金属结构上的涂漆铝夹心板组成。窗户和天井前方安装着色调相同的红色涂漆微孔板。连续而光滑的表皮在阳光下闪闪发光，随着日光的变化而变化。时不时露出的混凝土揭示了建筑结构的本质。

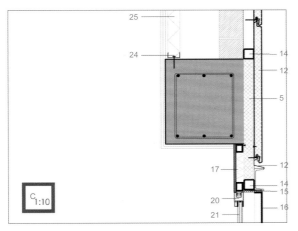

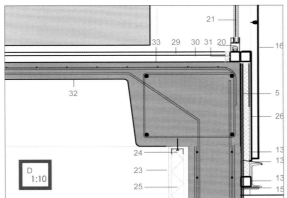

Façade detail
1. Drainage tile filtron R9
2. Feltemper V120 sheet
3. Rhenofol CG sheet
4. Feltemper V300 sheet
5. Extruded polystyrene insulation 6cm high density
6. Vapor barrier
7. Reticular slab – returnable coffered
8. Coping
9. Cleat
10. Plate for drip cap
11. Vertical substructure in steel 60x60x3 mm
12. Sandwich panel 50mm
13. UPN – 120
14. 60x60x3 mm steel beam
15. Steel plate 2cm
16. Micro – perforated and lacquered plate
17. Sandwich panel, lacquered
18. Steel plate
19. Perforated plate
20. Aluminium joinery w/ thermal break
21. Climalit 3+3- 12 – 4+4 glass
22. Dust cover plate
23. Double plasterboard 13+15mm
24. Steel substructure 70mm
25. Fiver glass insulation
26. Steel plate 2mm
27. Brick wall in "termoarcilla" 14cm
28. Brick wall in "termoarcilla" 19cm
29. Resin pavement
30. Mortar grip layer
31. Elastic band
32. Anti-dust surface treatment
33. Anti-impact layer

立面节点
1. 排水瓦管Filtron R9
2. Feltemper V120薄板
3. Rhenofol CG薄板
4. Feltemper V300薄板
5. 高密度挤塑聚苯乙烯隔热层6cm
6. 隔汽层
7. 网状板，可复位
8. 顶盖
9. 夹板
10. 滴水挑檐板
11. 垂直钢下层结构60×60×3mm
12. 夹层板50mm
13. UPN-120
14. 60×60×3mm钢梁
15. 钢板2cm
16. 微孔涂漆板
17. 涂漆夹层板
18. 钢板
19. 穿孔板
20. 铝接合处，带有断热功能
21. Climalit 3+3-12 – 4+4玻璃
22. 防尘罩
23. 双层石膏板13+15mm
24. 钢下层结构70mm
25. Fiver玻璃绝缘
26. 钢板2mm
27. 砖墙14cm
28. 砖墙19cm
29. 树脂铺装
30. 砂浆层
31. 橡皮筋
32. 防尘表面处理
33. 抗冲击层

School of Art and Design in Amposta

安波斯塔艺术设计学院

Lcoation/地点: Amposta, Spain/西班牙，安波斯塔
Architect/建筑师: David Sebastian + Gerard Puig Arquitectes.
Photos/摄影: Adrià Goula.
Completion date/竣工时间: 2013

Key materials: Façade – lacquer coated, perforated steel panel, solid brick
主要材料：立面——涂漆穿孔钢板、实心砖

Overview

This project was a proposal of a very different scale and with very different agendas: one was a teaching facility to relocate and improve the city's School of Art and Design and the other a fair pavilion to replace the existing one on the same site, currently very run down. Two adjacent sites are separated by a road. The first phase included the construction of the first building, the School of Art and Design.

A single structural system was used as a solution for the different program needed for the two buildings. A metallic structure initially formed by porticos with pillars every 10m which allows the architects to remove pillars where needed to form trusses. For each pillar they take out there are two diagonals, so the lights can reach spans of 30m. The resulting project is more a way of building the place, than a definitive answer.

Detail and Materials

The buildings are organised in sections 9m wide and of varying heights (+4m/+8m/+12m). The

new topography ranges from the more human scale of the porticos and lower parts to the scale of the building's height and skylights.

The architects propose a ventilated façade finishing installed with a metallic profile (Model Creta_Europerfil) of dimensions of 100x900cm, which includes a single piece all the width of the portico (900cm).

The wall consists of gero (structural ceramic brick) of 29x14X9cm, supported on the metallic structure, with stucco mortar for the outer face. On the wall of gero the architects install a substructure of metallic tubes of 80x80x4cm that will serve as the support for the metallic profile.

Among these metallic tubes there is based on extruded polystyrene thermal insulation of 5cm thick. Finally the architects mount the metallic profile on metal tubes, leaving a small air chamber of 3 cm between the thermal insulation the metal profile finish.

项目概况

项目的规模和日程安排都十分独特，分为两个部分：一是城市艺术设计学院的搬迁，另一个是对现有的贸易展厅进行重建。两个场地之间被一条公路隔开。一期工程师建造艺术设计学院。

项目为两座建筑选择了相同的结构系统。由间隔为10米立柱所组成的门廊形成的金属结构，让设计不再需要桁架立柱。每个立柱都有两条对角线，因此光线的到达范围可达30米。最终的项目更像是对场地的重塑。

细部与材料

建筑由9米宽和4米、8米、12米高的分段结构组成，从更符合人体尺寸的门廊和底层空间到建筑的顶部高度和天窗，错落有致。

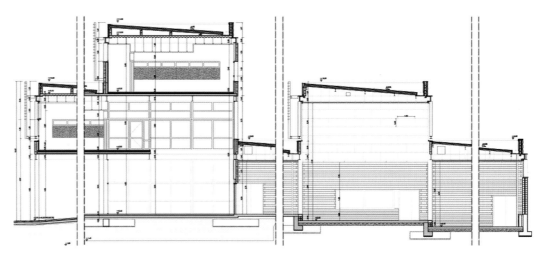

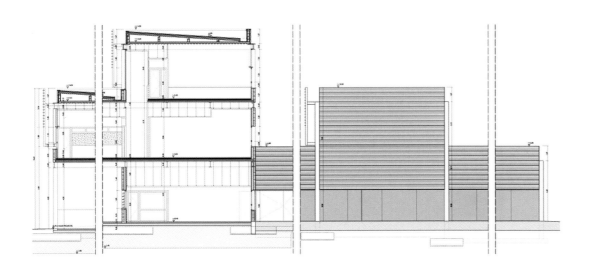

建筑师设计了一个通风立面系统，采用100x900厘米的金属型材（Model Creta_Europerfil），其中门廊结构处的单一金属板宽度为900厘米，横跨整个门廊。

结构陶瓷砖墙面由29x14x9厘米的gero砖块组成，下方有金属结构支撑，外部涂有灰泥砂浆。建筑师在墙面上安装了一个金属管（80x80x40厘米）组成的下层结构，作为金属型材的支撑。

在金属管之间安装着5厘米厚的挤塑聚苯乙烯隔热层。最后，建筑师将金属型材安装在金属管上，并且在隔热层和金属型材饰面之间留出了一个3厘米的空气层。

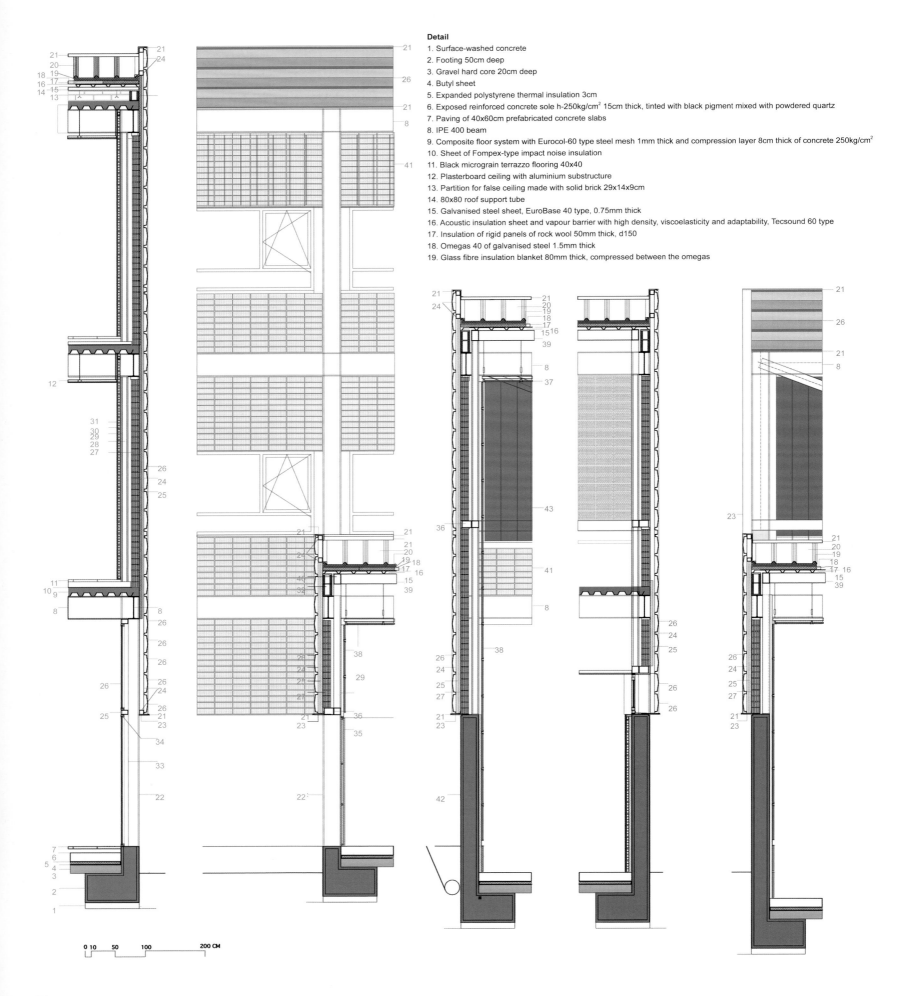

Detail

1. Surface-washed concrete
2. Footing 50cm deep
3. Gravel hard core 20cm deep
4. Butyl sheet
5. Expanded polystyrene thermal insulation 3cm
6. Exposed reinforced concrete sole h-250kg/cm^2 15cm thick, tinted with black pigment mixed with powdered quartz
7. Paving of 40x60cm prefabricated concrete slabs
8. IPE 400 beam
9. Composite floor system with Eurocol-60 type steel mesh 1mm thick and compression layer 8cm thick of concrete 250kg/cm^2
10. Sheet of Fompex-type impact noise insulation
11. Black micrograin terrazzo flooring 40x40
12. Plasterboard ceiling with aluminium substructure
13. Partition for false ceiling made with solid brick 29x14x9cm
14. 80x80 roof support tube
15. Galvanised steel sheet, EuroBase 40 type, 0.75mm thick
16. Acoustic insulation sheet and vapour barrier with high density, viscoelasticity and adaptability, Tecsound 60 type
17. Insulation of rigid panels of rock wool 50mm thick, d150
18. Omegas 40 of galvanised steel 1.5mm thick
19. Glass fibre insulation blanket 80mm thick, compressed between the omegas

20. Eurocobert 40n type roof sheeting stuck to omegas 40 metallisse grey 9006
21. Lacquered steel plate finish
22. HEB-300 metal pillar
23. Bracket 150x15mm
24. Painted galvanised steel tube 80x80x6mm
25. Insulation of expanded polystyrene 5cm thick
26. Creta-type perforated steel plate 0.75mm thick
27. 15cm ceramic sheet made with solid brick 29x14x9cm
28. Diagonal metal tube 80x80x6mm
29. Metal tube uprights 120x120x6mm
30. Partition wall with 5cm air cavity
31. Semi-smooth white-painted plaster render
32. Solid concrete block 40x20x15
33. Frame of galvanised steel tubes with 120x80x6
34. Perfrisa-type locks
35. Door with metal frame and 18mm birch veneer marine plywood panels
36. Crosspiece with 120x120 + 200x120 steel tube 8mm thick
37. Tubular steel upright 50x50mm
38. Exposed panels of Heraklith-type wood wool agglomerate with guide
39. IPE-180 band
40. Steel tube 180x120, 8mm thick
41. Exposed exterior ceramic sheet of striated solid brick with water-repellent finish
42. Wall of reinforced concrete 250kg/cm^2 30cm thick exposed on the exterior
43. U-glass-type reinforced structural glass with air cavity

节点
1. 表面冲洗混凝土
2. 基脚50cm
3. 碎石硬核20cm
4. 丁基薄板
5. 发泡聚苯乙烯隔热层3cm
6. 清水混凝土底座，h-250kg/cm^2，15cm厚，混合石英粉，黑色颜料染色
7. 40x60cm预制混凝土板铺装
8. IPE 400梁
9. 复合地板系统：1mm厚Eurocol-60型钢网+8cm厚250kg/cm^{2}混凝土压实层
10. Fompex型隔音板
11. 黑色微晶水磨石地面40x40
12. 石膏板吊顶，铝制支承结构
13. 假吊顶隔断，实心砖构成，29x14x9cm
14. 80x80屋顶支承管
15. 镀锌钢板，EuroBase 40型，0.75mm厚
16. 隔音板和隔汽层，具有高密度、黏弹性和适应性，Tecsound 60型
17. 刚性石棉板隔热层，50mm厚，d150
18. Ω40镀锌钢，1.5mm厚
19. 玻璃纤维隔热衬垫，80mm厚，压缩在Ω型材之间
20. Eurocobert 40n型屋顶板，粘在Ω40型材上，金属灰9006
21. 涂漆钢板饰面
22. HEB-300金属柱
23. 支架150x15mm
24. 涂漆镀锌钢管80x80x6mm
25. 发泡聚苯乙烯隔热层，5cm厚
26. Creta型穿孔钢板0.75mm厚
27. 15cm陶瓷板，由实心砖构成，29x14x9cm
28. 斜金属管80x80x6mm
29. 金属管支柱120x120x6mm
30. 隔断墙，带有5cm气腔
31. 半光滑白色石膏粉刷
32. 实心混凝土块40x20x15
33. 镀锌钢管框架120x80x6
34. Perfrisa锁
35. 门：金属框架+18mm桦木皮船用胶合板
36. 横档：120x120+200x120钢管，8mm厚
37. 钢管支柱50x50mm
38. 外露Heraklith型刨花板
39. IPE-180钢带
40. 钢管180x120，8mm厚
41. 外露陶瓷板：条纹实心砖配防水饰面
42. 钢筋混凝土墙，250kg/cm^2，30cm厚，外露
43. U形玻璃型强化结构玻璃，配气腔

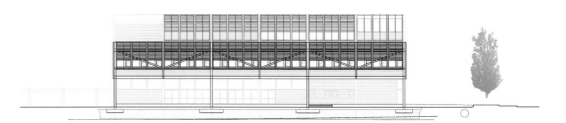

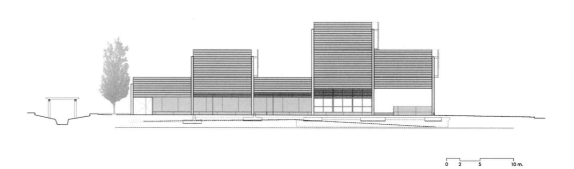

Social Cybercentre Macarena Tres Huertas

玛卡瑞纳特里斯胡尔塔斯社会数码中心

Location/地点: Sevilla, Spain/西班牙，塞维利亚
Architect/建筑师: MEDIOMUNDO arquitectos
Collaborator/合作设计: Mario Ortega Gómez (MOG-Arquitectos)
Photos/摄影: Fernado Alda
Area/面积: 410m²
Completion date/竣工时间: 2010

Key materials: Façade – perforated steel sheet Structure – brick wall
主要材料：立面——穿孔钢板
结构——砖墙

Overview

The district of Macarena Tres Huertas is characterised by the eight-floor dwelling building blocks supported by pilots that leave porches on the ground floor. This allows visual transparency and free circulation among the gardens that avoids the perception as a stationary and blurred space.

Therefore the new 'Macarena Social - CyberCentre' rests in this place by generating a visual and transit transversal in order to optimize the accessibility to the surrounding paths and open areas.

Moreover, ground floor releases the programmatic arrangement, by creating a wi-fi social plaza below the building with a small access garden. Cafeteria and wi-fi piazza are linked together through a porch and are connected with the multipurpose room. All of those spaces are offering a meeting point for the neighbours and a leisure space. On the top of those spaces arises a volume, which is lined with red lacquered sheet and accommodates the computer labs, workshops and offices.

The main idea is based to "raise to the power of

three" the former of the free spaces is now occupied by the building which follows the multiply n-times in the tangible spaces: on the 1st floor are located the wi-fi garden plaza; the multipurpose and connected spaces. Continuously the roof level is offered generously to the neighbours as a terrace to hold events and as river viewpoint.

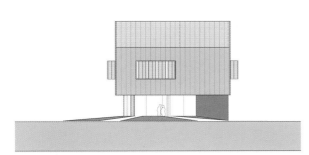

East elevation 东立面

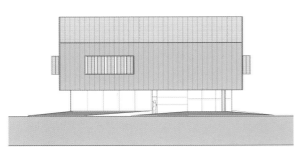

South elevation 南立面

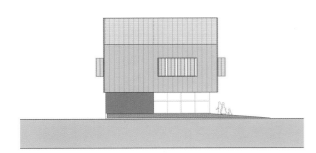

West elevation 西立面

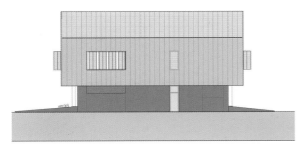

North elevation 北立面

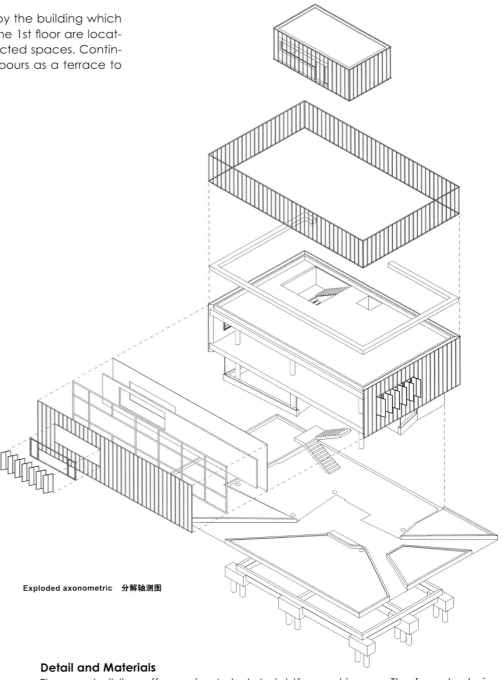

Exploded axonometric 分解轴测图

Detail and Materials

The new building offers a simple but straightforward image. The façade design had as a challenge the use of industrial materials into a small-scale building.

Subsequently the façade material has been selected to be a red-lacquered fold up metal sheet over the thermal insulation and the brick wall. Between the brick wall and the sheet layer there is a leaving a ventilated area for climate control. The steel sheet has different perforation densities that allow different levels of privacy and even security. There are several intimacy gradients managed by the 'gills' over the windows (vertical lama or banderols that make the building breath), orientated to free spaces, preserving the windows and views to the dwellings' privacy.

It is a statement on sustainability in terms of normalized construction, organized by structural units and standard module, with serial production process, controlled transport and executing time, that benefits the energy and emission control. The building follows passive construction on order to rationally deal with the extreme weather of Seville: make the most of thick isolation, natural ventilation and natural lightning.

项目概况

玛卡瑞纳特里斯胡尔塔斯区遍布着8层高的住宅楼，住宅楼的一楼大多以门廊为特色。这实现了透明的视觉感，让人们可以从花园里自由穿梭，避免了模糊静止的空间。

因此，新建的社会数码中心采取了类似的处理方式，优化了周边走道和开放区域的可达性和通透性。

此外，地面的室外空间还在建筑下方营造出一个无线网络社交广场和小花园。餐厅与无线网络广场通过门廊连接起来，共同与多功能厅相连。这些空间都为邻里提供了会面和休闲的场所。在这些空间上方，一个由红色油漆板包围的结构内设置着电脑机房、工作室和办公室。

项目的主要理念以"三合一功能"为基础，让建筑体现了多重性能：在一楼，无限网络花园广场、多功能厅和连接空间相互作用。建筑屋顶同样被利用了起来，能够为社区提供举办活动、观赏河景的平台。

细部与材料

建筑体现了简洁、直接的形象。立面设计的挑战在于如何将工业材料融入小型建筑之中。

建筑师选择将红漆折叠金属板安装在隔热层和砖墙之上，在砖墙和金属板之间留有一个通风层作为气候控制。根据不同的隐私和安全需求，各个部分钢板上的穿孔密度也各不相同。窗口的金属板像"鱼鳃"一样可以手动闭合，保证了建筑的通风。窗口朝向空地，既获得了风景，又不影响旁边住宅楼住户的隐私。

由结构组件和标准模块通过组装生产流程所形成的标准化建造方式有效控制了建筑的执行时间，有利于能源和废弃排放控制。为了应对塞维利亚的极端天气，建筑采用了被动式结构：充分利用了加厚隔热层、自然通风和自然采光。

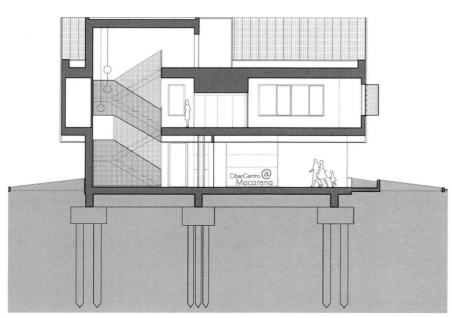

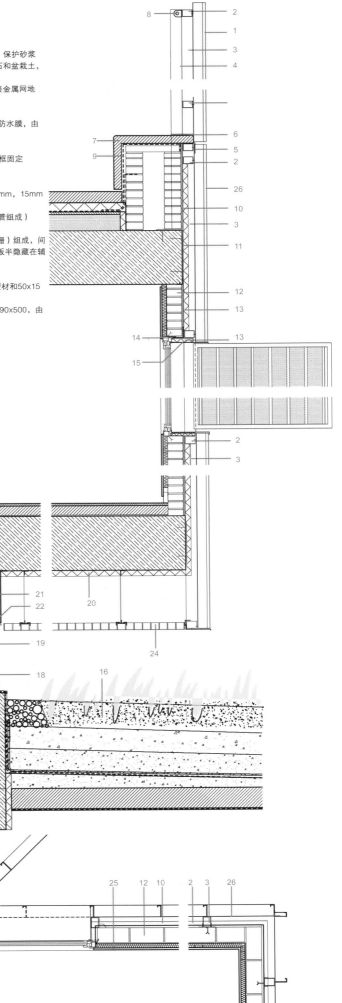

South façade section
1. Exterior finish
   Profiled sheet tray or type with 15% perforation lacquered in red coral, t=0.75mm placed vertically and attached to pillars with screws horizontal
2. Crosspiece 80.60.4 tubular profile welded to vertical upright
3. Vertical upright (1 every 3 mt) 2 tubular profiles 80.60.4 continuous weld
4. Edging covering strip
5. Bib bent sheet metal on crosspiece with recess for receiving pillars
6. Continuous foam sealing joint P
7. Prefabricated recess seat with riser
8. Fixture
9. Waterproof laminate
10. 20mm outdoor waterproof plaster
11. Upright tie plate
    Horizontal plate on a fixing manufactured system 200.250.10 attached with 2 screws 12mm diameter and welded to vertical plate 350.250.10 vertical shaft attached to 4 fixed screws with diameter 12mm
12. Siding on lintel consisting of:
    20mm outdoor waterproof plaster, made of 1/2 foot lad perf mineral wool 30mm, 15mm plasterboard panel
13. Insulation 30mm projected prior to fixation plate and lintel
14. Formation with NLP 70.70 lintel hook wrought by plate 50.5
15. Upper post of 2mm sheet metal lacquered red coral. Attached to auxiliary profile sheet folded in "C"
16. Planter waterproofing with variable height vapor barrier asphalt base layer of light weight concrete 10cm average thickness, layer regularization mortar in bitumen modifications membrane ibm-48, protective layer of mortar, anti-root material in the bordering area with build elements with rigid polyethylene beakers, gravel and sand and potting soil with an average thickness of 40cm, with weep holes for drainage
17. Floated in situ concrete interior with brass joints and welded steel mesh of flooring. Seated on extruded polystyrene panels
18. Reinforce concrete sill 20cm esp.
    50,150 pair receiving plate and cover carpentry drywall waterproofing polish membrane under sill composed of folded galvanised steel sheet forming Goteron lacquered.
    Outer sill, galvanised steel
19. Fixed security glazing 6 +6.6 b received by profile "or" stainless steel 40.30.2
20. Bottom face forged by 40mm projected polyurethane
21. Enclosure formed by:
    20mm drenched waterproof, made of 1/2 foot lad perfil 110mm insulation projected 40mm, 15mm plasterboard on 48mm perfil
22. Formation with NLP 50.150y, 8mm lintel plate forged by hook plated and steel pipes
23. Upper door post of bent sheet metal of lacquered galvanised steel. Lintel fixed to plate
24. Interior and exterior fake ceiling type by galvanised steel grating composed tramex decks every 50.3 rods 5mm 40mm and 80mm each. Clamp half hidded by auxiliary structure suspended ceiling
25. Inner cladding with 40mm SPF
    Double plasterboard 15mm on 70mm galvanised steel perfil rodapie profile and lacquered Aluminium I 50.15
26. Floor coating profiled sheet listing type Hacierba sr 1.90.500. by arcelormittal, lacquered red coral

南立面剖面
1. 外墙饰面
   压型钢板，15%穿孔，红珊瑚色油漆，t=0.75mm，垂直放置，以水平螺丝固定在立柱上
2. 横档80x60x4型材，焊接在垂直支柱上
3. 垂直支柱（间隔3米），双管型材80x60x4，连续焊接
4. 包边条
5. 横档上的弯曲金属板，凹槽用于安装立柱
6. 连续的泡沫密封接头
7. 预制凹槽底座
8. 固定装置
9. 防水层压板
10. 20mm户外防水抹灰
11. 支柱连接板
    水平板以2个12mm螺丝安装在200x250x10成套系统上，焊接在350x250x10垂直板上，通过4个12mm固定螺丝固定
12. 过梁上的护壁板，由以下结构组成：
    20mm户外防水抹灰
    石棉30mm，15mm石膏板
13. 30mm绝缘层，突出安装于面板和过梁上
14. NLP 70x70过梁挂钩和50x50面板组成的型材
15. 上门柱，2mm红珊瑚色穿孔金属板，固定在辅助型材板上
16. 种植花盆防水层：沥青底层隔汽层，轻质混凝土平均厚度10cm，砂浆层配有沥青调整膜ibm-48，保护砂浆层；防根系材料：聚乙烯材料、碎石、砂石和盆栽土，平均厚度40cm，带有排水孔
17. 就地浇筑混凝土地面，配有黄铜接头和焊接金属网地面；安装在挤塑聚苯乙烯板上
18. 钢筋混凝土门槛20cm
    门槛下方50x150成对接收板和木层干式墙防水膜，由折叠涂漆镀锌钢板组成
    外层门槛，镀锌钢
19. 固定安全玻璃6+6.6 b，由40x30x2不锈钢框固定
20. 地面锻造40mm聚氨酯
21. 外层围墙由以下结构组成：
    20mm浸湿防水层，由110mm突出绝缘40mm，15mm石膏板和48mm perfil聚乙烯纤维组成
22. 由NLP 50x150y、8mm过梁板（钩板和钢管组成）
23. 上门柱，涂漆镀锌钢，过梁安装在面板上
24. 室内外假吊顶，由tramex平板（镀锌钢格栅，间隔为50x3杆，5mm、40mm、80mm；夹板半隐藏在辅助结构吊顶上
25. 内包层，40mmSPF
    15mm双层石膏板，安装在70mm镀锌钢型材和50x15涂漆铝板上
26. 地面涂层，压型钢板，Hacierba sr型，1x90x500，由arcelormittal提供，涂有红珊瑚色漆

The "Coslada" Hybrid Complex
科斯拉达综合体

Location/地点: Madrid, Spain/西班牙，马德里
Architect/建筑师: Atxu Amann, Andrés Cánovas, Nicolás Maruri.
Photos/摄影: David Frutos
Completion date/竣工时间: 2012

Key materials: Façade – lacquered steel
Structure – reinforced concrete
主要材料：立面——涂漆钢
结构——钢筋混凝土

Overview
Located in Coslada, a recently redeveloped industrial town close to Madrid, the project is borne of the arrangement of different building uses at different heights: underground parking; street-level offices and commercial units; a raised public space; and residences above.

The volume, divided into four towers, is oriented around a raised public space that simultaneously acts as a cover for the entrances and commercial units below; as such the building becomes a visible urban joint connecting the city at a wider scale.

The street-level square is connected by an elevator and a hung metallic staircase to a square above that is very much the heart of the project. Elevated ten meters, it links the four towers horizontally whilst separating the diverse building uses vertically. Once equipped, it become a sensitive leisure space for collective socialization; a community square that serves the city.

Detail and Materials
The reinforced concrete structure is covered with a sheet skin of grey lacquered steel, acting as a ventilated façade those changes between opaque and perforated finishes depending upon the specific conditions of use. The resultant nuances in light and ventilation create an intense and complex building envelope.

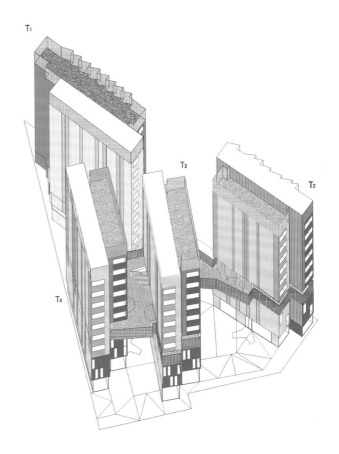

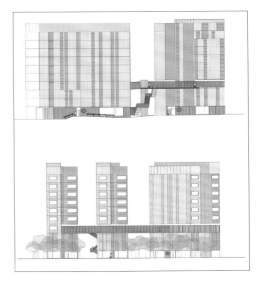

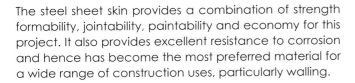

The steel sheet skin provides a combination of strength formability, jointability, paintability and economy for this project. It also provides excellent resistance to corrosion and hence has become the most preferred material for a wide range of construction uses, particularly walling.

Features
Excellent design fastening systems ensure the security and weather-light performance of steel walling during extreme weather.

The façade sheets are resistant to fire.
Excellent thermal properties that keep the building cooler in summer and warmer in winter.

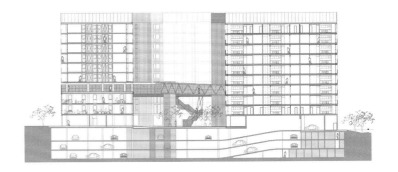

91

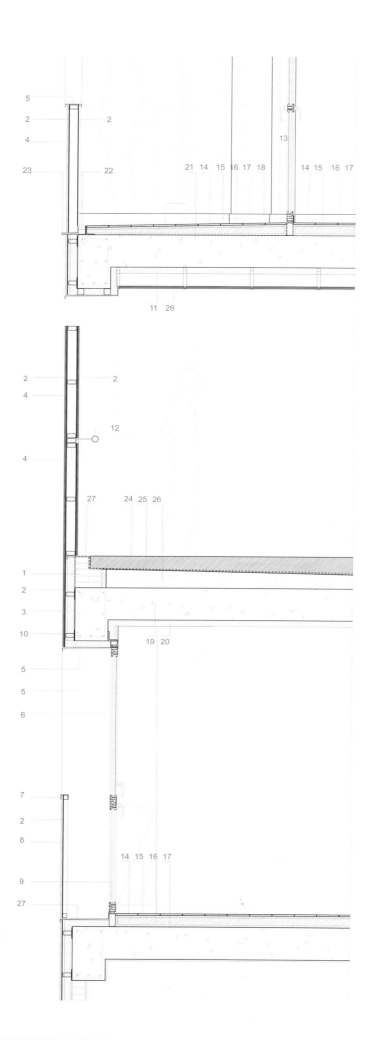

Façade construction detail

1. Concrete double hollow brick 1 foot
2. Corrugated and perforated steel model ARCELOR FREQUENCE 14.18c Corrugated steel
3. ARCELOR FREQUENCE model 14.18c
4. Painted galvanised steel tube 80×40mm color defined by D.F.
5. Folded painted galvanised steel , color to be determined by the D.F.
6. Carpinteria lacquered Aluminium, glass 8 + 10 +6 mm, folding top sheet, bottom sheet Swing
7. Painted galvanised steel sheet, color to be determined by LA D.F.
8. Steel tube 40 × 40mm
9. Galvanised steel tube 120×60mm fixing carpentry
10. Profile auxiliary fastening galvanised steel sheet
11. Glass and wool insulation e.60mm
12. Painted galvanised steel barrier tubes ø 60mm, color to be determined by LA D.F.
13. Terrace door, painted woodwork, Aluminium and glass 10+ 8 +6 mm
14. Stoneware 20×20mm e. 15mm, color defined by LA D.F. e.20mm
15. Gripping mortar e.20mm
16. Cement mortar e.65mm
17. Noise proofing insulation, polystyrene e.20mm
18. Painted wooden skirting, color to be determined by the DF e.15mm
19. Concrete slab e.30cm
20. Plaster and paint e.15mm
21. No adhesion double waterproof membrane
22. Folded galvanised sheet
23. Galvanised steel tube for drainage
24. Porous concrete pavement,"tennis-quick" painted, color to be determined by LA DF
25. No adhesion double waterproof membrane
26. Sloping lightweight concrete
27. Folded Painted galvanised steel, color to be determined by LA D.F

立面节点

1. 混凝土双层空心砖1英尺
2. 波纹穿孔钢ARCELOR FREQUENCE 14.18c型波纹钢
3. ARCELOR FREQUENCE 14.18c型钢
4. 涂漆镀锌钢管80x40mm
5. 折叠涂漆镀锌钢
6. Carpinteria涂漆铝,玻璃8+10+6mm,折叠式顶板,Swing下层板
7. 涂漆镀锌钢板
8. 钢管40x40mm
9. 镀锌钢管120x60mm,木工固定
10. 辅助固定镀锌钢板
11. 玻璃及石棉绝缘e=60mm
12. 涂漆镀锌钢护栏管ø=60mm
13. 平台门,涂漆木工+铝+玻璃10+8+6mm
14. 石件20x20mm e=15mm
15. 压层砂浆e=20mm
16. 灰泥砂浆e=65mm
17. 防噪绝缘,聚苯乙烯e=20mm
18. 涂漆木脚线e=15mm
19. 混凝土板e=30mm
20. 石膏+油漆e=15mm
21. 无胶双层防水膜
22. 折叠镀锌板
23. 镀锌排水钢管
24. 多孔混凝土铺装,tennis-quick涂装
25. 无胶双层防水膜
26. 斜坡轻质混凝土
27. 折叠涂漆镀锌钢

项目概况
项目位于科斯拉达———一座距马德里不远的新开发的工业小镇。项目将不同的功能设施分别设置在不同的楼层：地下是停车场，街面层为办公室和商铺，半空中设公共空间，最上面则是居民楼。

整个项目分为四个塔楼，环绕着一个空中公共空间展开。这个公共空间同时还为入口及商铺提供了顶盖。这座建筑在更广泛的层面上将城市空间连接了起来。

街面层的广场通过电梯和悬挂式金属楼梯与处在项目中心的空中广场相连。位于10米高空的广场将四座塔楼连接起来，同时也在垂直方向分割了不同的建筑功能。经过整装之后，它将成为一个社会化集体休闲空间，作为社区广场服务于城市。

细部与材料
钢筋混凝土结构上覆盖着一层灰色涂漆钢作为通风立面，立面根据功能条件的不同呈现为不透明或穿孔的形态。采光与通风的设计共同打造了一个紧凑而复杂的建筑外壳。

钢板表皮具有可成形性、可接合性、涂覆性和经济性，十分符合项目需求。此外，它还具有优秀的抗腐蚀性能，是大型墙面设计的最佳选择。

设计特色
优秀的固定系统保证了钢墙在极端天气中的安全性以及抗气候性能。
立面板材具有防火性能。
优秀的热性能保证了建筑冬暖夏凉。

151 Viviendas, Locales Comerciales y Garaje en Mieres

米耶雷斯151住宅、商店和车库

Location/地点: Mieres, Spain/西班牙，米耶雷斯
Architect/建筑师: Atxu Amann Alcocer, Andrés Cánovas Alcaraz, Nicolás Maruri González de Mendoza
Photos/摄影: David Frutos
Built area/建筑面积: 26,768.08m²
Completion date/竣工时间: 2012

Key materials: Façade – polycarbonate, lacquered steel
主要材料: 立面——聚碳酸酯、涂漆钢

Overview

This project is produced in a context where the volume of the building is previously strictly defined by urban regulations so the exterior form is not a problem to be considered. For this reason it is possible to concentrate the design energy in other conditions. In this case in the specific study of the functional variables of housing: type.

The proposal is a building protected by two galleries, that has an interior dwellings space released of the presence of the structure. The mechanical and wet spaces are concentrated around a "technical wall". This decision allows: cross ventilation, free disposition of storage and flexible spaces.

The galleries are configured as an intermediate space between the inner housing and contact with the city. They are warm in winter and cool in summer. They are a thermal insulation space for the building and a same time an in-between space for different uses. The dwelling is thought around natural regulated thermal condition, the space in-between defend from noise and seeks defuse light, and protect from exterior views, to provide the best living ambience. The use of galleries is very traditional in the North of Spain, where rain and clouds are common every day, and reduce the use of heating in the winter. The summers are very humid so easy cross ventilation is an important necessity. The interior patio is place for some trees that help to protect the views and the noise between different dwellings.

Detail and Materials

The urban façade is solved with a polycarbonate skin located on the front of the continuous terrace, on the first level, instead of polycarbonate; the façade is made of perforated steel sheet. This skin is opened by a sliding system, which solves the relationship of the dwellings with the outside. The second front of the gallery is lined with ceramic tiles of various colors. The system allows an intermediate location between the outside temperature and the internal working climate condition. The interior color of the ceramic is in contrast with the gray skies of the area.

项目概况

项目所在位置有着严格的城市规划，因此它的外观造型是固定的。因此，设计应将重点放在其他方面，即住宅类型的功能参数。

建筑由两条走廊环绕，其内部空间摆脱了建筑结构的束缚。机械和服务区围绕着一面"技术墙"展开。这种设计保证了交叉通风、自由的储藏空间和灵活的空间布局。

走廊被设计成住宅内部与外部城市之间的过渡地带。它们冬暖夏凉。建筑设有专门的隔热空间和多功能夹层空间。住宅采用自然控制热力状况，夹层空间保证了住宅不受噪声和直接光照的干扰，提供了最佳的居住环境。走廊的使用在西班牙北部的传统建筑中十分常见，因为那里多云多雨，并且走廊可以减少冬季的供暖需求。夏季十分潮湿，因此交叉通风十分必要。天井内种植了一些树木，有助于保护隐私、减少噪声。

细部与材料

建筑底层立面在连续的露台前方添加了一层聚碳酸酯表皮；二楼的立面则由穿孔钢板组成。表皮通过滑动系统打开，解决了住宅与外界的联系问题。走廊的另一面墙壁上铺满了各种颜色的瓷砖。这一系统在室外温度和室内气候条件之间形成了一个过渡区。内部瓷砖的色彩与灰色的天空形成了鲜明对比。

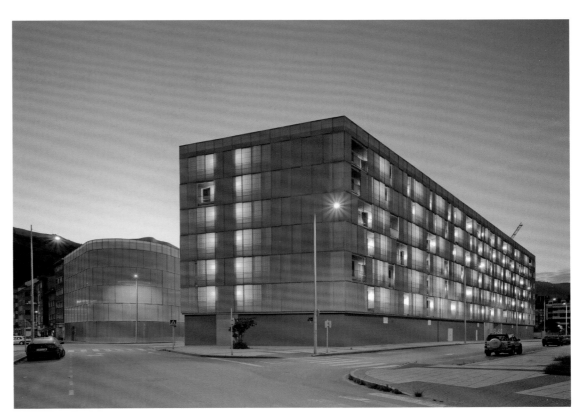

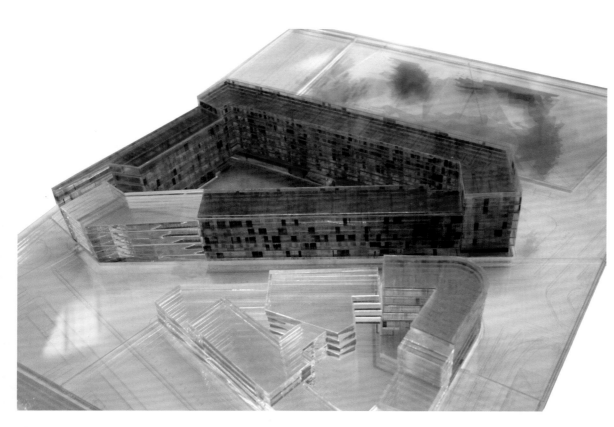

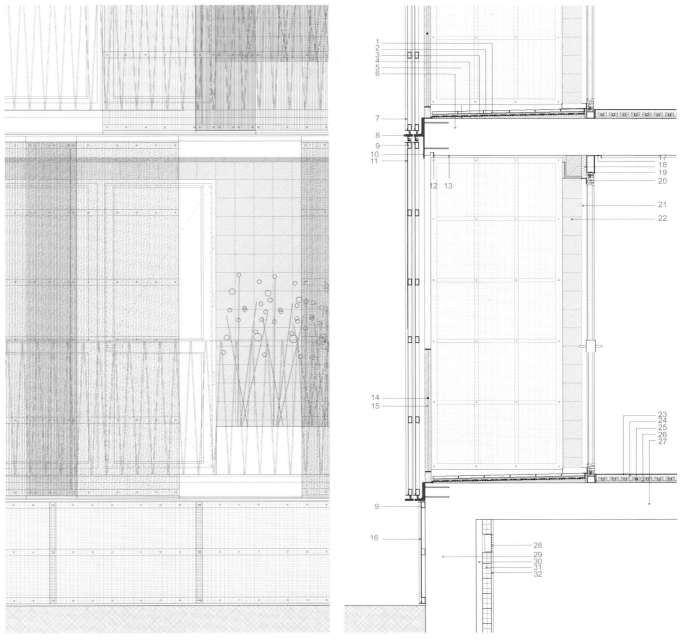

Façade construction detail
1. Ceramic flooring 20×20, color to be determined by LA D.F.(screed floor)
2. Gripping mortar 2cms
3. Separating layer of the synthetic weave
4. Double layer waterproofing sheet 4kg/m2
5. Concrete slope, 2%
6. Wrought, 32cm
7. Polyester sheets composed of glass fiber reinforcement impregnated with polyester resin and protected with a coating of gel mould, e=1.3mm, minimum
8. "L" painted and galvanised steel, color defined by LA D.F., 150×150×15m, each 1.2m, perforated each 40cms(radius=1cm)
9. Galvanised and lacquered steel tube, color defined by LA D.F., 50×30×2mm
10. Galvanised and lacquered steel sheet, color defined by LA D.F, forged, largest side 325×6, lower side 115×6
11. Corrugated and perforated steel sheet, ARCELOR FREQUENC MODEL 14.18c
12. Galvanised and lacquered steel sheet, color defined by LA DF, 50×2mm(cement cast)
13. Rough concrete cast and plastic painting, various colors defined by LA D.F, e=1.5mm
14. Circle ø 12mm. Smooth rail
15. Galvanised and lacquered steel sheet, color defined by LA D.F 1. 80×80×10 each 1.20m
16. Corrugated steel sheet, ARCELOR FREQUENC 14.18c
17. Galvanised and lacquered steel sheet, color defined by LA D.F, e=2mm, plaster equipped to tope
18. Extruded polystyrene, e=4cms
19. Carpentry door, galvanised and lacquered steel pipe, color defiined by LA D.F,140×60×3mm
20. Carpentry on sub-frame galvanised and coated steel, color defined by LA D.F
21. Galvanised and coated Aluminium plate, color defined by LA D.F
22. Ceramic finish 15*15cms (colors defined by LA D.F)
23. Linoleum flooring, e=2mm, color defined by LA D.F
24. Cement mortar, screed floor, e=4cms
25. Cross linked polyethylene tube, with anti- oxygen function, 16×1.8mm
26. Molded insulation, plasticized and dovetailing 20/45mm. Density 20kg/m3)
27. Wrought, 32cms
28. Aluminium ventilation grille, 15*15cms(each grid 20m2)
29. Reinforced concrete wall 40cms, variable width
30. camara bufa 5cms
31. Brick 7cms
32. Plastic plaster painting, e=1.5cms

立面构造节点
1. 瓷砖地面20x20
2. 抹平砂浆2cm
3. 合成编织隔断层
4. 双层防水板4kg/m²
5. 混凝土斜坡2%
6. 锻造层，32cm
7. 聚酯板，由玻璃纤维和聚酯树脂组成，带有胶膜涂层保护，e=1.3mm
8. L形涂漆镀锌钢，150x150x15mm，每条1.2m，每隔40cm穿孔（半径1cm）
9. 镀锌涂漆钢管50x30x2mm
10. 镀锌涂漆钢板，焊接，最大面325×6，下面115×6
11. 波纹穿孔钢板，ARCELOR FREQUENC MODEL 14.18c型
12. 镀锌涂漆钢板50x2mm（水泥浇筑）
13. 糙面混凝土浇筑和塑性涂料e=1.5mm
14. 圆环光滑围栏ø 12mm
15. 镀锌涂漆钢板80x80x10，1.20m长
16. 波纹钢板RCELOR FREQUENC 14.18c型
17. 镀锌突起钢板e=2mm，石膏盖顶
18. 挤塑聚苯乙烯e=4cm
19. 木门，镀锌涂漆钢管，140x60x3mm
20. 支架木工，镀锌涂层钢
21. 镀锌涂层铝板
22. 陶瓷饰面15x15cm
23. 油地毡地面=2mm
24. 水泥砂浆地面e=4cm
25. 交联聚乙烯管，带有抗氧化功能，16x1.8mm
26. 模制塑料绝缘，增塑+燕尾榫接合20/45mm，密度20kg/m³
27. 锻造层32cm
28. 铝通风格栅15x15cm（每个格栅20m²）
29. 钢筋混凝土墙40cm，宽度不定
30. 门厅
31. 砖7cm
32. 塑性石膏涂装e=1.5cm

Full of Triangles
三角楼

Location/地点: Fukuoka, Japan/日本，福冈
Architect/建筑师: Junichiro Ikeura (DABURA.i)
Photos/摄影: Satoshi Ikuma (Techni Staff)
Site area/占地面积: 148.52m²
Completion date/竣工时间: 2013

Key materials: Façade – Galvalume
主要材料：立面——铝锌涂层钢

Overview
It is a commercial building built in front of Fukuma Station that is the main station. In Fukuma Station, rebuilding is completed three years ago, and large-scale development business is developed in the former station backside (the station east) now. A house, an apartment complex and a doctor's office, commercial facilities are built in sequence and create an energic area. It is reverse to it, and, in contrast with the mood of the station back, old stores line up, and the station right side where this building is located cannot deny that stagnation has a slight it.

Originally, in the place where this building was located, the client did business in a commercial building for approximately 30 years. However, with the redevelopment around the station, front road is widened. A widening road interferes the building; of the site 1/3 will be quite sharpened. As a result, the old building was dismantled, and a wonderful triangular site was left.

It was difficult to plan the commerce space where functionality was demanded from this triangular land. It should have been rectangular to make it easy to cause a tenant if it is possible, but then the security of an effective area was difficult.

Detail and Materials
Finally the architect thought about these three angles of site shapes to give the building charm by constituting a building as a clue of the designs characteristically. Triangle was carried out thoroughly because the vertical plane daringly as makes a support of designing a triangular building. It renovates neighbouring environment by a characteristic building coming up and improves local charm. And it gave the value of the building and thought that it finally led to the commercial profit of the tenant.

The structure makes a cedar rail of 45*45 millimeter into a triangle form as the groundwork of the

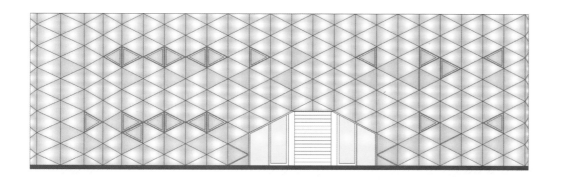

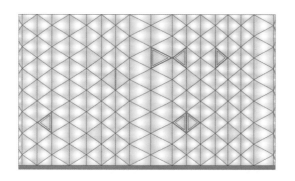

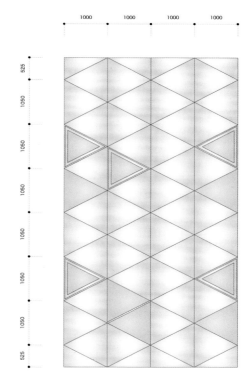

Facade

traditional method of construction + outer wall. Along the triangle lattice, the architect constitutes the appearance in the outer wall of 0.4 millimeters of gal barium colour steel sheets and an opening of the steel. Triangular acute angle individualises outer wall materials.

Particularly, when it is night, leaking light meets the people who arrived at the station from the triangular window shiningly and it is symbolic and exists at the station square.

项目概况

这是一座位于福冈车站前的商业楼。福冈站的重建于3年前完成,在原车站的后方(现车站东侧)进行了大规模商业开发,包括住宅、公寓楼、诊所和商业设施,使其变成了充满活力的区域。相反,与其形成对比的是,项目所在的车站右侧区域就显得过于萧条。

建筑的原址曾是一座拥有近30年历史的商业楼。但是随着车站的再开发,前方的道路得到了拓宽。拓宽的道路将建筑场地缩减了三分之一。因此,旧建筑被拆除,留下了一个奇妙的三角形场地。

在这个三角形地块上进行商业空间规划是一项困难的挑战。如果将场地切割成长方形或许更加简单,但是这样就难以保证有效地利用地块面积。

细部与材料

最终,建筑师决定利用三角造型为建筑增加独特的建筑魅力,并以三角为基础展开设计。三角形被贯彻始终,最终形成了一座三角建筑。它通过独特的设计为周边环境重新注入了活力,并且间接为商户带来了可观的商业价值。

建筑结构将45x45毫米的雪松围栏制成三角形,作为传统的地基和外墙。在三角格架外,建筑师用0.4毫米的铝锌涂层钢板构成外墙,并且形成了若干个钢结构窗口。三角形让建筑外墙个性十足。

最特别的是,夜晚,从三角形窗口渗透出来的光线打到行人身上,为车站增添了奇妙的光影效果。

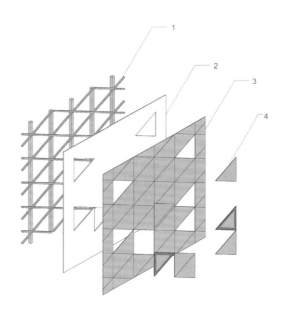

Façade diagram
1. Pillar: cedar 105x105
   Rail: cedar 45x45
2. Plasterboard t=12.5mm
   Conifer plywood t=12mm
3. Galvalume t=0.4mm
4. Steel sash

立面图解
1. 柱子:雪松105x105
   围栏:雪松45x45
2. 石膏板t=12.5mm
   针叶树胶合板t=12mm
3. 铝锌涂层t=0.4mm
4. 钢框格

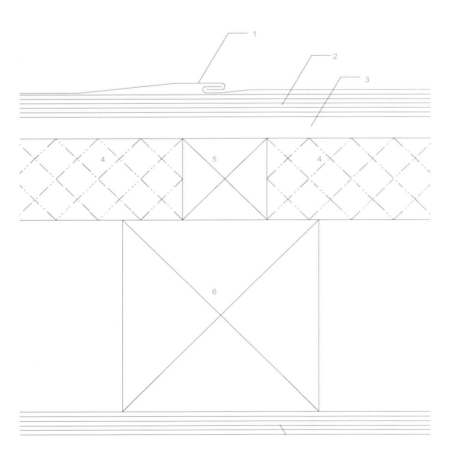

Detail 1
1. Galvalume t=0.4mm
2. Plasterboard t=12.5mm
3. Conifer plywood t=12m
4. Styrofoam t=45mm
5. Cedar 45x45
6. Cedar 105x105

节点1
1. 铝锌涂层t=0.4mm
2. 石膏板t=12.5mm
3. 针叶树胶合板t=12mm
4. 泡沫聚苯乙烯t=45mm
5. 雪松45x45
6. 雪松105x105

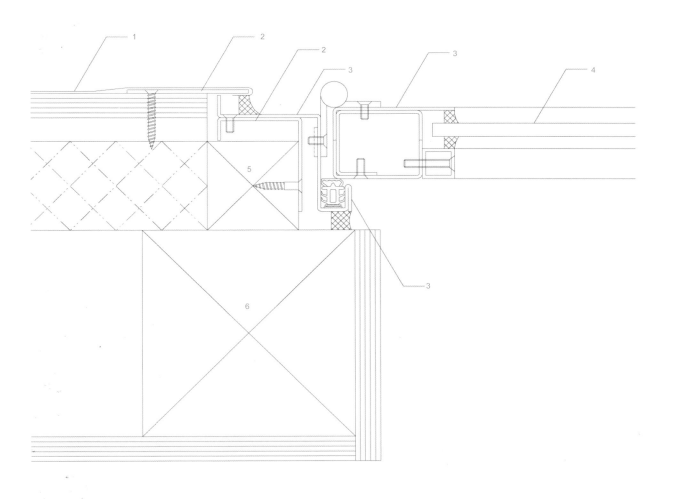
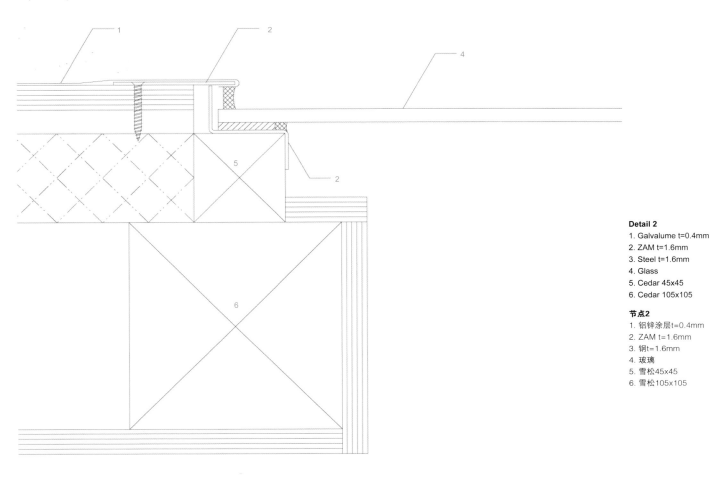

Detail 2
1. Galvalume t=0.4mm
2. ZAM t=1.6mm
3. Steel t=1.6mm
4. Glass
5. Cedar 45x45
6. Cedar 105x105

节点2
1. 铝锌涂层t=0.4mm
2. ZAM t=1.6mm
3. 钢t=1.6mm
4. 玻璃
5. 雪松45x45
6. 雪松105x105

KUKJE Gallery
库卡画廊

Location/地点: Seoul,Korea/韩国，首尔
Architect/建筑师: SO – IL
Photos/摄影: Iwan Baan
Site area/占地面积: 800m²
Gross floor area/总楼面面积: 1,260m²

Key materials: Façade – bead-blasted chainmail stainless steel mesh, curved low-iron laminated glass
Structure – steel, concrete
主要材料： 立面——链甲不锈钢网、弧形低铁夹层玻璃
结构——钢材、混凝土

Overview
The building contains a 16 x 9 x 6m gallery space, a 60 seat auditorium, project spaces, support and administrative functions for the gallery complex. The ground floor gallery is a single-story, column-free space optimized for large installations, performances and events. To light the art and keep a palpable relationship to the outside, daylight enters through a perimeter skylight. The skylight can be shaded or completely darkened, to create a black-box condition, which allows for light sensitive works or video. The first of two lower levels holds a 60 seat dark stained wood-clad auditorium, administrative areas, catering spaces, restrooms and mechanical spaces. The second basement holds storage and support spaces.

Detail and Materials
Considering the diagrammatic box geometry too rigid within the historic fabric, SO–IL enveloped the building in a mesh veil, creating a nebulous exterior that changes appearance as visitors move through the site. A custom stainless steel mesh produces a layer of diffusion around the structure, through a combination of reflections, openness, and moiré patterns produced through the interplay of its shadows. The mesh, made out of 510.000 individually welded rings, is strong yet pliable as it wraps around the building's irregular geometries. The result is an abstract 'fuzzy' object that accommodates a multiplicity of readings.

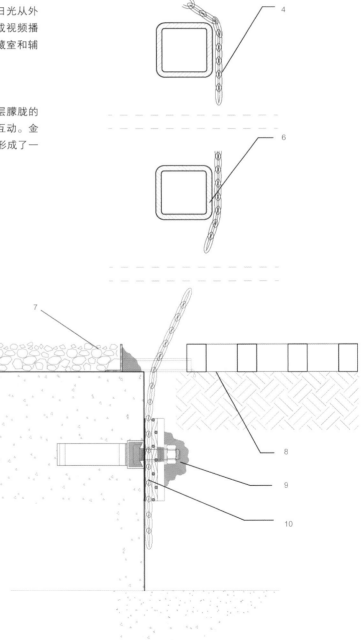

Façade detail
1. Threaded rod for tensioning mesh
2. Parapet mesh capture channel
3. Splice ring
4. Parapet coping
5. Parapet
6. Steel structure
7. gravel
8. Concrete pavers
9. Steel halfen channel
10. Steel clamp plate

立面节点
1. 张力网的螺纹杆
2. 护墙网捕捉槽
3. 接合环
4. 护墙压顶
5. 护墙
6. 钢结构
7. 碎石
8. 混凝土铺面
9. 钢槽
10. 钢夹板

项目概况

建筑包括一个16x9x6米的展览空间、一个可容纳60人的礼堂、项目空间以及辅助和行政空间。单层的展览空间没有立柱，适合举办大型装置展览、表演和活动。为了照亮艺术品并保持与外界的联系，日光从外围的天窗进入室内。天窗可以被局部或全部遮挡起来，形成黑盒的感觉，可以进行光敏作品展示或视频播放。地下一二层是拥有60个座位的礼堂、行政区、餐饮空间、洗手间和机械区。地下三层则是储藏室和辅助空间。

细部与材料

建筑师认为建筑的盒式造型在一片历史环境中过于突兀，因此为建筑披上了一层网纱，形成了一层朦胧的效果。定制的不锈钢网在建筑结构四周产生了漫射层，通过反射、开放、云纹实现了光影效果的互动。金属网由510,000个独立的焊接环制成，既坚固又柔韧，将建筑的不规则结构包裹起来。最终，建筑形成了一个模糊不清的物体，可以有多重解读方式。

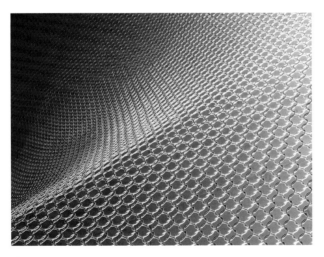

Vocational Education Centre
职业教育中心

Location/地点: Gordola, Switzerland/瑞士，戈尔多拉
Architect/建筑师: Durisch + Nolli Architetti Sagl, Lugano
Collaborators/合作设计: Dario Locher, Thomas Schlichting, Birgit Schwarz
Built area/建筑面积: 9,328m²
Completion date/竣工时间: 2010

Key materials: Façade – Inox by Arcelor Mittal/Montana
Platform – Concrete Platform with Cobiax® System for slim slabs and large spans
Inner walls – Duripanel®
主要材料: 立面——Inox不锈钢（Arcelor Mittal/Montana制造）
平台——Cobiax®系统混凝土平台，大跨度窄板
内墙——Duripanel®板

Overview
The project fits the given program in one, single building set at the edge of the plot of land. The building is a single volume, composed of a serial repetition of simple elements. The traditional typology is here characterised by the raising of the complex on a platform and the space below answers the necessity of parking spaces and different storage. The cover space is marked by the way to the parking entrance and, along the West side, by a second entrance for loading trucks and its flexibility for use is outstanding; it offers more outdoor cover spaces than originally requested, that could be used for many activities. The access to the platform is simple; three ramps are located next to the laboratory entrances that lead to the upper level. On the North side a freight lift for goods and disabled people can be found. The volume containing the workrooms and teaching rooms is designed to be simple, flexible, and functional. Somewhat like an industry building where students and teachers can experience a professional environment. The big north-facing sheds guarantee the perfect light for any training activity and the closed working spaces help capture the attention.

Detail and Materials
The reticular metal structure shaped as a Shed covers the whole width and develops along the length of the building, around 140 metres. The regular use of the same Shed element simplifies the construction and allows for the control of costs.

The project is characterised by the use of material and manufacturing that matches with the subjects taught in the centre. The raised platform is simple and rational in concrete. On top of this rough structure lays a light metal structure, in order to contain the weight on the platform. The outer coating skin is of a light steel inox. The internal coating of the walls and roof are realized using a simple metal box (Montanawall), where installation for acoustic insulation will be installed on the ceiling surface.

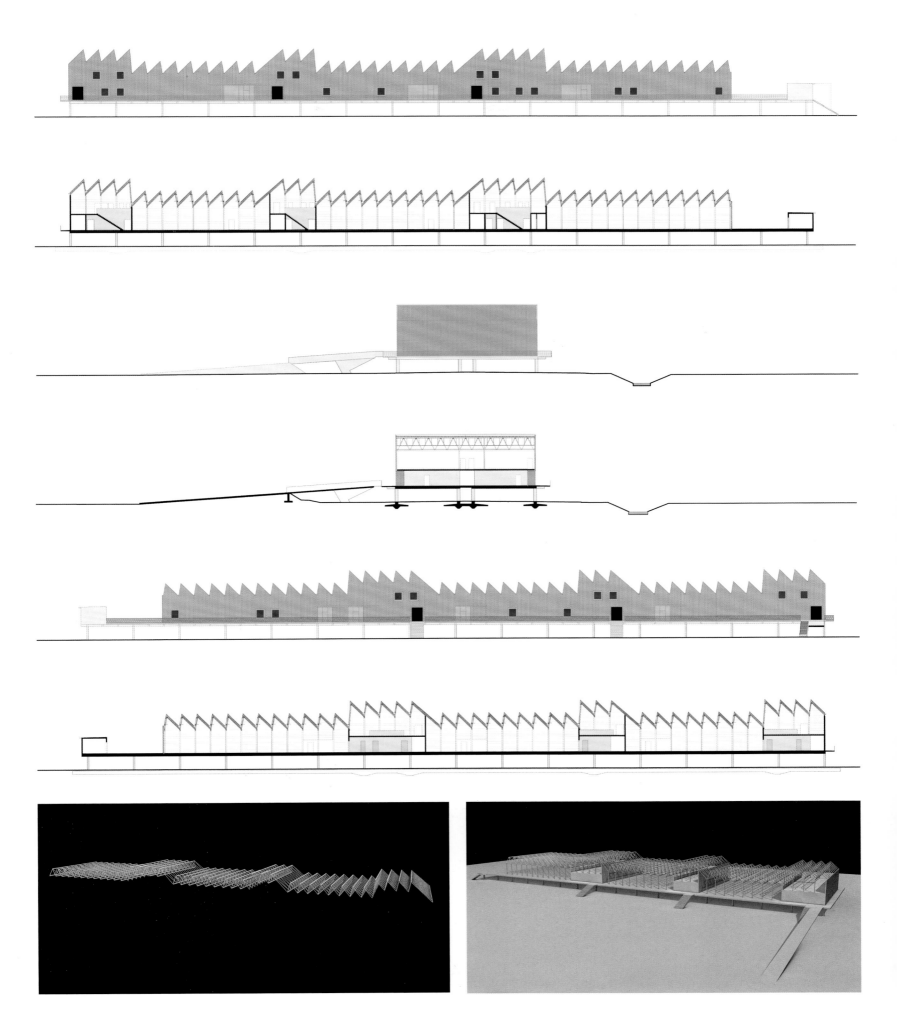

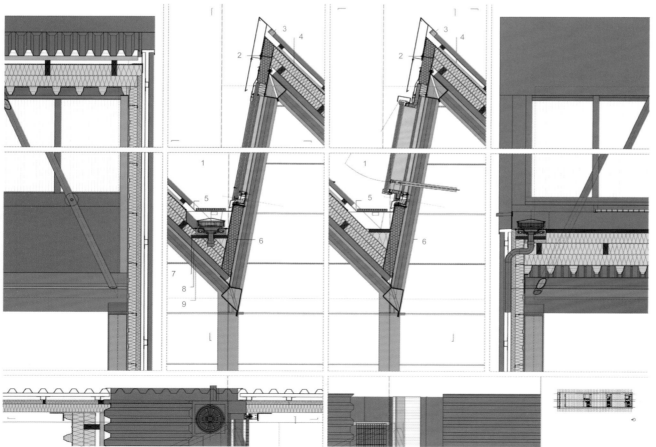

Façade detail
1. Single Sheet SP 45/900 / 0.8 mm 45 mm
   Substructure 60 mm
   Attic Sarnafil TU 111
   2 Layer Insulation Boards Type Flumroc PARA 2x80 mm
   Vapor barrier Sarnavap 1000E
   Sound-absorbing inserts for sheet SP59A
   Montana sheet SP59 59/900 / 1:00 A 59 mm
   Carpenteria Metallica
2. Fixing with glue rivet +
3. Profile of Closure
4. Attic Sarnafil TU 111
5. Bracket for Grid
   Grid walkway
6. Type Isopanel
7. Vapor barrier
   Type Sarnavap P1000E
8. sheet SP 59 A / 1mm
9. Carpenteria Metallica

立面节点
1. 单层板SP 45/900/0.8 mm，45 mm
   下层结构60mm
   Sarnafil阁楼TU 111
   双层隔热板FLUMROC PARA型2×80mm
   隔汽层SARNAVAP 1000 E
   插入隔音层SP59A
   Montana板SP59/900/1.00 A 59mm
   Carpenteria Metallica金属板
2. 铆钉安装
3. 包层型材
4. Sarnafil阁楼TU 111
5. 栅格支架
   栅格走道
6. Isopanel板
7. 隔汽层
   Sarnavap P1000E型
8. 薄板SP 59 A/1mm
9. Carpenteria Metallica金属板

Detail
1. MOUNTING BRACKET
2. COCKPIT PLUVIA
3. TABLE FOR LAYING CHANNEL
4. 2 LAYER INSULATION TYPE FLUMROC PARA 2x80 mm
   VAPOR BARRIER Sarnavap 1000 E
   INSERTS FOR ACOUSTIC SP59A
   SHEET MONTANA SP59/900/1.00 A 59 mm
5. PILLAR HEA 200   190 mm
   BOX TYPE MONTANA MK 80/500 80 mm
   TERM ISOLATION. TYPE ISOVER CLADISOL 120 mm
   SUBSTRUCTURE / OMEGA PROFILE 2X40 mm
   CORRUGATED SHEET SP 45/900/0.8 mm, 45 mm

节点
1. 安装支架
2. PLUVIA底盘
3. 槽体安装面
4. 双层隔热板FLUMROC PARA型2x80mm
   隔汽层SARNAVAP 1000 e
   插入隔音层SP59A
   Montana板SP59/900/1.00 A 59mm
5. 丙烯酸羟乙酯柱200，190mm
   箱式Montana MK80/500, 80mm
   绝缘层，ISOVER CLADISOL型120mm
   下层结构/OMEGA型材2x40mm
   波纹板SP 45/900/0.8mm, 45mm

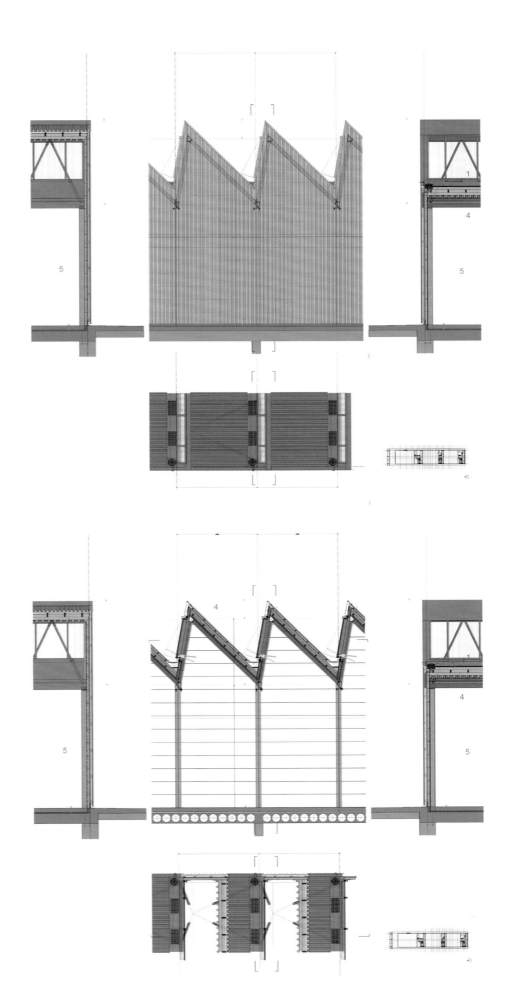

项目概况

项目将所有功能合并在一座位于场地边缘的单体建筑中，由一系列的简单元素重叠而成。建筑建在一个平台上，下方空间满足了停车和仓储的需求。

下方的覆顶空间以停车入口通道和西侧的装货卡车入口为特色，极具灵活性。建筑所提供的覆顶空间比最初的要求要多，可以进行许多额外的活动。

平台的入口十分简单：三条坡道紧邻通往上层空间的实验室入口。北侧设有载货电梯（也可供残障人士使用）。建筑内部的工厂间和教室设计十分简单、灵活、实用，看起来更像是一座供师生体验职业环境的工业建筑。向北的朝向为教学活动提供了完美的光照，而封闭的工作区则有助于集中注意力。

细部与材料

棚屋造型的网状金属结构覆盖了建筑的跨度，其长度达140米。棚屋结构元素的运用简化了施工流程，控制了项目成本。

项目的材料和建造方式富有特色，与内部教学的科目相互匹配。抬高的平台由混凝土建成，简洁而合理。在这个粗糙的结构上方设置着一个金属结构，保证了平台的承重力。建筑外层表皮采用轻质Inox不锈钢。内层墙面和屋顶则使用了简单的金属盒结构（Montanawall），天花板上安装了隔音设备。

Yapı Kredi ACCR
亚比信贷银行ACCR大楼

Location/地点: Gebze, Turkey/土耳其，盖布泽
Architect/建筑师: Mehmet Kütükçüoğlu, Ertuğ Uçar
Photos/摄影: Cemal Emden, www.cemalemden.com
Site area/占地面积: 198,000m²
Built area/建筑面积: 15,500m²
Completion date/竣工时间: 2011
Estimated cost/预算造价: 9,210,000Euro/欧元

Key materials: Façade – stainless steel panel
Structure – steel
主要材料: 立面——不锈钢板
结构——钢材

Overview
The ACCR Building has been built to the campus of Yapi Kredi Bank Operational Centre in Gebze. The building -named by its programs (Archive) on basement floor, (Call Centre) ground floor and (Restaurant) first floor- is a prism of 75x30 meters. The dimensions of the building and the gap between the existing campus are coming from the interior street's grids of the existing building. Its total area is approximately 15.000 sqm. Ten percent of this area consists of corridors, bridges, halls and entrance areas -for connecting to the existing system tightly- and sports Centre, public spaces and courtyards on that program. This new building makes the campus reach the slope on south side of the site. The service way, service courtyard, garbage collecting amenity, technical Centre and mailing units have been revised in the scope of the project. Besides that a conceptual design has been studied for a future hotel project on the campus.

The building with its dimming contours open and become transparent at the corner where the new entrance is located. At the same time, this opening is the beginning of the movement separating the restaurant canopy from the prism.

Detail and Materials
ACCR building's plain prismatic construction is covered with stainless steel panels. The façade integrates with the uninterrupted grass park on the ground and the sky on the eaves by those panels.

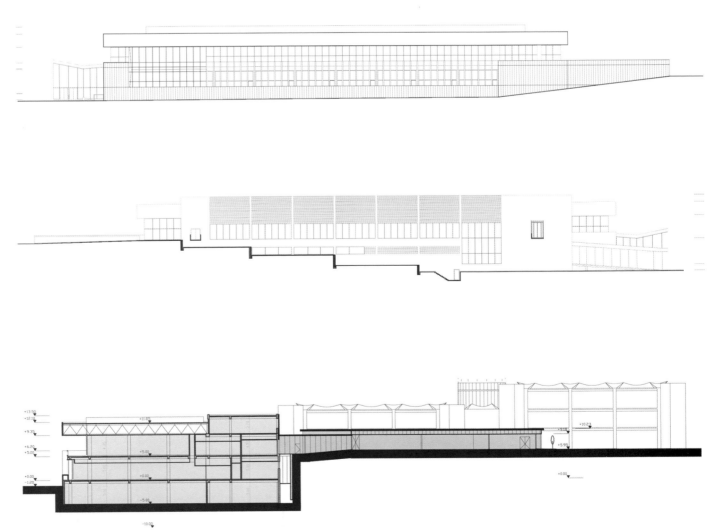

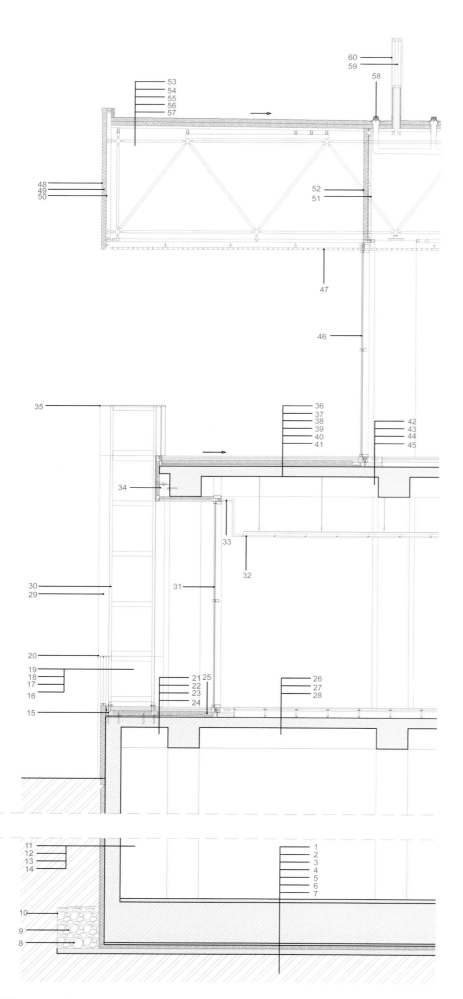
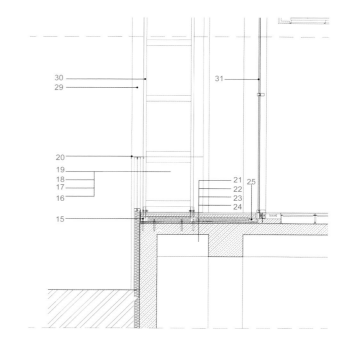

Detail
List of Materials

1. 7cm screed with wax polish
2. 80cm reinforced concrete foundation
3. 4cm protective concrete
4. 7cm rigid thermal insulation
5. Waterproofing membrane
6. 10cm blinding concrete
7. Compacted soil
8. 200mm perforated foundation drain
9. Pebbles
10. Geotextile
11. 30cm load bearing concrete wall
12. Two layer waterproofing membrane
13. 8mm rigid thermal insulation
14. Draining sheet
15. Anchorage set
16. 0.15cm galvanised steel sheet
17. Steel mullion, 5cm. U shape
18. Steel reinforcing band, U profile
19. 0.15cm galvanised steel sheet
20. Parapet: 1mm steel sheet
21. Epoxy flooring
22. 2cm self levelling
23. 8cm rigid thermal insulation
24. Two layer waterproofing membrane
25. Drainage canal
26. Carpet flooring
27. 15cm raised floor
28. 20cm reinforced concrete slab
29. 1.5cm galvanised steel column cover
30. Steel column, square profile 50x100x5mm
31. Transparent façade module:
    Float glass extraclear, 6mm
    HP neutral 61 (coating)
    16mm / air %100
    Float glass clear, 6mm
    U-value= 1.4 W/m²K
    SC= 0.45
32. Metal suspended ceiling with acoustic insulation: 60x60
33. Plasterboard suspended ceiling
34. Anchorage set
35. Parapet: 1mm steel sheet
36. 2cm stone flooring
37. 2cm mortar bed
38. 7cm rigid thermal insulation
39. Two layer waterproofing membrane
40. 2cm levelling concrete
41. 20cm reinforced concrete slab
42. 2cm stone flooring
43. 6cm mortar bed
44. 10cm levelling concrete
45. 20cm reinforced concrete slab
46. Transparent façade module:
    Float glass extraclear, 6mm
    HP neutral 61 (coating)
    16 mm / air %100
    Float glass clear, 6mm
    U-value= 1.4 w/m²k
    SC= 0.45
47. Metal streched suspended ceiling: 60x120 panels
48. 0.07cm galvanised steel cladding
49. Rockwool thermal insulation
50. 8cm Betopan cladding
51. 2cm OSB panel
52. 10cm plasterboard sandwich panel
53. 0.12cm membrane
54. 15cm rockwool thermal insulation
55. Vapour barrier
56. Trapozoidal metal sheet
57. Space frame steel structure
58. Water filter
59. Betopan cladding
60. Steel construction

材料节点

1. 7cm砂浆层，上蜡抛光
2. 80cm钢筋混凝土地基
3. 4cm防护混凝土
4. 7cm刚性隔热层
5. 防水膜
6. 10c盖面混凝土
7. 压实土
8. 200mm穿孔基础排水
9. 卵石
10. 土工布
11. 30cm承重混凝土墙
12. 双层防水膜
13. 8mm刚性隔热层
14. 排水板
15. 锚固定组件
16. 0.15cm镀锌钢板
17. 钢框，5cm，U形
18. 钢加强带，U形剖面
19. 0.15cm镀锌钢板
20. 护墙：1mm钢板
21. 环氧树脂地面
22. 2cm自流平
23. 8cm刚性隔热层
24. 双层防水膜
25. 排水槽
26. 地毯地面
27. 15cm抬高地面

28. 20cm钢筋混凝土板
29. 1.5cm镀锌钢柱顶
30. 钢柱，方型材50x100x5mm
31. 透明立面模块：
    超清浮法玻璃，6mm
    高性能涂层61
    16mm空气层
    清晰浮法玻璃，6mm
    传热系数1.4 W/m²K
    SC= 0.45
32. 金属吊顶，配有隔音板：60x60
33. 石膏吊顶
34. 锚固定组件
35. 护墙：1mm钢板
36. 2cm石地面
37. 2cm砂浆层
38. 7cm刚性隔热层
39. 双层防水膜
40. 2cm找平混凝土
41. 20cm钢筋混凝土板
42. 2cm石地面
43. 6cm砂浆层
44. 10cm找平混凝土
45. 20cm钢筋混凝土板
46. 透明立面模块：
    超清浮法玻璃，6mm
    高性能涂层61
    16mm空气层
    清晰浮法玻璃，6mm
    传热系数1.4 W/m²K
    SC= 0.45
47. 金属吊顶：60x120板
48. 0.07cm镀锌钢包层
49. 石棉隔热层
50. 8cm Betopan包层
51. 2cm定向刨花板
52. 10cm石膏夹层板
53. 0.12cm膜
54. 15cm石棉隔热层
55. 隔汽层
56. Trapozoidal金属板
57. 空间框架钢结构
58. 滤水层
59. Betopan包层
60. 钢结构

项目概况

ACCR大楼位于盖布泽的亚比信贷银行运营中心园区。建筑名称来自于建筑功能区首字母的缩写：地下室的档案室（Archive）、一楼的呼叫中心（Call Centre）和二楼的餐厅（Restaurant），是一座75x35米高的棱柱建筑。建筑总面积约15,000平方米，其中10%的空间为走廊、连接桥、大厅和入口区（与原有园区建筑紧密结合）以及体育中心、公共空间和庭院。新建筑让园区延伸到场地南侧的斜坡。服务通道、服务庭院和垃圾收集设施、技术中心、收发室全部进行了重新翻修。此外，银行还研究了未来在园区中建造一家酒店。

建筑调光立面对外开放，在入口转角处变得透明。同时，这一开口将餐厅天棚与棱柱建筑分隔开来。

细部与材料

ACCR大楼朴素的棱柱结构上覆盖着不锈钢板。立面通过这些面板与绵延不断的草地和头顶的蓝天紧密地结合起来。

Neiman Marcus at Natick Collection
纳蒂克内曼·马库斯百货商店

Location/地点: Natick, USA/美国，纳蒂克
Architect/建筑师: Elkus Manfredi Architects
Photos/摄影: Bruce Martin
Gross floor area/总楼面面积: 9,290m²

Key materials: Façade – stainless steel panel
Structure – steel
主要材料：立面——不锈钢板
结构——钢材

Overview
The Neiman Marcus specialty store in Natick is the most unusual and unique store for a company whose corporate policy prescribes that each store be tailored to its business location. Furthermore, Neiman Marcus requires that the architecture respond to its primarily female customer base. Elkus Manfredi Architects designed the undulating patterned stainless steel exterior of the store to evoke a sophisticated silk dress or scarf from the Neiman Marcus couture line billowing in the coastal breezes of New England. The surrounding landscape recalls the sea grass of tidal marshes, the traditional stone walls, and birch forests of the region.

Detail and Materials
The most striking element of the project is an undulating two-story high stainless steel exterior that is meant to represent a silk scarf or dress in multiple hues of bronze, champagne and silver. To create the image of the fabric, Elkus Manfredi explored not only the form but the color, texture, and pattern of the façade. The colors of the metal are timeless — bronze, champagne, and silver. The pattern follows the form and enhances moments within the building, such as the entry and signage.

Coordination of structural steel with multi-colored architectural stainless steel panels was a major challenge; this required extensive vendor and subcontractor management.

Early coordination with the panel vendor assured seamless harmonization with the structural steel vendor. Three-dimensional spatial coordinates were furnished in electronic format, and then the structural steel fabricator used this information to fabricate bearing seats.

Proactive planning for the interface between the structural members and stainless steel panels meant that the different colored ribbons appeared flawless.

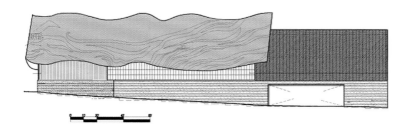

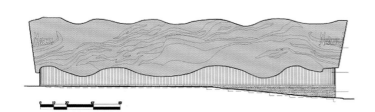

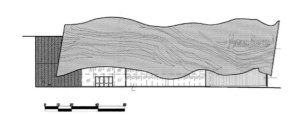

Section at entry
1. Roof vents for rain screen cavity
2. Metal rain screen
3. Metal panels
4. Stainless steel soffit panel
5. Glazed Aluminium curtain wall
6. Stainless steel entrance doors
7. Entrance mat
8. Stone pavers
9. Lighting
10. Partition entry vestibule

入口剖面
1. 雨幕空洞处屋顶通风口
2. 金属雨幕
3. 金属板
4. 不锈钢底板
5. 玻璃铝幕墙
6. 不锈钢入口门
7. 入口地垫
8. 石铺面
9. 照明
10. 分隔入口门廊

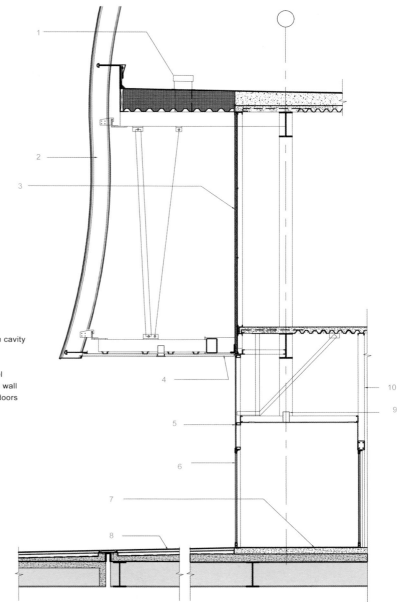

项目概况

纳蒂克内曼·马库斯百货商店充分体现了该品牌因地制宜的形象策略。此外，内曼·马库斯还要求建筑能够反映他们的女性客户基础。Elkus Manfredi建筑事务所为商店设计了波浪式的不锈钢外立面，像是由内曼·马库斯高端定制的精致的丝绸连衣裙或丝巾，在新英格兰的海风中迎风飘动。建筑周边的景观令人联想起潮汐沼泽的海草、传统的石墙和桦树林。

细部与材料

项目最引人注目的元素就是双层楼高的波浪形不锈钢外立面，它以古铜色、香槟色和银色呈现出丝巾或连衣裙的效果。为了实现织物的效果，建筑师不仅将重点放在造型上，还从色彩、纹理、图案等多方面进行了考虑。金属的色彩十分经典，呈现为古铜色、香槟色和银色。图案与造型相互映衬，并且提升了入口、招牌等建筑细节。

如何将结构钢材与多彩的不锈钢板结合起来是项目面临的首要挑战，要求许多供应商和承包商通力合作。

与不锈钢板供应商的初期协调保证了钢板与结构钢材和谐统一。项目首先进行计算机三维空间建模，然后结构钢材制造商跟着这一信息制造轴承座。结构元件和不锈钢板的前期规划保证了各种颜色的彩带都呈现出完美的效果。

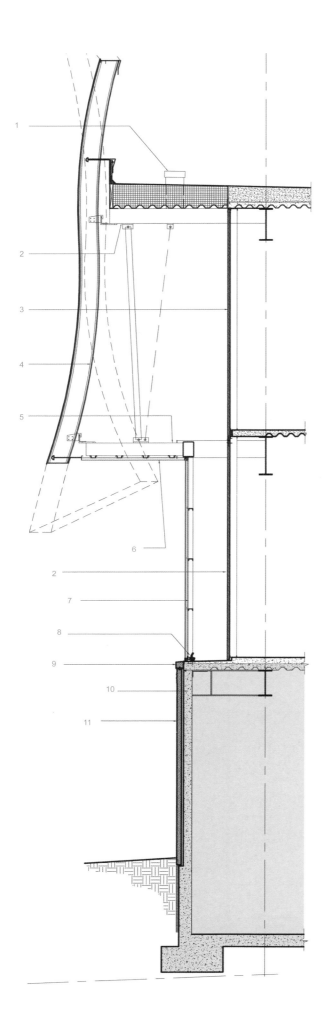

Polycarbonate wall

1. Roof vents for rain screen cavity
2. Steel outriggers-armature flushed framed at top of seel
3. Metal panel
4. Architectural metal rain screen and framing system
5. Galvanised steel outriggers
6. Stainless steel perforated soffit
7. Polycarbonate wall system
8. Light fixture
9. Granite silt
10. Concrete foundation wall
11. 4" granite wall

聚碳酸酯墙面

1. 挡雨板空洞处屋顶通风口
2. 钢板顶部钢框支架
3. 金属板
4. 金属雨幕和框架系统
5. 镀锌钢支架
6. 穿孔不锈钢底板
7. 聚碳酸酯墙面系统
8. 灯具
9. 花岗岩渣
10. 混凝土地基墙
11. 4"花岗岩墙

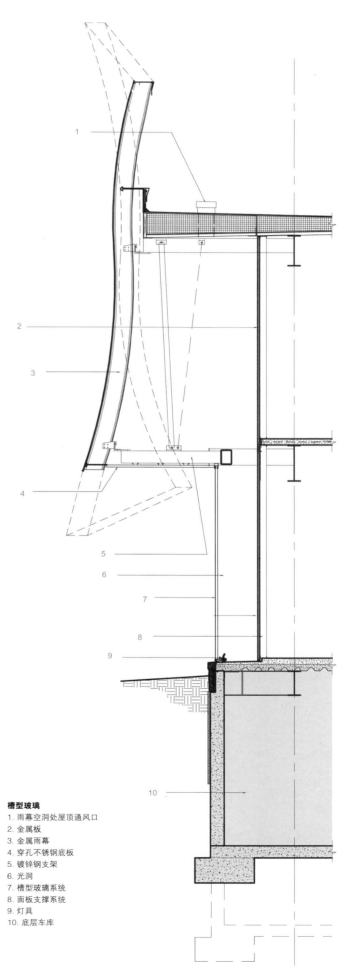

Channel glass

1. Roof vents for rain screen cavity
2. Metal panel
3. Metal rain screen
4. Stainless steel perforated soffit
5. Galvanised steel outrigger
6. Light cavity
7. Channel glass system
8. Panel support system
9. Light fixture
10. Lower level garage

槽型玻璃

1. 雨幕空洞处屋顶通风口
2. 金属板
3. 金属雨幕
4. 穿孔不锈钢底板
5. 镀锌钢支架
6. 光洞
7. 槽型玻璃系统
8. 面板支撑系统
9. 灯具
10. 底层车库

FRIEM Headquarters

弗利姆总部

Location/地点: Milan, Italy/意大利，米兰
Architect/建筑师: Angelo Lunati, Luca Varesi
Design team/设计团队: Thilo de Gregorio, Rossella Locatelli, Francesco Marilli, Mariana Sendas, Giulia Vrespa, Roberto Lamanna
Photos/摄影: Helene Binet
Completion date/竣工时间: 2010

Key materials: Façade – stainless steel, glass
主要材料：立面——不锈钢、玻璃

Overview

The main purpose of the project is the overall re-organisation of an area used for the production of electrical converters, located in an industrial compound along the Cassanese state highway, east of Milan.

The main idea is to define the boundaries of the production site by building a continuous wall to create a feeling of an enclosed urban block. The building "thickens" this wall near the angle, separating the internal space from the external one, with an internal garden and public spaces along the street.

The project foresees research spaces on the ground floor, withdrawn and facing the internal garden and offices on the first floor. The volume folds around the angle and stretches out with two bodies, characterised by a continuous glass façade: the first body, facing the birch garden, is designed to be used for directors' offices, while the second one, facing the street, will be used for the operating offices. On the far west side, in direction of the state road, the volume becomes a vertical body, with the tower hosting the technical stations.

Energy efficiency is achieved through several strategies: a high performance covering system for both opaque and transparent portions, external shielding systems against solar radiation, internal curtains, technical systems based on heat pumps and the recovery of rain water for irrigation.

Detail and Materials

The metal skin, constituted by shaped and

drilled sheet metal in opaque stainless steel, envelopes the building entirely like a curtain, thus conferring to the volume a balanced character, where the opaque and transparent portions are ambiguously defined. The skin filters the light into the building at different opening degrees, depending on the exposure to the sun and the degree of intimacy desired in the interior.

As compared to indistinctly open spaces, generally used for the buildings for the tertiary sector or to the anonymous forms standing in the background of large coloured signs, which are typically of the suburban areas of European cities, the general perception is of a substantially homogenous building, whose unique character derives from the complexity of the covering. In this sense, the building tries to find a balance between the identity purpose of FRIEM and a

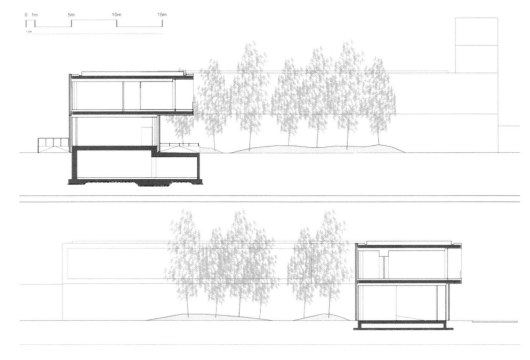

general urban attitude defining rules and hierarchies.

The texture of the metal facing was developed in several phases, starting with standard industrial parts, like elongated, drilled and bent plates, in order to create a solution specifically designed to convey the idea of a varied covering both at a layout and visual level. The self-supporting elements with mechanical attachment at the ends have different widths and height varying from 4 to 4.5m. The selected sheet metal is AISI 304 stainless steel with a thickness of 2mm, press-formed with custom openings cut by lasers with fine brushed finish.

The contrast between the apparent monolithic purpose and its detailed organisation, participates to the abstraction effort on which all the project is based and confers to it a particular and recognisable identity.

项目概况

项目的主要目的是在米兰东部卡萨尼斯州际高速公路旁的工业区打造一个电子复律器制造总部。

设计的主要想法是利用连续的墙壁来界定生产区，形成一座封闭的城市建筑。建筑将这面"墙"在转角处增厚，隔离了内外空间，形成了内部花园和沿街的公共空间。

项目将研究空间设在一楼，朝向内部花园，而将办公室设在二楼。建筑沿着转角折叠，形成两个结构，采用相同的连续玻璃立面。第一个结构朝向桦树花园，被用作主管办公室；第二个结构临街，被用作运营办公室。在西侧，朝向州际公路的方向，建筑呈直立状态，在塔楼内设置着技术站。

项目通过以下策略实现了节能设计：高性能的覆盖系统（分为不透明和透明两部分）、外部遮阳系统、内部窗帘、热泵技术系统以及雨水收集灌溉系统。

细部与材料

由穿孔不锈钢所制成的金属表皮像幕帘一样将整个建筑包围起来，赋予了建筑均衡的特色，模糊了透明与不透明部分的界限。表皮以不同的开放度过滤进入室内的阳光，开放程度取决于空间的曝光度和室内空间的隐私需求。

这种朦胧的开放空间通常应用于第三产业建筑或位于巨大的彩色标牌背景下的无名建筑（这在欧洲城市的近郊十分常见）。该项目打造了一座具有可持续性的建筑，其特色在于立面的复杂性。建筑视图在弗利姆公司的形象和城市规划准则之间找到一个平衡点。

金属立面纹理的开发经过了若干个阶段：从拉伸、钻孔、弯曲等标准工业制作开始，目标是在布局和视觉上实现建筑独有的立面设计。自支撑元件在两端设有机械连接点，宽度和高度在4~4.5米之间。项目选用AISI 304型不锈钢板，厚度为2毫米，被压制出各种造型，以激光开口，进行精刷面处理。

建筑的整体感与精致的布局之间形成了鲜明的对比，奠定了项目的设计基础并展现出独特而具有辨识度的形象。

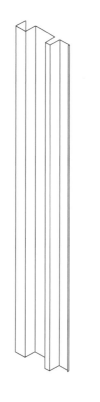 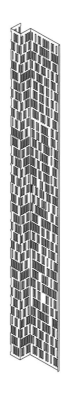 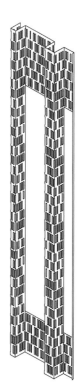

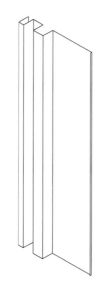 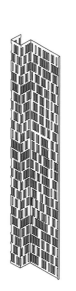 

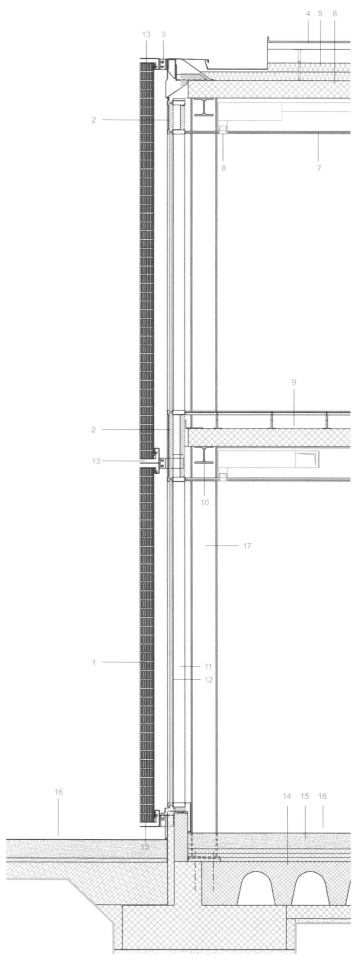

Section
1. Screen, shake and drilled stainless steel aisi 304
2. Painted Aluminium sheet
3. Support for shaped sheet metal stainless steel
4. Shaped Aluminium metal cover made
5. Rock wool insulation
6. Vapor barrier
7. False ceiling with sound insulation
8. Linear diffuser for primary air supply
9. Reinforced concrete floor
10. Heb 200 beam of connection between steel portals
11. Painted Aluminium upright curtain wall
12. Glass: 8mm / 16 mm with argon gas 4+4 mm
13. Stainless steel string-course
14. Ventilated crawl space
15. Polystyrene insulating layer
16. Industrial floor
17. HEA 300 steel pillar

建筑剖面
1. 幕墙，定型穿孔不锈钢板AISI 304
2. 涂漆铝板
3. 定型穿孔不锈钢板支架
4. 定型铝金属面
5. 石棉隔热层
6. 隔汽层
7. 隔音假吊顶
8. 条形散流器
9. 钢筋混凝土地面
10. HEB 200钢门连接梁
11. 涂漆铝幕墙
12. 玻璃；8mm/16mm，氢气夹层4+4mm
13. 不锈钢弦
14. 通风空隙层
15. 聚苯乙烯隔热层
16. 工业地面
17. HEA 300钢柱

119

Office Building, Ravezies

莱夫西斯办公楼

Location/地点: Bordeaux, France/法国，波尔多
Architect/建筑师: FLINT, architect
Photos/摄影: Arthur Pequin
Built area/建筑面积: 3,250m²
Completion date/竣工时间: 2010

Key materials: Façade – corrugated steel sheet, Aluminium
Structure – concrete
主要材料：立面——波纹钢板、铝材
结构——混凝土

Overview
This office building is set in the north of Bordeaux, in the heart of a new district.

The developer, ICADE, is one of the most important in France. For this project, their idea was to build it for themselves with the intention of having a high degree level of sustainable conditions.

They focused our design on a generic form, considering that a cube would meet many of the aimed targets. However we also tried to get away from the well known image of an office building that has its windows on a line between 1 m and 2.5m from the floor of each level.

So this generic volume would have 3 different programmatic strata, the parking lot would be located beneath the ground, the ground floor would be a public showroom, and the 4 upper levels would be office space.

Detail and Materials
The ground floor is fully glazed to allow easy identification of what is visible inside to passersby.

The office block façade is clad with corrugated steel sheets reflecting shades ranging from silver to gold. The windows are laid out every 2.7m so that the interior space can be freely planned. The influence of the abstraction of this skin came

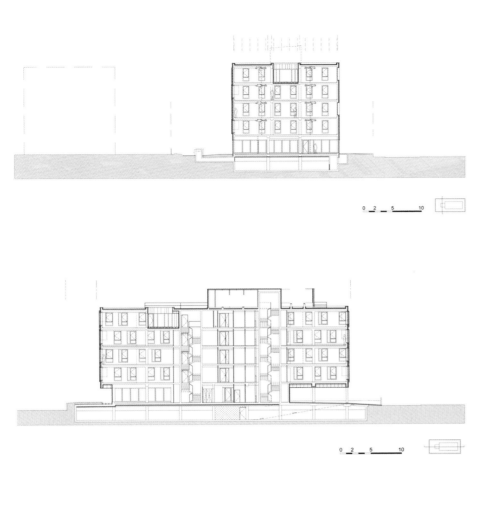

from the work of Donald Judd and his work on simple geometrical forms and specifically the box.

To provide shadow, the windows facing east, south and west have manually sliding shutters with a very simple, specifically designed mechanism.

The shutters are made of stainless steel sheets perforated with a maximum of 30% so that even when closed there is still enough daylight.

The difference in appearance between the materials, whether they reflect or absorb light, gives to the façades depth and vibrations. Around the building, and because of urban rules, the left over space is turned into a garden.

项目概况

这座办公楼位于波尔多南部莱夫西斯新区的中心。开发商ICADE公司是行业中的佼佼者，他们希望将这座建筑作为自己的办公楼，同时希望在设计中加入大量的可持续元素。

设计选择了通用造型，因为立方体可以满足多方面的要求。然而，建筑师决定摒弃办公楼的一般形象，并没有让各个楼层的窗口离楼面1~2.5米呈直线排列。

立方体建筑分为三个功能区：停车场设在地下，一楼是公共展览室，而上面四层才是普通的办公空间。

细部与材料

一楼全面采用玻璃幕墙围起，让过往的行人可以清晰地看到室内的景象，增加了建筑的辨识度。上层办公空间的立面采用波纹钢板覆盖，反射出金银的色彩。窗户的间隔达2.7米，可以灵活安排室内空间。抽象的表皮设计的灵感来自唐纳德·贾德，他的作品以简单的几何造型（特别是盒子造型）为特色。

为了遮阳，朝向东、西、南三面的窗户设有手动百叶窗。百叶窗的机械系统十分简单，由30%镂空的不锈钢板制成，即使全面封闭也能提供充足的日光。

多样化的材料外观为建筑立面带来了层次感和变化感。为了满足城市规划需求，建筑四周的空地被改造成了花园。

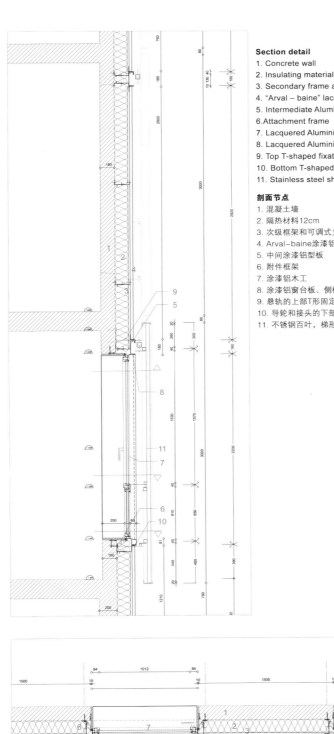

Section detail
1. Concrete wall
2. Insulating material 12cm
3. Secondary frame and adjustable bracket
4. "Arval – baine" lacquered Aluminium cladding
5. Intermediate Aluminium lacquered shaped plate
6. Attachment frame
7. Lacquered Aluminium carpentry
8. Lacquered Aluminium sill pans, sides and lintel
9. Top T-shaped fixation for suspension rail
10. Bottom T-shaped fixation for guiding wheel & abutment
11. Stainless steel shutter trapezoid profiles in frame

剖面节点
1. 混凝土墙
2. 隔热材料12cm
3. 次级框架和可调式支架
4. Arval-baine涂漆铝包层
5. 中间涂漆铝型板
6. 附件框架
7. 涂漆铝木工
8. 涂漆铝窗台板、侧板和过梁
9. 悬轨的上部T形固定
10. 导轮和接头的下部T形固定
11. 不锈钢百叶，梯形剖面框架

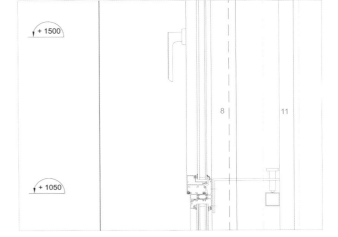

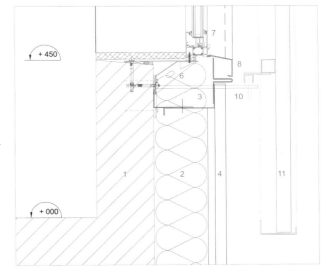

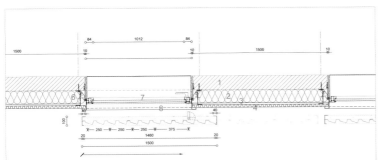

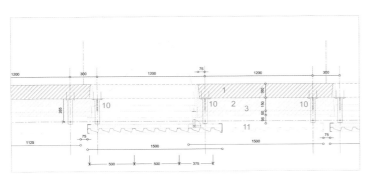

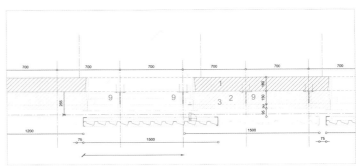

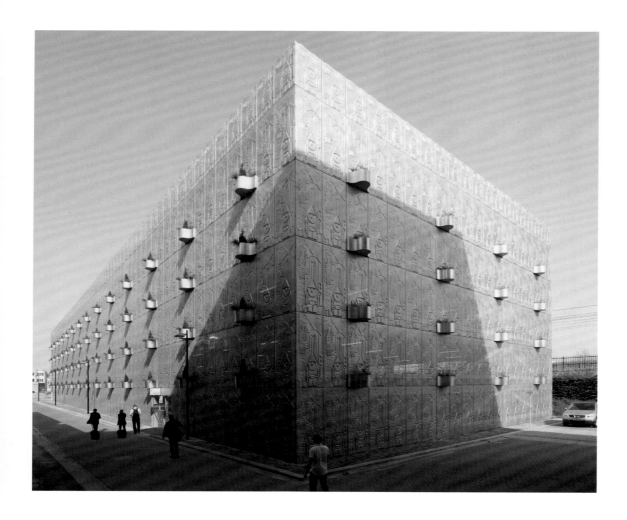

Gnome Parking Garage
小矮人停车楼

Location/地点: Almere, The Netherlands/荷兰, 阿尔梅勒
Architect/建筑师: Mei architects and planners stedenbouwers
Team/团队: Robert Winkel, Menno van der Woude, Hennie Dankers, Robert Platje, Pepijn Berghout, Maurice de Ruijter, Nars Broekharst
Photos/摄影: Jeroen Musch, Rotterdam
Construction area/建筑面积: 14,500m² (413 parking places)（413个停车位）
Completion date/竣工时间: 2011
Green supplies/景观设计:
Copijn garden and landscape architects
Copijn garden and landscape architects takes care of the plants on the façade. On each façade there are specific plants based on the orientation of the façade.
Copijn花园景观建筑事务所负责立面上的植物设计。根据不同的朝向，每个立面上都栽种有不同的植物。
Building costs/建筑成本: 5,400,000 Euro/欧元

Key materials: perforated stainless steel panel
主要材料：穿孔不锈钢板

Façade material producer:
外墙立面材料生产商：
The Voest Alpine (automotive company)

Overview
The development of the Gnome parking garage, with 413 parking spaces, is part of the expansion of the city centre of Almere-Buiten. Features of this plan are a lot of green and a clear urban structure. At the architectural level many experiments with innovative materialisation and new constructions have been made, as also in the Gnome parking garage.

The garage is covered with perforated stainless-steel panels that convey the green character of Almere. All new buildings in Almere Buiten must convey the green character of Almere. They usually do this with green roofs and terraces. While exploring the green character of Almere, we came across front gardens with gnomes and windmills, and fringes of reeds in the surrounding polders, and birdhouses in municipal trees. And since the green is not always present in the winter, we decided in consultation with the client (municipality of Almere) to cover the car park with figuration depicting the everyday greenery of Almere Buiten. Such a lavishly adorned building seldom appears, and in Almere Buiten the building looks like an old-fashioned tin tobacco box between the flat brick and rubble walls that are common there. The figuration and narrative skin make this a welcome change in the somewhat dreary centre of Almere Buiten.

Detail and Materials
The Voest Alpine automotive company in Spakenburg, previously known as Polynorm, made the stainless-steel panels. These panels are vacuum formed with the help of 3D computer techniques, special sheet-metal perforating machines, and moulds milled from cast steel. The small square perforations are specially rounded because of the danger that the corners of the perforations would tear when the sheets are being 3D vacuumed. Square-shaped perforations were chosen because the entire steel structure is composed of square tube profiles. Material development has therefore been conducted

in almost molecular fashion, and the stainless-steel skin acquires a relation with the supporting steel structure. This will be the only building in the Netherlands with such panels, given that Voest Alpine Austria has decided to confine its activities to the automotive industry. The delicacy of the material and its treatment with the accompanying production methods mark a return to the idea of crafted products in unique fashion.

The stainless-steel panels are designed in such a way that their air permeability is 30% on account of the ventilation requirements for car parks. They are also detailed for safety to prevent people from falling through them, and are also almost entirely transparent during daylight from inside the garage and, as a result, ensure a socially safe space.

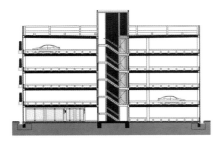

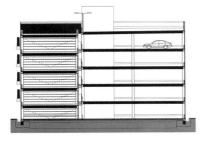

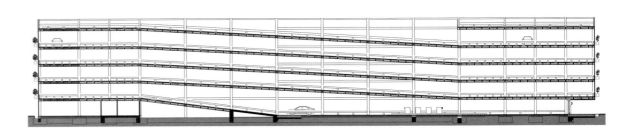

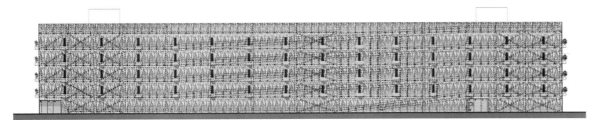

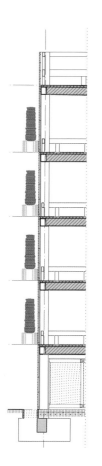
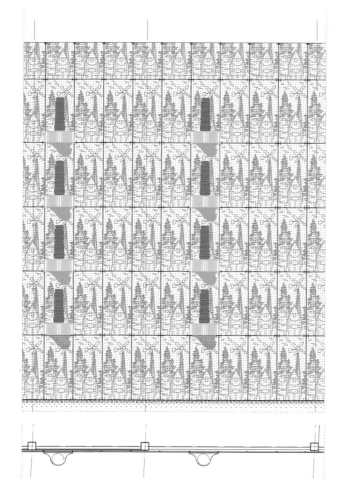

项目概况

小矮人停车楼拥有413个停车位，是阿尔梅勒布顿市中心扩建的工程的一部分。项目特色在于绿色景观和清晰的城市结构。从建筑层面上讲，小矮人停车楼进行了多种创新材料和构造技术的试验。

停车楼以穿孔不锈钢板覆盖，体现了阿尔梅勒的绿色特质。阿尔梅勒布顿的所有新建筑都必须体现该城的绿色特质。大多数建筑选择采用绿色屋顶和露台。在研究阿尔梅勒的绿色特质时，建筑师发现了带有小矮人和风车的前庭花园、围垦地里的芦苇和市政树木上的鸟舍。因为大多数植物不能保持常青，建筑师与市政当局磋商，决定用图形来描绘出城市的绿色特质。带有这样复杂装饰的建筑十分少见，在一片砖石围墙的建筑中，停车楼像一个旧式的锡铁烟盒。带有纷繁图案的建筑表皮让停车楼在略显单调的市中心深受欢迎。

细部与材料

斯巴肯堡的奥钢联汽车公司为停车楼制造了不锈钢板。这些板材利用三维计算机技术、特制的金属穿孔机和铸钢建模实现真空成形。为了避免三维真空成形时撕裂，小方孔特别采用了圆角。方形穿孔的选择是因为整个钢结构都由方形管材组成，这让不锈钢表皮与支撑钢结构实现了联系。这将是荷兰唯一一座使用此种面板的建筑，因为奥钢联公司现在仅承接汽车工业项目。精致的材料和处理方式象征着传统手工艺的回归。

不锈钢板的设计让透气率达到了30%，满足了停车场的通风需求。立面还保证了人们不会撞墙跌倒。白天，停车楼几乎是全透明的，形成了一个富有社会安全感的空间。

Façade detail
1. Coupling panels
   Stainless steel bolt & nut with M10 connection
   A plastic spacer sleeve between
2. Guardrail structure
   Uprights: Steel tube
   Beams: steel tube
3. Composite steel bracket d = 4.00mm
   Stainless steel pins around 10mm
   TBV mounting panels
4. Shims serving align
5. Stud(stud welding)
   M12x40mm, 2x, module size 1238mm
6. Line façade grid
7. Steel façade panel
   40% perforated
   Secured to steel floor beam
   Mutually coupled
8. THQ steel girder
   Connection
   Kluft connection by means
   Kluft connection
   Fire-resistant coating
   Declared constructor
9. Steel column 350x350
   Fire-resistant coating
10. Composite panels mounting bracket
11. Infill
    Power floated low pressure
    Hollow core slab

立面节点
1. 连接板
   M10连接不锈钢螺栓和螺母
   中间带有塑料挡套
2. 护栏结构
   支柱：钢管
   横梁：钢管
3. 复合钢支架d=4.00mm
   不锈钢栓，约10mm
   TBV安装板
4. 垫片
5. 螺柱（螺柱焊接）
   M12x40mm，2x，模块尺寸1238mm
6. 立面网格线
7. 立面钢板
   40%穿孔
   固定在钢板地面梁上
   双向接合
8. THQ钢梁
   连接
   缝隙连接方式
   缝隙连接
   耐火涂层
   公开建造商
9. 钢柱350x350
   耐火涂层
10. 复合板安装支架
11. 填充层
    低压机动抹板
    空心板

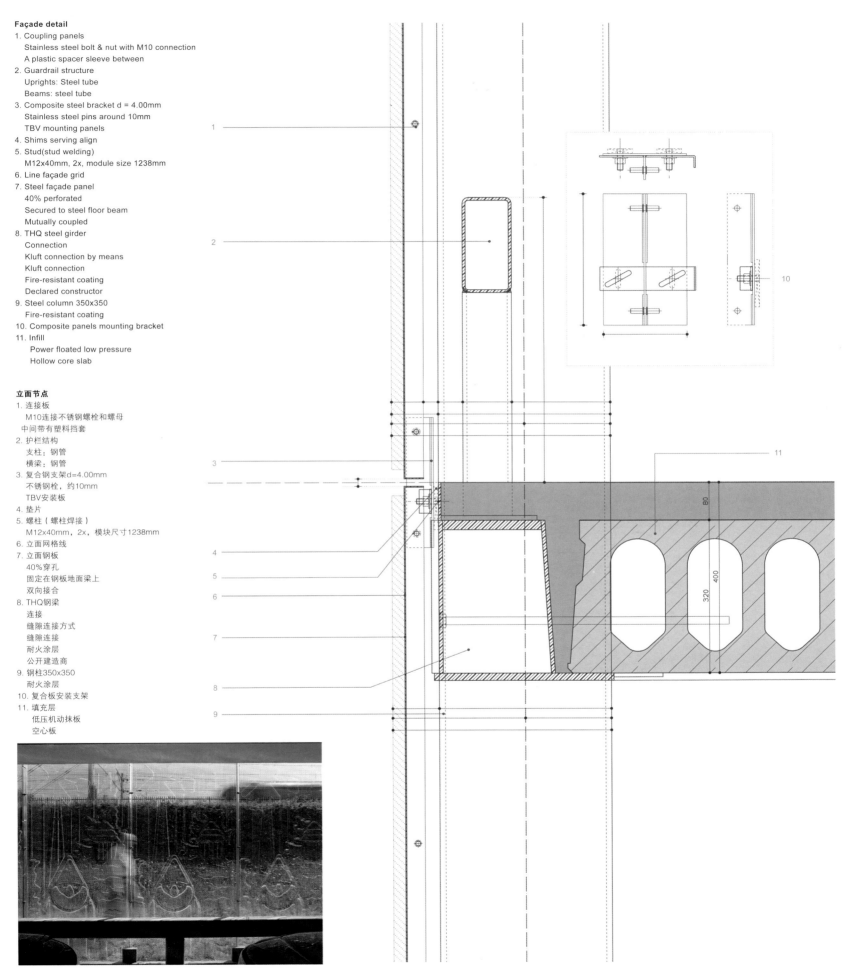

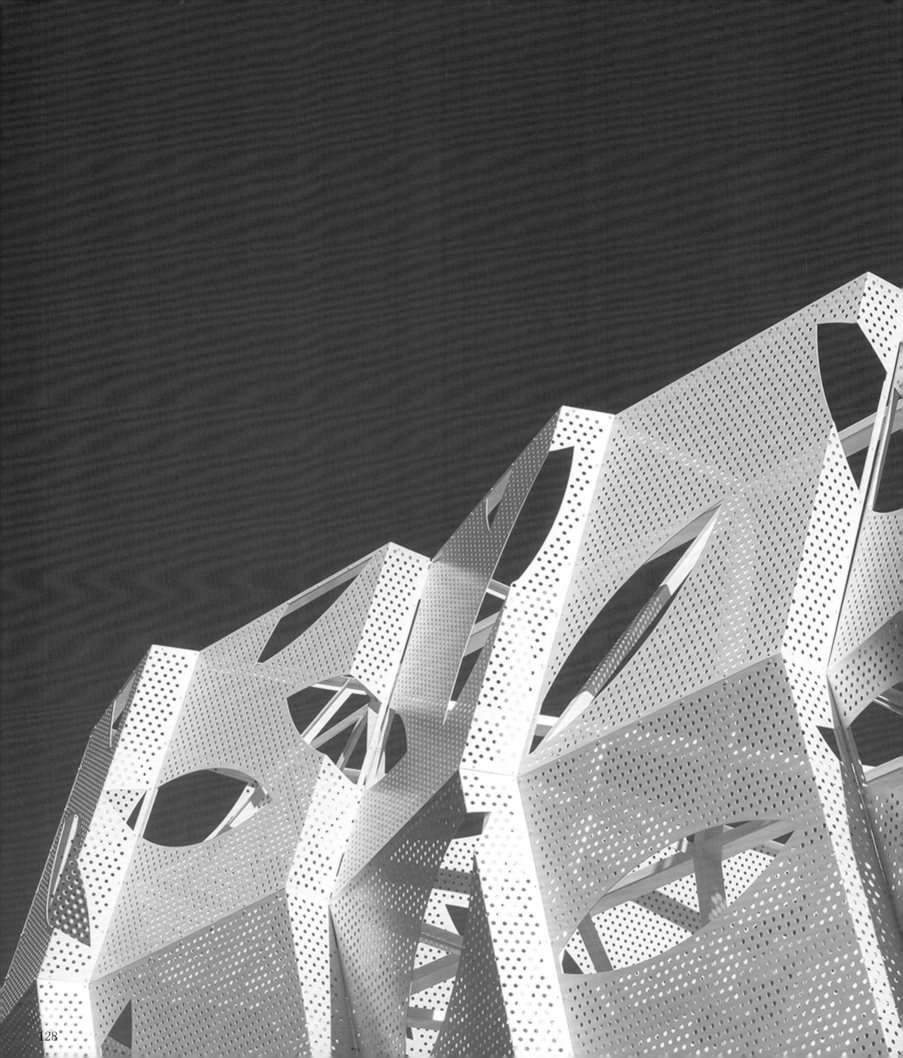

Chapter 2
Basic Information of Aluminium

第二章 铝

Aluminium (or Aluminum) is a chemical element in the boron group with symbol Al and atomic number 13. It is a silvery white, soft, ductile metal. Aluminium is the third most abundant element (after oxygen and silicon), and the most abundant metal, in the Earth's crust. Aluminium is remarkable for the metal's low density and for its ability to resist corrosion due to the phenomenon of passivation. Structural components made from aluminium and its alloys are vital to the aerospace industry and are important in other areas of transportation and structural materials.

Aluminium use in buildings covers a wide range of applications, including roofing, foil insulation, windows, cladding, doors, and architectural hardware and so on.

铝是一种化学元素，属于硼族元素，其化学符号是Al，原子序数是13。铝是一种较软、易延展的银白色金属。铝是地壳中第三大丰度的元素（仅次于氧和硅），也是丰度最大的金属。铝因其低密度以及耐腐蚀（由于钝化现象）而受到重视。利用铝及其合金制造的结构件不仅在航空航天工业中非常关键，在交通和结构材料领域也非常重要。

铝在建筑中有广泛的应用，其中包括屋顶、箔绝缘、窗户、覆盖层、门、建筑五金等。

Aluminium's corrosion resistance can be excellent due to a thin surface layer of aluminium oxide that forms when the metal is exposed to air, effectively preventing further oxidation. Experimental results show that aluminium far exceeds general steel in corrosion resistance. Still, the oxidised "skin" of aluminium is thin and soft, and can be damaged by external forces or bumps. In general condition, untreated aluminium sheet's service life can be about 20 years, without an ideal corrosion resistance. Special treatments such as chemical oxidation or electrochemical oxidation can produce an oxide layer of 0.025 to 0.05mm (or even thicker). PVDF coated aluminium sheet's service life is long as 50 years or more.

Aluminium also has several properties: strength to ratio, electrical and thermal conductivity, light and heat reflectivity, toxicity, recycling.

Aluminium alloy and Aluminium composite panels are prevalently used in architectural industries. (See Figure 2.1, Figure 2.2)

2.1 Aluminium Alloys
Definition

Aluminium alloys (or Aluminium alloys) are alloys in which aluminium (Al) is the predominant metal. The typical alloying elements are copper, magnesium, manganese, silicon and zinc.

There are two principal classifications, namely casting alloys and wrought alloys. About 85% of aluminium is used

铝在空气中氧化之后，表面会生成一层牢固而致密的氧化膜，这层防腐外衣可以防止内部被继续氧化。实验证明，铝的抗腐蚀性远远超过一般的钢材。尽管如此，铝的自然氧化"外衣"依然存在薄、软的缺点，容易因外力摩擦、磕碰等而遭到破坏。一般情况下，未经特殊处理的铝板使用寿命多在20年左右，其防腐性能还未达到理想要求。通过特殊的处理方式，如化学氧化或电化学氧化的方式可以在其表面生成一层0.025~0.05毫米（甚至更厚）的氧化膜，如经氟碳漆（PVDF）涂饰的铝板，其使用寿命长达50年或更长。铝还具有一些其他属性：高强度比、导电性、导热性、重量轻、高反射率、毒性和可回收性。

铝合金与复合铝板被普遍应用于建筑工业。（见图2.1、图2.2）

2.1 铝合金
定义

铝合金是以铝为主要金属的合金，其主要合成元素为铜、镁、锰、硅和锌。

Figure 2.1, Figure 2.2 Buildings enveloped by Aluminium **图2.1、图2.2** 铝外壳建筑

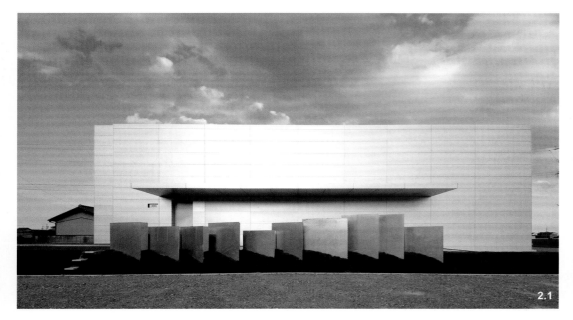

for wrought products, for example rolled plate, foils and extrusions. Cast aluminium alloys yield cost-effective products due to the low melting point, although they generally have lower tensile strengths than wrought alloys. The most important cast aluminium alloy system is Al-Si, where the high levels of silicon (4.0-13%) contribute to give good casting characteristics. Aluminium alloys are widely used in engineering structures and components where light weight or corrosion resistance is required.[1]

Properties
Aluminium alloy surfaces will formulate a white, protective layer of corrosion aluminium oxide if left unprotected by anodising and/or correct painting procedures. In a wet environment, galvanic corrosion can occur when an aluminium alloy is placed in electrical contact with other metals with more negative corrosion potentials than aluminium, and an electrolyte is present that allows ion exchange. Referred to as dissimilar metal corrosion this process can occur as exfoliation or intergranular corrosion. Aluminium alloys can be improperly heat treated. This causes internal element separation and the metal corrodes from the inside out.

Another important property of aluminium alloys is their sensitivity to heat. Workshop procedures involving heating are complicated by the fact that aluminium, unlike steel, melts without first glowing red. Forming operations where a blow torch is used therefore require some expertise, since no visual signs reveal how close the material is to melting. Aluminium alloys, like all structural alloys, also are subject to internal stresses following heating operations such as welding and casting. The problem with aluminium alloys in this regard is their low melting point, which make them more susceptible to distortions from thermally induced stress relief. Controlled stress relief can be done during manufacturing by heat-treating the parts in an oven, followed by gradual cooling – in effect annealing the stresses.

Aluminium Alloys VS Types of Steel
Compared to steel, aluminium alloys have higher specific strengths. The densities of most aluminium alloys are about 1/3 of that of steel. In fact, the strength ratio of Type 6000 aluminium alloy commonly used in architecture is higher than that of commonly-used carbon steel. Therefore, the superstructure made of aluminium alloys instead of steel or concrete will reduce the self weight of structure significantly, thus lowering construction intensity, shortening construction period, as well as expanding the structure's extreme span and height.

The prices of most aluminium alloys are 7 to 10 times to that of steel. As the densities are commonly 1/3 of that of

铝合金主要分为两类：铸造用铝合金和锻造用合金。大约85%的铝被用于锻造用品，如轧制铝板、铝箔和挤制铝板。尽管铸造用铝合金的拉伸强度较低，但是由于它的熔点低，铸造用铝合金不能制造出经济划算的产品。铸造用铝合金中最重要一种是铝–硅合金，高含量的硅（4.0%~13%）使其具有良好的铸造特性。铝合金被广泛应用于要求重量轻、耐腐蚀的工程结构和组件中。[1]

属性
如果不经过阳极氧化或涂层处理，铝合金表面会形成白色氧化膜。在潮湿的环境中，当铝合金与其他更耐腐蚀的金属相接触或由电解液实现离子交换时，会发生电偶腐蚀。这种过程又被称为异种金属腐蚀，可能以剥落或晶间腐蚀的形式出现。不当的热处理会使铝合金导致内部元素分离和由内至外的金属腐蚀。

铝合金另一个重要的属性是热敏感性。与钢不同，铝的熔化没有发红的过程，因此涉及到加热的加工程序更为复杂。因为没有明显的熔化信号，使用焊接用喷灯的成型操作要求专业人员来进行。与所有结构合金一样，铝合金也受焊接、铸造等加热操作所形成内在应力支配。由于铝合金熔点较低，它们对热感应应力所导致的扭曲更为明显。在制造过程中，可以通过在烤炉中进行加热处理来实现可控的应力消除，然后再进行逐渐冷却。

铝合金与钢材的比较
与钢材相比，铝合金的最大优势在于它的比强度高。铝合金的密度约为钢材的1/3，建筑中常用的6000系列的铝合金的强度比一般常用的碳素钢的强度还要大。因此，采用铝合金代替钢材或者混凝土建造房屋建筑的上部结构，可以大大减少结构的自重，从而减小施工强度，缩短施工周期等，同时还可以拓展结构的极限跨度和高度。

但相对于钢材而言，铝合金型材的价格约为钢型材价格的7~10倍，而考虑到其密度只有钢材的1/3，铝合金使用的直接价格约为钢材的3倍左右。如若考虑下部结构、施工的便捷与周期和维护费用，铝合金结构还有

Figure 2.3, Figure 2.4 Aluminium alloy encloses the building 　图2.3、图2.4 铝合金包层建筑

steel, the direct prices of aluminium alloys are 3 times to steel. With considerations of factors such as substructure, convenient construction, short construction period and maintenance cost, aluminium alloy structure is price competitive, especially in the construction of large-span space structure.

Alluminum Alloys in Architectural Field

In architectural field, as structural and finishing materials, aluminium alloys are extensively used. For example, aluminium alloy of Type 3000 Al-Mn-Mg is commonly used in roofing and walls due to its good ductility, corrosion resistance and processability. In addition, aluminium alloys can be extruded as various profiles to be used as light-weighted structural beam columns, door and window frames, curtain wall brackets, finishing profiles, etc.(See Figure 2.3, Figure 2.4)

2.2 Aluminium Composite Panel
Definition

Aluminium Composite Panel (ACP) also Aluminium Composite Material, (ACM) is a type of flat panel that consists of two thin aluminium sheets bonded to a non-aluminium core. Many people use Dibond or sandwich panel as a generic name (genericized trademark) for Aluminium composite panels that are typically 0.3 to 0.5mm thick Aluminium sheets covering a solid polyethylene core with a total thickness of 3mm or

其价格竞争优势，尤其在能充分发挥其优势的大跨空间结构方面。

铝合金在建筑中的应用

在建筑中，铝合金作为结构和装饰材料具有重要的用途，是建筑装饰领域最为广泛使用的材料之一。其中，以铝锰镁合金为主的3000系列板材因其良好的延展性、抗腐蚀性能、加工性能等被广泛用于建筑屋面和墙面等结构。此外，铝合金还可以通过挤压法生产出各种断面的型材，用作轻型结构梁柱、门窗框架、幕墙结构架及装饰型材等。（见图2.3、图2.4）

2.2 复合铝板
定义

复合铝板又称复合铝材，是一种由两层薄铝板和其他材料的夹层所形成的平板材料。许多人以商品名Dibond或夹心板来称呼由0.3~0.5mm铝板和聚乙烯夹层所组成的3mm复合铝板或类似的复合铝板。[2]

more, or similar panels from any manufacturer.[2]

Properties

Aluminium sheets can be coated with PVDF, fluoropolymer resins (FEVE) or Polyester paint. Aluminium can be painted in any kind of colour, and ACPs are produced in a wide range of metallic and non-metallic colours as well as patterns that imitate other materials, such as wood or marble. The core is commonly low density polyethylene, or a mix of low density polyethylene and mineral material to exhibit fire retardant properties.

Aluminium Composite Panel in Architectural Field

ACP is mainly used for external and internal architectural cladding or partitions, false ceilings, signage, machine coverings, container construction, etc. Applications of ACP are not limited to external building cladding, but can also be used in any form of cladding such as partitions, false ceilings, etc. ACP is also widely used within the signage industry as an alternative to heavier, more expensive substrates. (See Figure 2.5, Figure 2.6)

Reference / 参考文献
1. J. Polmear, Light Alloys, Arnold, 1995
2. "The original aluminium composite is Dibond". Retrieved 2013-01-31

属性

铝板可以附加聚偏二氟乙烯（PVDF）、氟聚合物树脂（FEVE）或聚酯涂料涂层。铝可以被漆成任意色彩，聚合铝板可以呈现出金属感和非金属感等多种色彩，还可以添加模仿其他材料（如木材或大理石）的纹理图案。复合铝板的夹层通常采用低密度聚乙烯，有时还会添加具有阻燃属性的矿物材料。

复合铝板在建筑中的应用

复合铝板主要用于建筑的内外墙壁包层或隔断、假吊顶、引导标识、机器覆盖层、集装箱建筑等。复合铝板的应用不仅局限于建筑外墙，还可以用作多种包层，例如隔断、假吊顶等。复合铝板还广泛应用于标识产业，用于替代厚重且价格高昂的底板。（见图2.5、图2.6）

Figure 2.5, Figure 2.6 Aluminium composite panels cover the building surface

图2.5、图2.6 复合铝板覆盖的建筑表面

Cluj Arena
克鲁日体育场

Location/地点: Cluj-Napoca, Romania/罗马尼亚，克鲁日
Architect/建筑师: Dico si Tiganas Design Office
Photos/摄影: Cosmin Dragomir
Completion date/竣工时间: 2011
Site area/占地面积: 19,600m²
Construction area/建筑面积: 8,800m²

Key materials: Façade – aluminium perforated sheets, façade cassettes, perforated tape woven into the façade;
Structure – steel

主要材料: 立面——穿孔铝板、立面暗盒、交织穿孔带
结构——钢

Overview
In 1911 the first athletic and soccer stadium was built in Cluj, at the edge of the Central Garden, close to the river. A wooden tribune gave shelter to the spectators. After 50 years, in 1961, a new concrete stadium was built in replacement. It had a horse shoe shape, and received after some time the name of an athlete. After another 50 years, following more than a decade of debates and proposals, the new Cluj Arena was opened in 2011. It is the most ambitious construction in the City finalized in Cluj in the last more than 30 years. The design principles are targeting a special connection of the building with the surrounding public space. It offers from far views a silhouette which shapes the skyline. Closer you may discover a big volume, breathing through its' skin and reflecting the colours of the clouds, the sunset and the river. Even closer you may circle it walking on the high promenade and looking inside. This visual connection is one of the key drivers of the concept. From inside you are focused on the playground but still connected to the landscape. The scale of the building is diminished by the artificial hill, covering the underground spaces for training. The grass covered planes are continuing the soccer ground to outside and the alleys surrounding the whole building are somehow like the inside athletic tracks. The public called the new stadium the UFO landed in town. At night it becomes the biggest lamp in town, spreading light and shining through its envelope. A tower was proposed to counterbalance the composition, connected to the stadium and hosting a hotel, conference centre and offices. The verticality of the tower was supposed to rise to be perceived along the river and to announce the presence of the stadium as churches and important public palaces are doing. The tower was designed to be plugged-in the stadium ring via the conference hall, marking the main entrance. The stadium roof was supposed to be developed vertically up to the top of the tower, sheltering a belvedere platform. The times changed and the tower was never built. The Arena waits a new companion, a multi sports indoor hall, under construction, right beside, for a new architectural dialogue.

Detail and Materials
The façades are translucent trough the perforated steel plates, keeping the views accessible to the public from the circulation areas under the tribunes. The slight asymmetry of the four articulated sectors of the stadium was born out of the constrictions of the site for accommodating the requested number of 30,000 spectators. Curves, waves and inclined planes are responding to the movement of athletes, players, people, sun, water and seasons. The façades are made of Aluminium perforated sheets and consists of 5,000m² façade cassettes and 2,000m² of perforated tape woven into the façade.

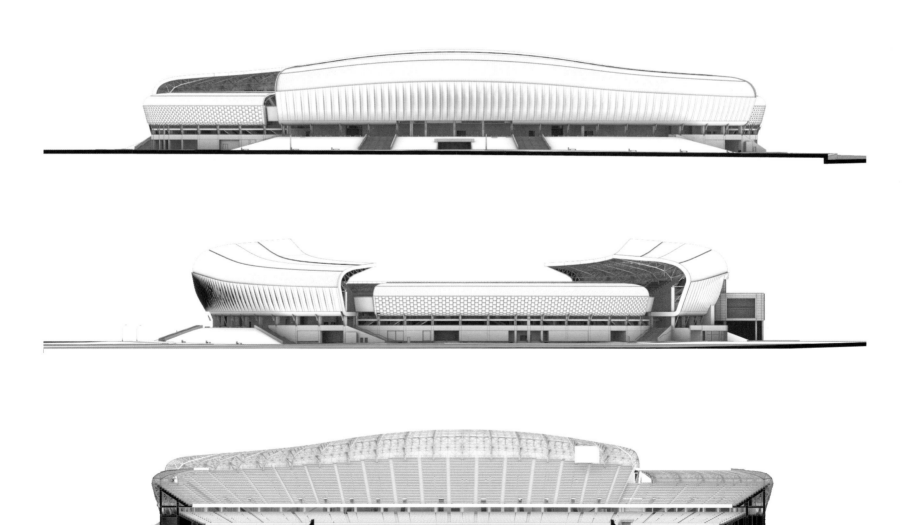

项目概况

1911年，克鲁日在河畔中央花园的边缘建造了它的第一座田径足球体育场，体育场的看台是木制的。50年后的1961年，一座混凝土建成的体育场替代了原体育场，呈马蹄铁形，以一位田径运动员的名字命名。又过了50年，新克鲁日体育场横空出世。它是克鲁日近30多年来最宏伟的建设项目。

设计的目标是在建筑与周边公共环境之间建立起独特的联系。远远看去，体育场在天际线上划出了柔和的轮廓。在近处，你可以看出体育场通透的表皮反射出云彩、夕阳和河流的景色。你可以在高高的长廊上绕着体育场漫步，看到内部的景象。这种视觉连接是设计概念的主要出发点。虽然置身场内，但是仍然与景观相连。体育场旁的人造山体下方是训练场，弱化了体育场庞大的体量感。地面上的绿草将主球场延伸到了体育场外面，而环绕体育场的走道则类似于田径跑道。公众把新体育场叫做"城市中的UFO（不明飞行物）"。夜晚，体育场会变成城中最大电灯，透过表皮熠熠发光。

建筑师曾规划体育场旁将建造一座大厦，内设酒店、会议中心和办公设施。大厦将凸显体育场的存在感，使其像教堂等重要的公共空间一样。大厦通过会议厅与体育场相连，标志出主入口。体育场的屋顶将上升到大厦的顶部，形成观景平台。由于多种原因，大厦一直没有建成。体育场边上正在规划建造一个

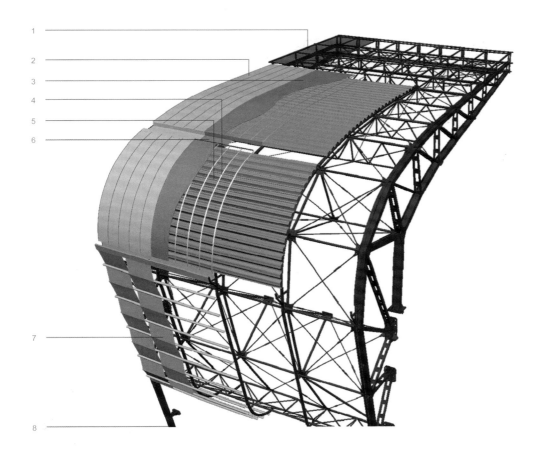

Façade detail
1. Ppolycarbonate sheet
2. Steel standing seam (Onyx White Pearl Effect Gold PVDF coating)
3. Rigid mineral wool
4. Omega steel profiles
5. Corrugated perforated steel sheets
6. Steel chute with Pluvia
7. Braided aluminium sheet with perforations (Onyx White Pearl Effect Gold PVDF coating)
8. Steel façade structure

立面节点
1. 聚碳酸酯板
2. 直立锁边钢板（珠光白效果氟碳烤漆）
3. 刚性矿物棉
4. 欧米茄钢型材
5. 波纹穿孔铝板
6. 钢斜槽
7. 编织穿孔铝板（珠光白效果氟碳烤漆）
8. 钢立面结构

多功能室内体育馆，期待与体育场形成全新的建筑组合。

细部与材料

穿孔金属板让建筑立面呈现为半透明的状态，公众可以在看台下方的通道看到外面的景象。体育场外围相互铰接的四块不对称结构是根据实际需求形成的：建筑场地有限，体育场却要容纳30,000名观众。曲线、波浪和斜面都呼应着运动员、球员、人、太阳、水和四季的运动。建筑立面由穿孔铝板组成，其中包括5,000平方米的立面暗盒和2,000平方米的交织穿孔带。

Maison des Sciences De l'Homme de Dijon, Université de Bourgogne

勃艮第大学人类科学楼

Location/地点: Dijon, France/法国，第戎
Architect/建筑师: Nicolas Guillot architect
Photos/摄影: ©Fabrice Ferrer, ©Erick Saillet
Site area/占地面积: 3,426m²

Key materials: Façade – perforated aluminium tube
Structure – wood, concrete, stone
主要材料： 立面——穿孔铝板
结构——木材、混凝土、石材

Overview
A building in the image of knowledge: both rigorous and unpredictable.

The Maison des Sciences de l'Homme is a platform for interdisciplinary work in the social sciences – a centre for research and the propagation of knowledge. As a hub of the Université de Bourgogne and the CNRS (the French national scientific research organisation), it provides tech¬nical support for researchers and doctoral students.

Architecturally speaking, the building is simple in its form and configuration, and rational in its construction. Like a set of building blocks, its components are superimposed crosswise. Along with individual workspaces, there are places for encounters and social interaction.

Detail and Materials
The forms are recursive, without superfluity, and the overall effect is compact. The range of materials is restricted: there is concrete, Burgundy stone for the facings, wood for the structure, and perforated metal for some of the external surfaces.

There is also a quality of unexpectedness about the building. The metal skin has several functions: that of sunshades, railings and walkways for maintenance purposes. It duplicates the

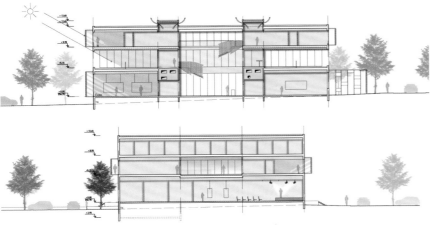

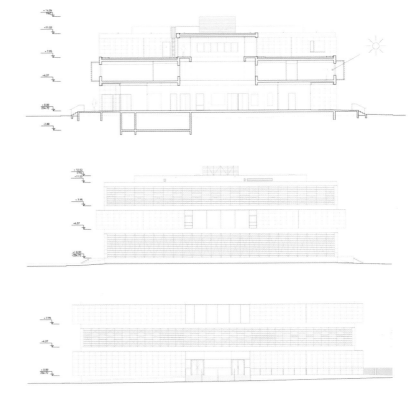

façades, which are open onto the outside world along their entire length. The 90° angle between levels produces a stratified effect. The basic modules of solar protection, with their laser-cut motifs, are used in different ways and combinations, with their presence or absence producing effects of alternation. And there is a progression from the base, which is stable and full, to the top, which is most open, as if to illustrate the fluidity of intellectual processes.

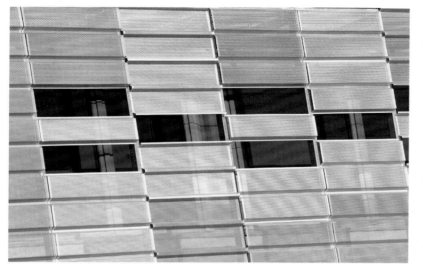

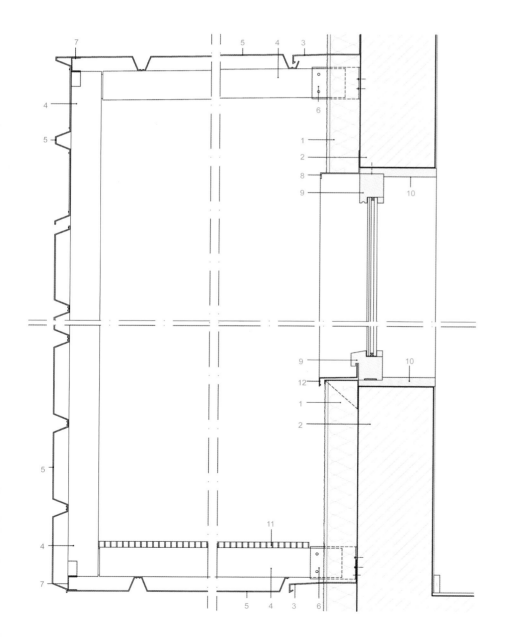

Façade detail
1. Insulation and coating
2. Concrete
3. Bent metal sheeting, for water runoff
4. Galvanised steel tube
5. Folded, perforated aluminium tube
6. Mineral wool-filled, structure-reinforcing sleeve
7. Galvanised right-angle sheeging support
8. Painted insulation-stop flap
9. Untreated larch window frame
10. Three-ply untreated larch slab
11. Metal flooring
12. Painted water runoff flap

立面节点
1. 保温层和涂层
2. 混凝土
3. 弯曲金属板，用于排水
4. 镀锌钢管
5. 折叠穿孔铝板
6. 填充矿物棉的结构加强套
7. 镀锌直角挡板支架
8. 涂漆绝缘翻板
9. 落叶松原木窗框
10. 三层清水松木板
11. 金属地面
12. 涂漆排水翻板

项目概况

一座代表知识的建筑：严谨而出人意料。

人类科学楼是一个社会科学的跨学科平台，是研究和传播知识的中心。作为勃艮第大学和法国国家科学研究组织的核心机构，它为研究人员和博士研究生提供了技术支持。

从建筑方面来讲，建筑造型简洁，结构合理。建筑看起来像积木横向叠加起来一样。除了独立工作区之外，楼内还设有社交和会面空间。

细部与材料

建筑造型反复循环，没有多余的装饰，整体效果紧凑简洁。建筑所采用的材料十分有限：贴面装饰采用混凝土、勃艮第石材，基础结构采用木材、部分外立面采用穿孔铝材。

建筑也有出人意料的地方。金属表皮有多重功能，可用作遮阳、围栏和维护通道。金属表皮复制了整块朝向外界的立面。不同楼层之间的90度夹角产生了层次感。基本遮阳模块上带有激光切割图案，以各种方式组合应用，形成了变化的效果。从建筑底部到顶部有一种渐进效果：底部坚固而封闭，顶部基本完全开放，展示了学术研究的流动性。

Public Library in Ceuta

休达公共图书馆

Location/地点: Ceuta, Spain/西班牙，休达
Architect/建筑师: Ángela García de Paredes, Ignacio Pedrosa / Paredes Pedrosa, arquitectos
Technical control/技术总监: Juan Antonio Zoido, Luis Calvo
Structure/结构设计: Alfonso Gómez Gaite. GOGAITE, S.L.
Mechanical engineers/机械工程: JG Ingenieros
Photos/摄影: Fernando Alda
Area/面积: 6,159m²
Completion date/竣工时间: 2013

Key materials: Façade – perforated aluminium
主要材料： 立面——穿孔铝材

Façade material producer:
外墙立面材料生产商：
Jofebar: www.jofebar.com

Overview

The new Library in Ceuta is conditioned by the steep topography of the plot and by the Arab Marinid archaeological excavation of the XIV century that determine all interior spaces of the Library.

The Library is conceived as a compact volume that preserves the archaeological area as the core of the public spaces, creating a sense of openness and transparency between reading spaces and visitors to the Marinid centre. The library is organized in terraces placed on the slope that embrace the remains of the past. The lecture rooms are stacked in several levels overlooking the void where groups of hanging triangular lamps with peaks in both geometries are set over the archaeological centre. Two different entrances in two levels, one to the Library and other to the visitors centre, are placed linking the inside to the nearby streets.

Seven triangular concrete pillars support the building with a program organized vertically. The third floor with the general book display is placed over the concrete structure that covers the archaeological site. Over it a light steel structure in six levels stacks the program being the highest one the book depot, archives and offices. A concrete plied basement runs along the steep streets and several concrete structural voids are cut up in the double façade of the Library as viewpoints towards the city. On the terrace in the roof level an open reading room is placed, shaded by the aluminium-perforated skin that wraps up the building that filters sun and open views towards both seas, Europe and Africa.

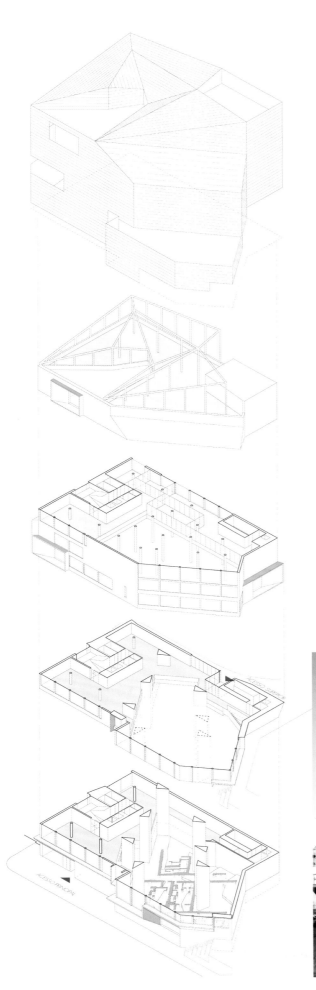
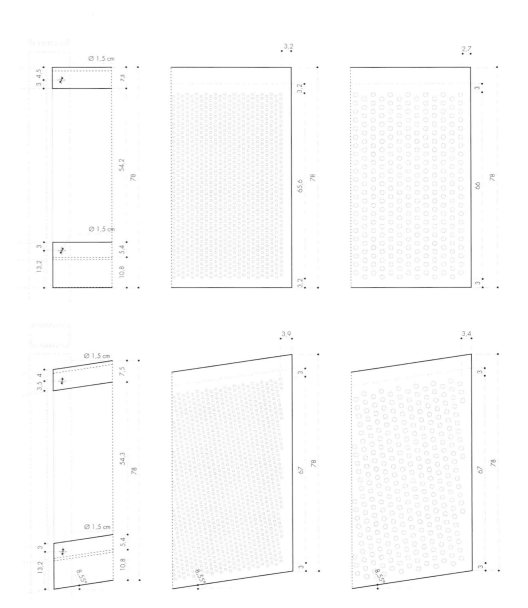

Detail and Materials

The compact folded volume is entirely wrapped up in an aluminium-perforated skin that reduces glare and solar gain and maximises the use of natural daylight reducing long-term energy costs. The mesh mitigates the sometimes-harsh qualities of daylight thus minimising the use of artificial light to avoid contrast and helping to illuminate the depth of the space. The final façade includes different glass-metal layers, energy efficient: an interior glass one and an outer metal one, as a veil, that interplay with the changing light conditions protecting the inside from the sun and heat. Slight variations in the make-up of the panels, for different orientations, provide the library with a differentiated yet uniform skin, emphasising the faceted shape of the building. Between them a gallery permits easy maintenance of glass openings and simple installations.

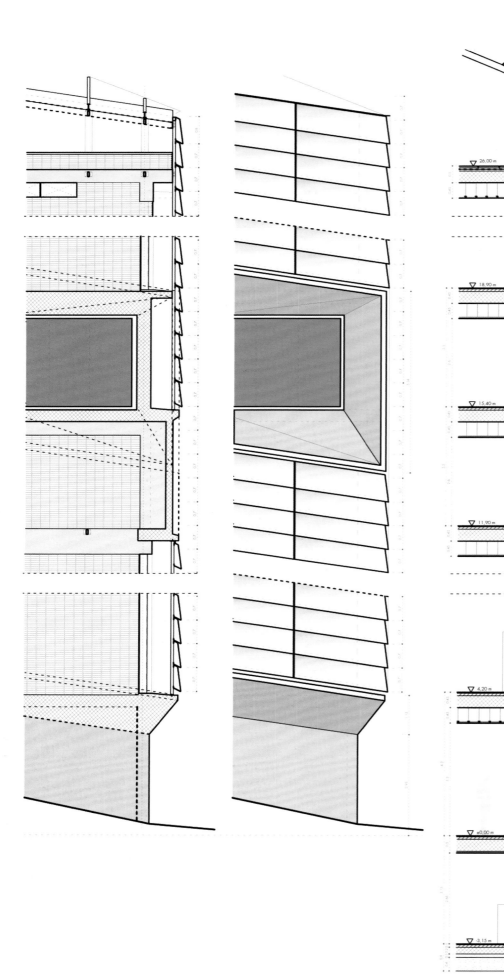

Façade detail

1. Folded and microperforated aluminium sheet plate (t=4mm)
2. TRAMEX 40.30.30 galvanised steel
3. Passable inverted roof; draining floor tile (t=3cm) on XPS (t=5cm) double layer asphalt cloth, geotextile sheet and insulating concrete for slope formation
4. Folded and microperforated aluminium sheet plate (t=4mm)
5. UPN 200
6. Galvanised steel bracket: tubular section 140.80.5
7. Galvanised steel: tubular section 140.80.5
8. Air conditioning return.
9. Exterior Windows of White in situ concrete
10. circular tube fired to the concerete
11. Aluminium curtain wall
12. Galvanised steel sheet plate
13. Galvanised steel: tubular section 50.50.4
14. 1/2 foot brick wall.
15. Pine strips 60x100 mm
16. Sandwich panels
17. Linoleum flooring: Colorette Amstrong (T=2mm)
18. Acustic insulation (t=5mm)
19. Canal Edge galvanised steel conduit wiring embedded in the ground
20. Perforated and laminated plaster suspended ceiling. Battens each 0.3 m
21. Flooring parts gray phyllite ridge
22. Acustic insulation (t=5mm)
23. Exterior Windows of White in situ concrete
24. White concrete wall formwork tablet placed vertically. e = 30cm
25. Veneer wall lobby with pieces of gray phyllite ridge
26. Half foot brick perforated chamber forming rough for comic
27. Flooring parts gray phyllite ridge
28. Acustic insulation (t=5mm)
29. Pile wall. D 45cm
30. Projected cement mortar with mesh e = 6cm d8/15cm and ground anchors d12
31. Vertical drainage sheet
32. Half foot brick perforated chamber forming rough for comic
    Plaster and painted with plastic paint
    Ventilation grilles stainless steel
    Mate placed on the top and bottom alternately every 2m
33. Flooring of changing rooms
34. Porcelain tile pressing, unglazed, slip 20x10cm
35. Horizontal drainage sheet
36. Reinforced concrete floor. mesh d8/15
37. Sandy river bed e = 10cm
38. Pitching draining compacted e = 40cm

立面节点

1. 微穿孔折叠铝板（t=4mm）
2. TRAMEX 40.30.30镀锌钢
3. 可通行倒置屋顶；排水地砖（t=3cm）XPS（t=5cm）双层沥青布，土工布和绝缘混凝土，用于构建坡体
4. 微穿孔折叠铝板（t=4mm）
5. UPN 200
6. 镀锌钢架：管状剖面140.80.5
7. 镀锌钢：管状剖面140.80.5
8. 空调回风口
9. 外窗，白色现浇混凝土
10. 固定在混凝土上的圆管
11. 铝幕墙
12. 镀锌钢板
13. 镀锌钢：管状剖面50.50.4
14. 1/2'砖墙
15. 松木条60x100 mm
16. 夹层板
17. 油地毡地面：Colorette Amstrong（t=2mm）
18. 隔音板（t=5mm）

19. 管道边缘镀锌布线钢管，嵌入地面
20. 穿孔夹层石膏吊顶，板条间隔0.3m
21. 地面部分灰色千枚岩脊
22. 隔音板（t=5mm）
23. 外窗，白色现浇混凝土
24. 白色混凝土墙模片板，垂直排列，间距30cm
25. 外墙走廊，配有灰色千枚岩脊
26. 1/2'穿孔砖
27. 地面部分灰色千枚岩脊
28. 隔音板（t=5mm）
29. 桩墙（d=45cm）
30. 凸出的水泥砂浆，带有网眼，e= 6cm，d=8/15cm；地锚桩钉d=12
31. 垂直排水板
32. 1/2'穿孔砖
   石膏，涂有塑性涂料
   不锈钢通风格栅
   上下搭接，间隔2cm
33. 更衣室地面
34. 瓷砖压面，无光泽，规格20x10cm
35. 水平排水板
36. 钢筋混凝土地面，网眼d=8/15cm
37. 沙河床，e= 10cm
38. 斜面排水，e= 40cm

Façade detail
1. Steel section HEB200 with welded platens (t=2cm)
2. Flame resistant panels RF-180
3. Reinforced concrete slab
4. White reinforced concrete bracket. Formwork with splints in vertical position. Drip underneath.
5. White concreate hanging beam. Formwork with splints in vertical position
6. Screed and slope formation with waterproof mortar.
7. Slope formation mortar for pavement placement.
8. Terrazo flooring (t=2cm)
9. Linoleum flooring: Colorette Amstrong (T=2mm)
10. Galvanised steel edge gutter for wiring and installations.
11. Tube 60.60.40
12. Anodized aluminium curtain wall profile of 25 microns. Mullions and transoms of 100x52mm
13. Glass 3+3/12/6
14. Steel sheet culmination (t=2mm)
15. 1/2 foot brick wall.
16. Sandwich panel, Escanplac type, formed by: exterior coretech board (t=16mm), XPS nucleus (T=80mm) and interior plasterboard (t=13mm)
17. Pine transom substructure 100x60 mm for sandwich panel fixation. Mullions at 1.2 m intervals
18. Thermal insulation (t=4cm)
19. Perforated and laminated plaster suspended ceiling. Battens each 0.3m
20. Galvanised steel profile section L40.4
21. Vertical Galvanised steel tube 140.80.5. Substructure for the exterior skin
22. Screwed galvanised steel bars.
23. Screwed microperforated and folded aluminium sheet (t=4mm)
24. Air conditioning return

立面节点
1. 型钢HEB200，配有焊接压板（t=2cm）
2. 防火板RF-180
3. 钢筋混凝土板
4. 白色钢筋混凝土支架，垂直位置有夹板模板，下方有滴水槽
5. 白色混凝土吊梁，垂直位置有夹板模板
6. 砂浆层和斜坡构造，带有防水灰浆
7. 斜坡构造灰浆，用于布置铺装
8. 水磨石地面（t=2cm）
9. 油地毡地面：Colorette Amstrong（t=2mm）
10. 镀锌钢边槽，用于布线和安装
11. 管60.60.40
12. 氧化铝幕墙（25micron），竖挡和横梁尺寸为100x52mm
13. 玻璃3+3/12/6
14. 钢板顶点（t=2mm）
15. 1/2'砖墙
16. 夹层板，Escanplac型，由外层coretech板（t=16mm）、XPS内核（t=80mm）和内层石膏板（t=13mm）组成
17. 松木横梁下层结构100x60mm，用于固定夹层板，竖挡间距1.2m
18. 保温层（t=4cm）
19. 穿孔夹层石膏吊顶，条板间隔0.3米
20. 镀锌型钢L40.4
21. 垂直镀锌钢管140.80.5，外表皮的下层结构
22. 螺丝镀锌钢条
23. 螺丝微穿孔折叠铝板（t=4mm）
24. 空调回风口

项目概况

休达的新图书馆位于一处斜坡地形上，其室内设计以14世纪的阿拉伯马林王朝考古发掘为主题。

图书馆被设计成一个紧凑的空间，保留的考古遗址被用作核心公共空间，在阅读空间和马林王朝遗址之间营造出一种开放感和透明感。图书馆呈阶梯形式沿坡度展开，将考古遗址环绕起来。教室分布在不同的楼层里，俯瞰着中空空间。在这个中空空间顶部，三叉形吊灯照亮了考古中心。建筑在两层分别有两个入口，一个通往图书馆，另一个通往游客中心。

7根三角形混凝土柱将建筑支撑起来。图书馆三楼的普通图书展示就设置在考古遗址的上方。6层高的轻质钢书架上摆放着图书和档案文件。由混凝土建成的地下室沿着陡峭的街道延伸，图书馆的双层立面上被切割出若干个开口，展现了城市的景象。屋顶平台上设有一个露天阅览室，由穿孔铝材制成的建筑表皮不仅为它提供了遮阳，还呈现出跨越欧非大陆的海洋美景。

细部与材料

紧凑的建筑空间被包裹在一层穿孔铝材表皮内，既削弱了眩光和太阳热增量，又最大限度地利用了自然采光，从而缩减了能源成本。网眼表皮缓和了日光刺眼的效果，同时又缩减了人造光的使用，避免了明暗反差，使空间深处也能得到照明。建筑立面由不同的节能玻璃金属层构成：内层为玻璃，外层为金属，像面纱一样与光照条件相互作用，为建筑内部遮阳遮阴。金属板排列方向的微妙变化为图书馆提供了分化而统一的表皮，突出了建筑的切面造型。两层立面之间的走廊让玻璃窗的安装和维护变得更简单。

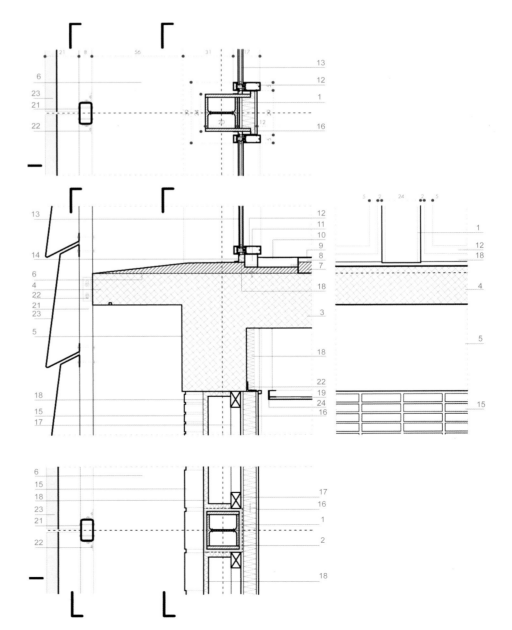

TIZIANO 32
蒂奇亚诺32

Location/地点: Milan, Italy/意大利，米兰
Architect/建筑师: Park Associati
Photos/摄影: Leo Torri
Site area/占地面积: 9,700m²
Construction area/建筑面积: 1,030m²

Key materials: Façade – aluminium and expanded metal panels, micro-perforated aluminium sheet
主要材料： 外立面——铝材和多孔金属板、微穿孔铝板

Overview

Located close to Milan's historic trade fair area and dating back to the late 1950s, this office block was one of a number of service buildings designed by some of the most important architects of the day that were erected in a middle-class residential district which had developed on uniform lots after the Second World War. The Generali insurance company, in particular, opted for a headquarters designed by the architect Cesare Donini which had the dual function of providing office spaces and residential units for its employees. Park Associati and Zucchi & Partners have been assigned the task of resystematizing and upgrading the 'common' areas of the volume on via Tiziano, which was the main element in a staged development arranged around a semi-indoor garden. This consists in a seven-floor block, an imposing metal canopy and a thin walled garden.

A key and necessary feature of the new project, which specifically addresses this high volume and is exclusively concerned with the upgrading of its common areas, has been the recuperation of the basement area. The entrance to the building has been moved down to this level, accessed via a gentle flight of steps, so as to be able to adequately redistribute a number of spaces that were previously unused and, at the same time, to bestow a new contemporary image upon the whole building, but one which respects the undeniable quality of the original architectural design. This solution involved the addition of a new entrance canopy built from concrete, steel and glass which starts from the building's exterior.

Detail and Materials

The vertical and horizontal surfaces of the canopy are covered in a skin of aluminium and expanded metal panels, and are arranged in such a way as to stagger the light and dark elements. The structure is a sort of 'accent' - a strong sign visible from the street that indicates the presence of the renovation project. But it is also above all an element that leads the visitor towards the heart of the building. As a result of the arrangement, the relationship with the external garden surrounding the building also becomes more intense: while the canopy itself seems to open up unexpectedly, like a telluric yet orderly movement on the smooth green lawn, the project has also carefully considered the redesign of the exterior fencing of the lot. Thus, metal pillars maintain the rhythm of the bays of the building, and transparent glass panes have been mounted in between them so as to create a hanging, almost floating boundary that does not separate the building from the rest of the city. The colour theme, one of the key elements of the project, is immediately visible on the exterior of the building in the large window pane of the former conference room, now transformed into the reception of one of the office sectors, where acid yellow glass panels have been added.

Finally, the general refurbishment of the volume's frontage is now enhanced by a series of thin continuous vertical wings made from micro-perforated aluminium sheets, which seek to stress the three-dimensionality of the façade, in a game of light and dark that varies according to the amount of sunlight.

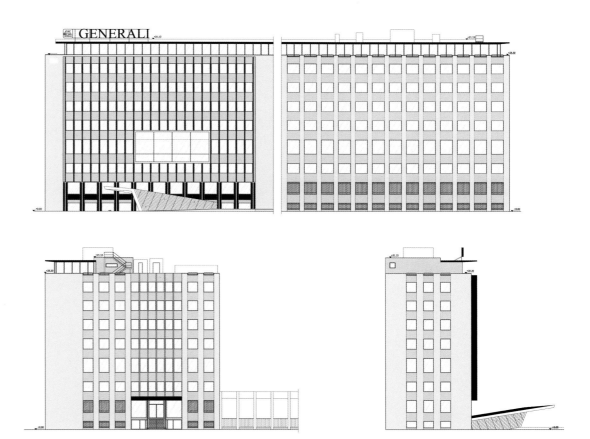

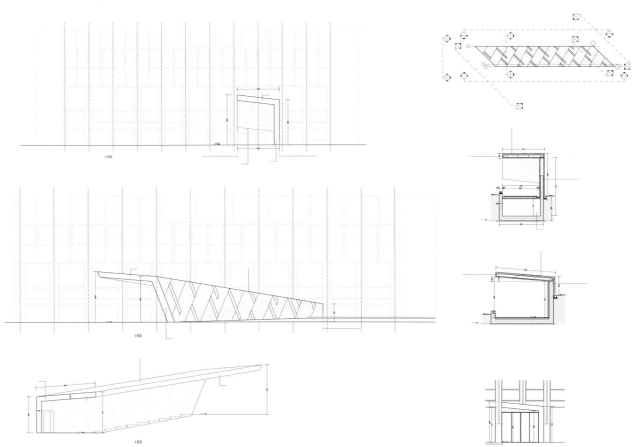

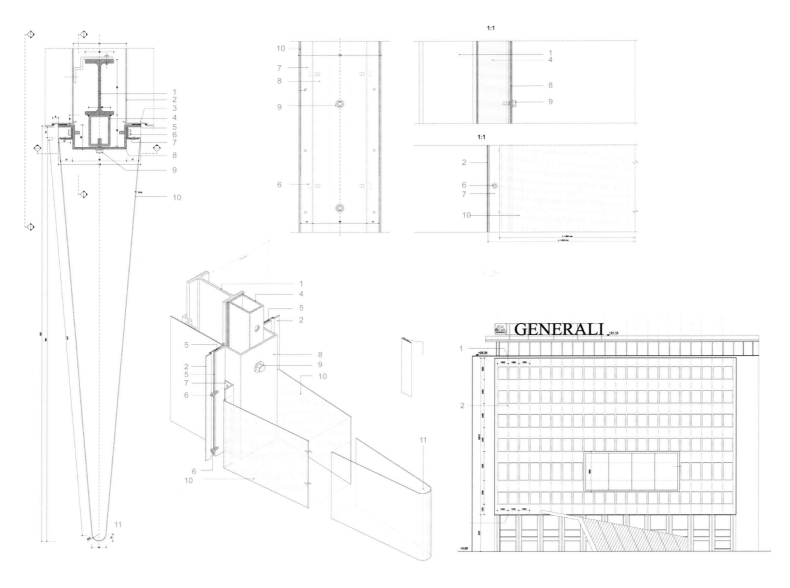

Detail of the façade of the blade mounting horizontal section
1. Existing IPE 80 post
2. Aluminium frame of the existing façade
3. Maximum size of the "blade"
4. 45x30 steel box welded to the existing IPE 80 post sp.3mm
5. Neoprene gasket
6. Allen screws securing the blade to the metal structure
7. Profile pre-painted steel silver soldered and riveted sp.2mm AL 17x17 workshop in the perforated plate steel
8. Continuous profile of galvanised steel bent sp.3mm
9. Screws securing the continuous profile at the box-shaped steel
10. "Blade" steel sheet SP. 8/10 perforated 3mm pitch to 4.5mm pre-painted silver
11. Radius of curvature of the bending of the steel sheet: 10mm

New statement of 'building with an indication of "lame" façade
1. The top end cap sheet steel SP.8/10 welded to the "blade"
2. "Blade" steel sheet SP. 8/10 perforated 3mm pitch to 4.5mm pre-painted silver
Note the length of the "lame" and measured by aligning the outer edge of the window façade

安装叶片的立面节点
1. 原有的IPE 80杆
2. 原有立面的铝框
3. 叶片的最大尺寸
4. 45x30钢框，与原有的IPE 80杆焊接起来，sp.3mm
5. 氯丁橡胶垫片
6. 六角固定螺丝，将叶片固定在金属结构上
7. 预涂漆型钢，采用银焊和铆钉固定，sp.2mm
8. 连续的镀锌弯型钢，sp.3mm
9. 螺丝，将连续的型钢固定在钢框上
10. 叶片穿孔钢板，sp. 8/10，倾斜3~4.5mm，预涂银色漆
11. 钢板弯曲半径 10mm

建筑立面叶片的节点设计
1. 顶端钢盖板sp.8/10，与叶片焊接
2. 叶片穿孔钢板，sp. 8/10，倾斜3~4.5mm，预涂银色漆
（注意叶片的长度，对准窗立面的外围进行测量）

项目概况

这座办公楼的位置靠近米兰的传统贸易区，在周围建于二战后的统一模式的中产阶级住宅建筑中显得与众不同。忠利保险公司选择了由建筑师切萨雷·多尼尼设计的总部大楼。大楼兼具办公楼和住宅楼的功能，为员工提供了住宅单元。公司委托Park建筑事务所和Zucchi事务所对建筑实施改造，项目由7层高的办公楼、壮观的金属遮篷和一个小围墙花园组成。

地下室的修复是项目的关键，是公共区域升级改造的重点。建筑的入口被挪到了地下，通过一排缓和的台阶进入。这种设计让空间得到了重新配置，为建筑带来了全新的现代形象感，同时又不会削弱建筑原来的建筑风格。新增的入口遮篷是该设计的重点，遮篷由混凝土、钢和玻璃建成，从建筑外部一直向下延伸。

细部与材料

遮篷的垂直表面和水平表面上覆盖着铝材和多孔金属铝板，将明暗元素交错并置。这一结构是建筑的特征之一，让人们从远处的街道上就能辨识出来。当然，它最重要的功能是引领访客进入建筑的中心。这种布局让环绕建筑的室外花园设计也具有了强烈的设计感。遮篷以出人意料的方式打开，在光滑的草坪上拔地而起；与此同时，项目对围墙也进行了重新设计。金属柱保持了建筑凹进处的韵律感，并且在之间安装上了玻璃板，形成了轻盈、悬浮的边界，保持了建筑与城市不被分隔开。项目的色彩主题明确地呈现在建筑宽大的玻璃窗上，柠檬黄的玻璃板让窗口从建筑立面中跳脱出来。

最后，项目的临街面被装饰了一排薄薄的垂直微穿孔铝板，突出了建筑立面的立体感，实现了独特的光影效果。

Orthogonal section on cantilever roof
1. Metal mesh
2. Extraclear satin glass, anodized Aluminium window frame
3. Gutter
4. Anodized Aluminium plate 20/10 thick
5. Beam 220 primary structure
6. Aluminium mesh
7. RC wall 25cm thick
8. Downpipe Φ60mm
9. Aluminium plate box
10. Linear lamp disano lunar FL 1x58E (IP657) l=690mm
11. Gutter with recessed greed
12. Wallwasher fluorescent lamp
13. Slide covered with slabs of Matraia stone
14. Stairs cladding: Matraia stone
15. Painted Aluminium plate
16. Wall with light box

悬臂式顶板的垂直剖面
1. 金属网
2. 超清丝光玻璃，阳极氧化铝窗框
3. 排水槽
4. 阳极氧化铝板，20/10厚
5. 主结构横梁220
6. 铝网
7. 钢筋混凝土墙，25cm厚
8. 落水管，直径60mm
9. 铝板箱
10. 线形灯FL 1x58E (IP657) l=690mm
11. 嵌入式排水沟
12. 荧光洗墙灯
13. 滑动盖板，Matraia石板
14. 台阶覆层：Matraia石板
15. 涂漆铝板
16. 墙上灯箱

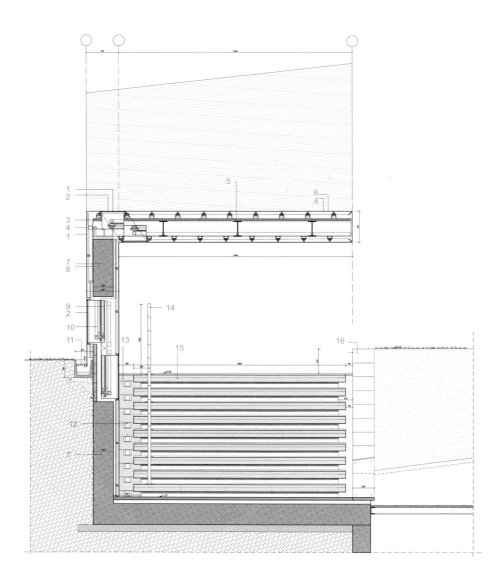

Centre for Manufacturing Innovation, Metalsa CIDeVeC

曼特沙公司工业创新中心

Location/地点: Apodaca, Mexico/墨西哥，阿波达卡
Architect/建筑师: Brooks + Scarpa
Area/面积: 1,750m² (phase I), 3,500m² (phase II)
一期工程1,750m²；二期工程3,500m²

Key materials: Façade – perforated Aluminium, glass
主要材料: 立面——穿孔铝材、玻璃

Overview

Industrial buildings of this type are rarely a model for workplace innovation. They are typically a direct, and often nefarious programmatic response to the function inside with little consideration for the occupants needs. The approach to this project was to preserve the integrity of a high bay industrial facility and program, while providing a model environment for the users and visitors.

A saw-toothed roof draws from the geometry of old factories and the surrounding Monterrey Mountains. The angled elements of the roof provide abundant natural daylight to the spaces below at the building's northernmost elevations. By modulating space and light thru a fractured roof geometry, the building is able to maintain a rational plan to meet the rigorous requirements of the program, while providing a strong connection to the landscape both visually and metaphorically.

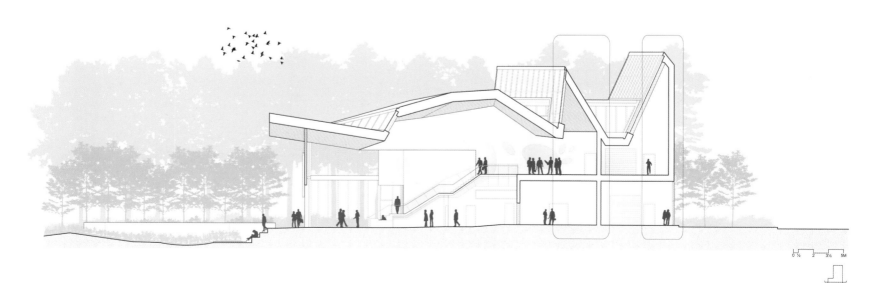

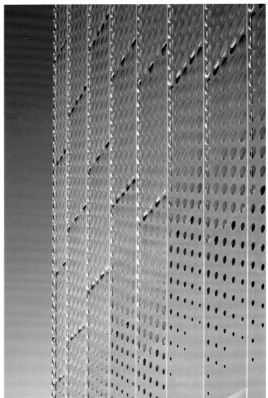

Programmatically, the building is divided into two volumes – warehouse/labs and offices functions. The upper storey of the offices cantilevers over the lower storey to the west and is clad in a highly perforated metal skin and is the main entry façade. The lower storey is mainly glazed and open to reveal portions of the research laboratory, machine room and other industrial functions not requiring visually security. From the exterior, the warehouse appears to float lightly over the mechanical and intellectual heart of the program, reversing the notion that an industrial building should be solid and protected. Rather, the building seems very open and is intended to feel vulnerable revealing parts of its inner program to public view.

Detail and Materials

A major feature of the building is the perforated metal skin that clads the entire façade. The custom Aluminium skin is both perforated and etched. It incorporates interplay of solid and void, orchestrating areas of both light and shadow, while limiting views into the research areas, necessary to protect proprietary trade secrets. Thus, the industrial program has been transformed from a black box environment to a light filled space with a strong visual connection to the outside.

Each of these strategies and materials, exploit the potential for performance and sensibility while achieving a rich and interesting sensory and aesthetic experience.

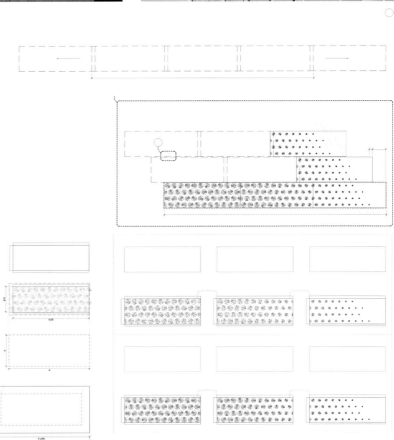

South façade detail
1. 20.GA.GSM ridge cap with 18 GA.GSM clips @400mm(16")O.C., TYP
2. Cont' silicon caulking W/backer rod over W.P. membrane and flashing
3. Provide HSS extention and cont' STL. bent plate at skylight head BLK support – PTD.All exposed STL. white – TYP.
4. Provide cont' 14 GA. Clip@ end of netting below STL purlins
5. SIM., no insulation
6. Perforated vertical Aluminium panel
7. Steel purlin per structural
8. Cont, L-bent plate @ conc, deck edge
9. Acoustical isolation JT. By construction specialties INC.
10. Cont' purlins @ storefront mullion header
11. Provide con't 10mm, STL plate @ B.O purlin
12. W/low-E coating
13. Exterior slab on grade TYP
14. Conc, grade beam/foundation per structural
    600mm isolated mat slab @ hydraulic room
15. Cont' flexibly rubber membrane W/mineral wool infill at isolation joint
16. 100mm sand with 10 mil VISQUEEN @ mid-depth
17. Acoustical isolation JT. By construction specialties INC.-HB-400
18. Baseplate and drypack per structural

West façade detail
1. Cont' 10mm bent plate
2. Perforated vertical Aluminium panel
3. Steel purlin, per structural
4. 35mm multi-wall polycarbonate panel
5. Align B.O. GSM flashing /cover
6. Cont. sealant at channel base, TYP.
7. Flush mount wood cherry HAVC floor register- Whittington #215464, TYP.
8. HVAC ducts per mech.
9. Steel post, hang from STL. beam above
10. W/Low-E coating

南立面节点
1. 20.GA.石膏脊盖，配有18 GA.GSM夹片，@400mm(16")O.C.
2. 防水膜和防水板上的硅嵌缝/泡沫条
3. 在天窗上顶部支撑结构上安装高速钢架和钢曲板，所有裸露的钢结构均为白色
4. 在钢檩条下方网眼的两端安装14 GA.夹片
5. 接缝，无隔热
6. 穿孔垂直铝板
7. 钢檩条
8. 平台边缘的L形曲板
9. 隔音层
10. 正面窗框顶部的檩条
11. 在檩条上安装10mm钢板
12. 低辐射涂层
13. 外层斜坡板
14. 混凝土地基梁/地基
    液压泵室的600mm绝缘垫楼板
15. 弹性橡胶薄膜/隔离接缝以矿物棉填充
16. 100mm砂层，中间带有10mil VISQUEEN材料
17. 隔音层
18. 基板和干燥混合料

西立面节点
1. 10mm曲板
2. 穿孔垂直铝板
3. 钢檩条
4. 35mm多层壁聚碳酸酯板
5. 石膏防水板/顶盖
6. 渠道底座密封剂
7. 嵌入式安装樱桃木空调地面出风口
8. 空调管道
9. 钢杆，悬挂于上方钢梁
10. 低辐射涂层

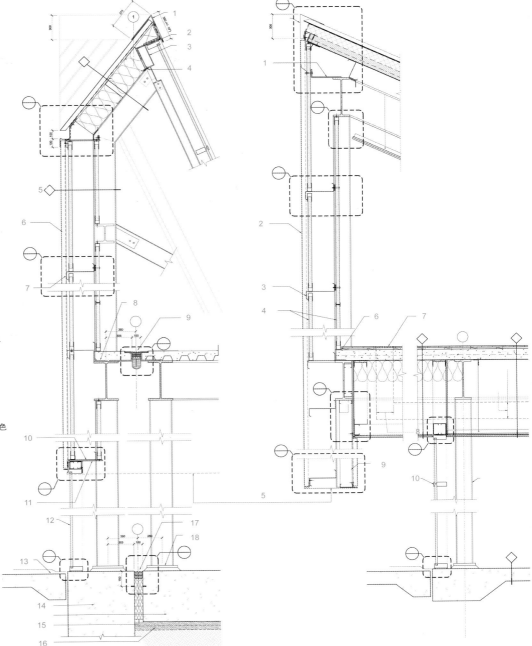

项目概况

这种类型的工业建筑很少能呈现出创新设计。它们通常以简单粗暴的方式实现内部的功能性，很少考虑到使用者的需求。本项目的设计方案既考虑了工业设施的功能价值，又注重了使用者和访客的环境体验，堪称典范。

锯齿形屋顶从旧工厂和周边的蒙特雷山中获取了灵感。屋顶的棱角结构为室内空间提供了充足的自然采光。"断裂"的屋顶结构调整了空间和光线，让建筑能够保持合理的规划来满足严格的功能要求，同时又与周边的景观环境巧妙的联系起来。

从功能上讲，建筑分为两个部分：仓库/实验室和办公区。上层办公区在建筑西面悬挑出来，以镂空的金属板包裹，是建筑的主入口立面。下层楼体主要是玻璃幕墙，展现了部分研究实验室、机房和其他无需保密的工业空间。从建筑外面看，仓库看起来像是悬浮在机械创新区域之上，颠覆了工业建筑封闭的形象。建筑看起来十分开放，将一部分内部功能直接展现在公共视野之中。

细部与材料

包裹着整个立面的穿孔金属表皮是建筑的主要特征之一。定制的铝制表皮采用了穿孔和蚀刻技术，实现了虚与实、光与影的灵活搭配，同时也将部分研究区域遮挡起来，保护了商业机密。这样一来，建筑将工业项目从封闭的黑匣子带到了明亮的空间中，与周边环境建立起了紧密的联系。

这些技术与材料开拓了项目的潜力，实现了丰富的质感和良好的美学体验。

Castle of Skywalkers
天行者城堡

Location/地点: Chungcheongnam-do, Korea/韩国，忠清南道
Architect/建筑师: Doojin Hwang(Doojin Hwang Architects)
Photos/摄影: Youngchae Park+Suyeon Yun
Site area/占地面积: 21,955m²
Gross floor area/总楼面面积: 9,355.67m²
Design date/设计时间: 2011
Completion date/竣工时间: 2013

Key materials: Façade – Aluminium expanded metal
Structure – SRC+RC
主要材料：立面——铝金属网
结构——型钢混凝土+钢筋混凝土

Overview

Castle of Skywalkers is a training complex for the Hyundai Capital Skywalkers, a leading professional volleyball team in Korea. Located on a gentle, rolling hill in the suburbs of the city of Cheonan, this 4-storey, 9,355m² building commands a panoramic view toward the surroundings.

At the heart of the building is an atrium, with various training facilities and dorm rooms centred on a volleyball court at the centre. The plan, based on a rectangle, a circle and diagonals, is an exercise in manipulating layered geometry that corresponds to different programmes at different levels.

Separated from the main building is an annex, whose serpentine lines are shared with the surrounding landscape. Conceived as a small "party house", the annex is a place for amusement and relaxation for professional sportsmen going though intense training routine.

Castle of Skywalkers showcases a new trend in sports facilities design seeking to create both an efficient training environment and a pleasant and comfortable place of living for a highly competitive athletic team.

Detail and Materials

The external walls are conceived as a "functional skin"; the outer layer of expanded Aluminium panels, separated 250mm from the inner layers of structure and insulation, keeps the high summer sun out and lets in the low winter sun, reducing the long-term operational cost of the building at a reasonable initial cost. The highly porous cladding also generates delicate and versatile aesthetics that constantly changes depending on weather, sun's direction and viewers' locations.

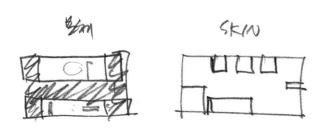

项目概况

天行者城堡是现代资本天行者排球队的训练楼,天行者是一支韩国著名专业排球队。坐落在韩国天安市郊区一个舒缓的山坡上,这座四层总面积9,355平方米的建筑享有周边自然环境优美的风景。

建筑的中央是中庭,各种训练设施和宿舍环绕着中央排球场展开。楼面设计以长方形、圆形和对角线为基础,是一个多层几何空间,分别应对了不同楼层的不同功能设施。

脱离主楼的附属建筑蜿蜒起伏,与周边景观融为一体。作为一间小型"派对房",附属建筑为专业运动员在高强度的训练之余提供了休闲放松的空间。

天行者城堡展示了体育设施设计的新趋势,力求为一支高水平运动队打造高效训练环境与舒适生活空间的混合体。

细部与材料

外墙被设计成"功能表皮",外层铝网距离内层结构和隔热层250毫米,既能阻挡夏季的高角度日照,又能引入冬季的低角度日照,将建筑的长期运营费用降到了一个相对合理的数字。多孔的外墙包层还形成了精致多变的美观效果,会随着气候、日照角度和观察者的位置而不断变换。

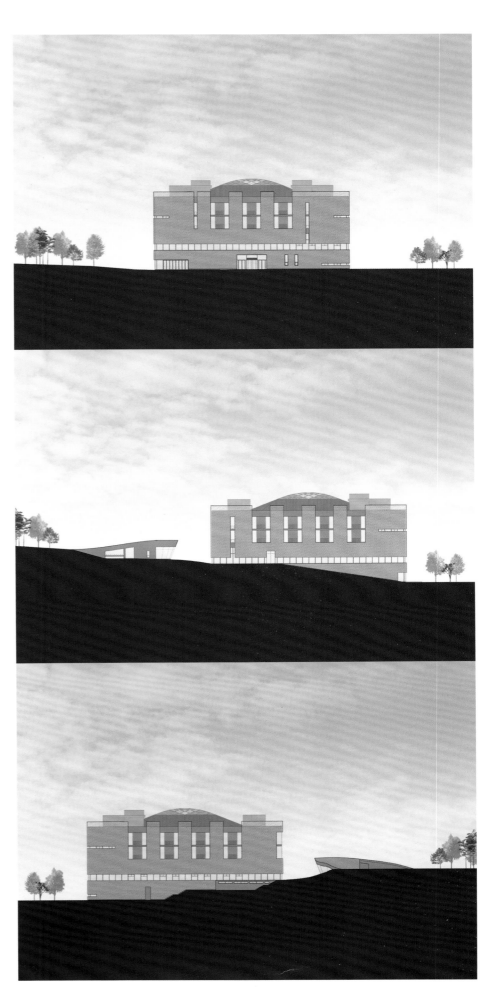

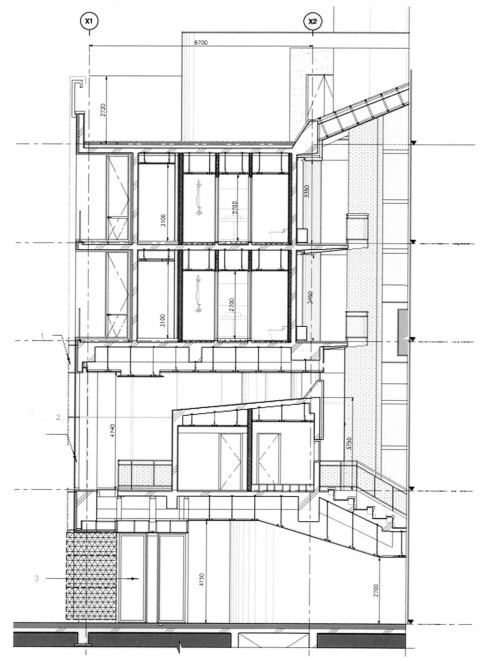
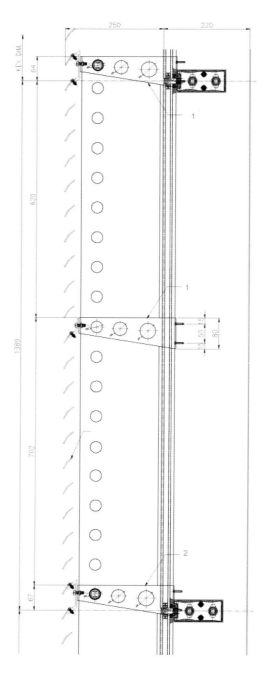

Section detail 1
1. Anodizing on aluminium
2. Thk 24 transparent pair glass
3. Auto revolving door

剖面节点1
1. 阳极氧化铝
2. 24厚透明双层玻璃
3. 自动旋转门

Section detail 2
1. Aluminium extrude bar ex metal bracket
2. Ex metal/anodizing (W188, H66, Thk 2)

剖面节点2
1. 金属支架外的铝条
2. 外层金属/阳极氧化铝（宽188，高66，厚2）

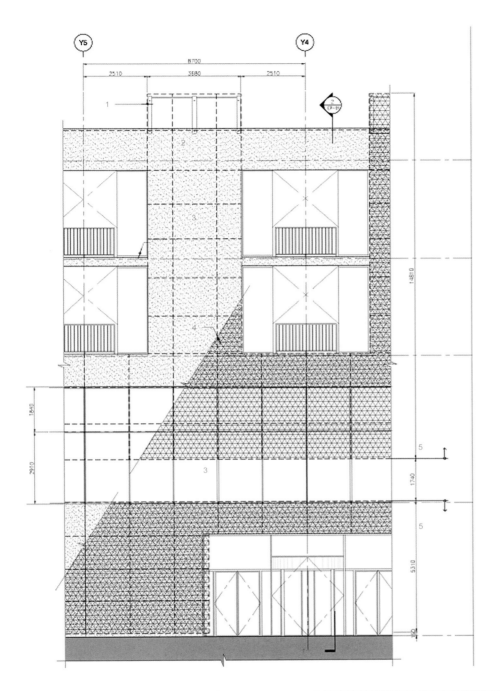

Elevation detail
1. Outline of exterior metal plate
2. Custom coloured drivit
3. Thk 24 transparent pair glass
4. Anodizing on aluminium
5. Exterior expanded metal reference point

立面节点
1. 外层金属板轮廓
2. 定制彩色保温层
3. 24厚透明双层玻璃
4. 阳极氧化铝
5. 外层金属网参考点

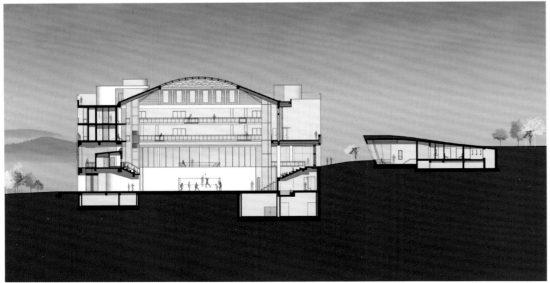

Parametric Fragment

参数碎片住宅

Location/地点: Matsusaka, Japan/日本，松阪
Architect/建筑师: Takashi Yamaguchi (TAKASHI YAMAGUCHI & ASSOCIATES)
Associate/助理建筑师: Jun Murata (TAKASHI YAMAGUCHI & ASSOCIATES)
Completion date/竣工时间: 2011
Photos/摄影: TAKASHI YAMAGUCHI & ASSOCIATES
Site area/占地面积: 667.84 m²
Architectural area/建筑面积: 189.15m²
Total Floor area/总楼面面积: 293.19m²

Key materials: Façade – aluminium panel
Structure – plaster board
主要材料: 立面——铝板
结构——石膏板

Overview

The house is located in a rural area near Ise Bay. There are vast rice fields around the site. During rice planting season, the fields are bathed in water and in autumn, harvesting season gives fields a golden glow. The area is also subject to high rainfall and frequent flooding.

The volume is elevated on a mound and rests on a green foundation. The site demanded a plan in which this foundation could protect the building from flooding and also allow the view of countryside to blend in harmoniously with green grass planted on the foundation.

Detail and Materials

The structure consists of double skin system of the main frame and a surface of Aluminium panels. To reduce the amount of radiant heat in summer, west and south sides are designed without openings and the width of space in the double skin of walls and roof on west and south are wider. This system controls the circulation of warm air so that it rises through the inside of double skin and extracted through louvers.

What the architects express here is a non-segmented form, in which the parts blend seamlessly with the whole by changing dimensions and angles parametrically for the same material. The external Aluminium panes covering the volume vary to the louvers according to changes in size and gradual fragmentation. The angles, widths and lengths of panels are adjusted and louver aperture rations are determined according to the location of each room opening, loft opening, flue window and ventilation and air-conditioning outlets.

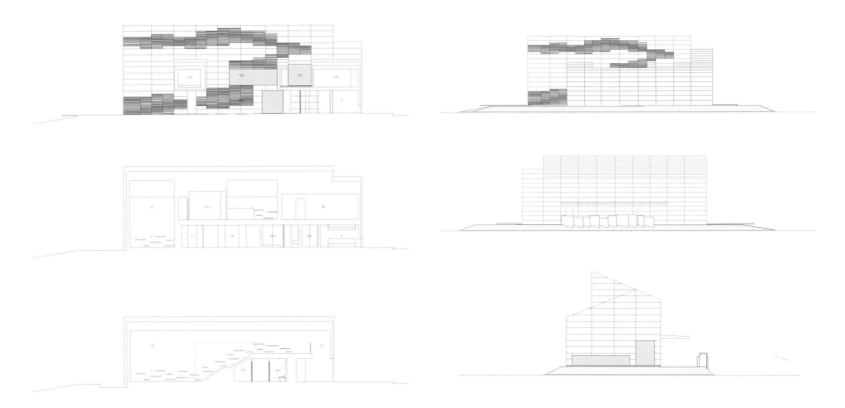

项目概况

住宅坐落在靠近伊势湾的乡村，四周环绕着大片的稻田。在水稻种植的季节，稻田里充满了水；秋天，收获的季节为稻田染上了金色的光晕。该地区雨量充沛，常遇洪水。

住宅位于一个土堤上，下方是绿色的地基。场地条件要求地基能够保护建筑不受洪水淹没，同时又让乡村景观与地基上的绿草和谐地融合起来。

细部与材料

建筑结构由双层表皮系统组成，包括一个主框架和一层铝板。为了减少夏季的热辐射量，住宅的西、南两侧没有开窗，且双层墙面和屋顶的宽度都设计得更宽。这个外墙系统控制了热空气的流通，使其从双层表皮的内部上升，然后从百叶窗排出。

建筑师采用了分割式造型，采用相同材料的各个部分以变化的尺寸和角度融为一体。外层铝板的尺寸根据百叶窗而变化，形成了渐变的碎片。铝板的角度、宽度和长度得到了调整，百叶窗口的比例由各个房间的窗口位置、阁楼天窗、烟道和通风口的位置所决定。

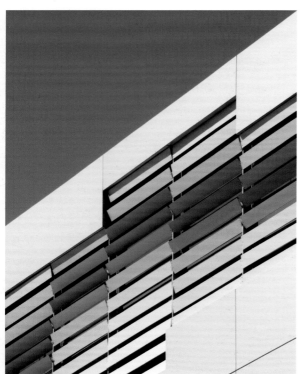

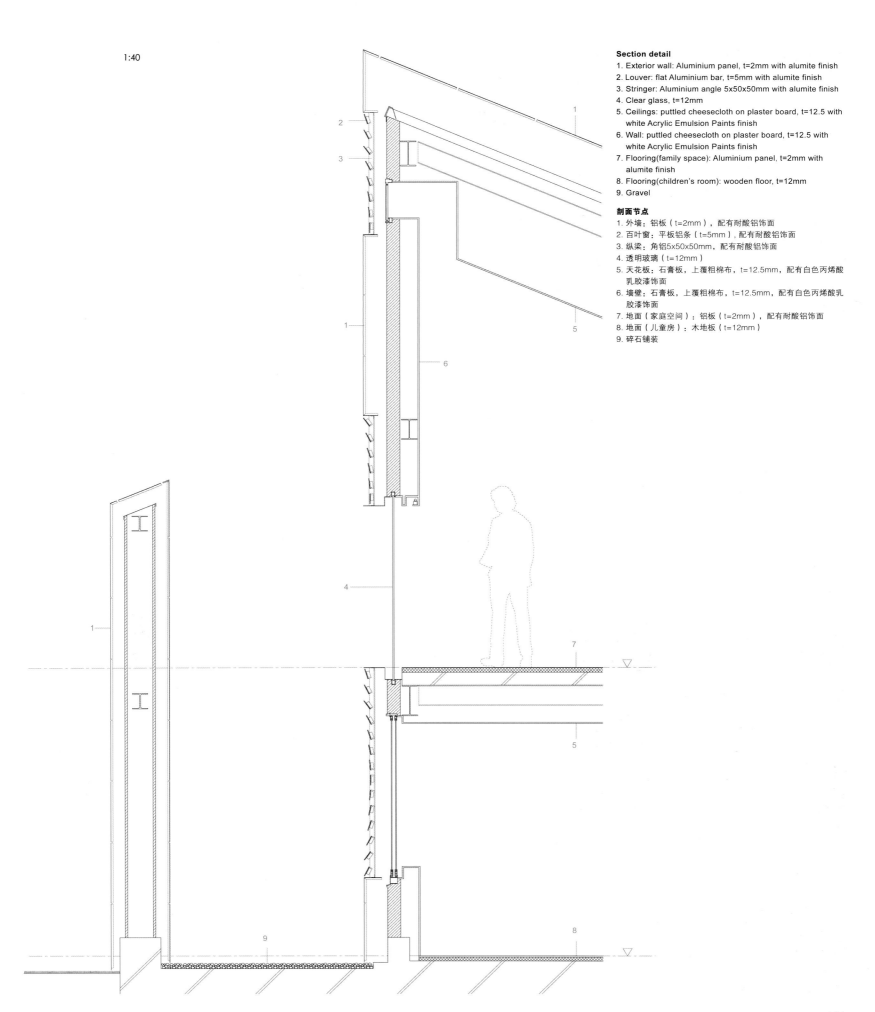

Office Building in Barcelona

巴塞罗那办公楼

Location/地点: Barcelona, Spain/西班牙，巴塞罗那
Architect/建筑师: BCQ arquitectura Barcelona. David Baena, Toni Casamor, Manel Peribáñez, Maria Taltavull.
Collaborating architects/合作建筑师: Catarina Crespo, Marta Cid, Kim Hamers, Mika Iitomi (landscaping)
Photos/摄影: ©Pegenaute
Completion date/竣工时间: 2011
Built area/建筑面积: 6,291m²

Key materials: Façade – Aluminium slat
Structure – Steel reinforced brick wall
主要材料：立面——铝板条
结构——钢筋砖墙

Overview

The building is situated in Ciutat Vella district, Barcelona. Its overall appearance is that of a glass box protected by a translucent metallic filter, which screens the light and the views.

The building is organised in three overlapping volumes whose rotation among them has been suggested by the geometry of the buildable lot: the transparent volume of the ground floor adopts the geometry of the site. This floor establishes a continuing relationship with the public space in which it is situated. The volume of first to fourth floors regains the alignment of the street l'Om, and creates, with the projection originated by this shift, a large porch that frames and protects the building entrance. This shift enables the offices to have a rectangular floor plan absolutely regular, which facilitates distribution and allows maximum flexibility of use. At the same time it creates more space on the side square. The third volume of the fifth and sixth floors is guided by a turn in between the two previous volumes. The fact that this third volume is smaller than the lower one, generates a terrace accessible to the users of the building from the fifth floor.

Detail and Materials

The need for natural light which is determined by the administrative use of the building is resolved with the large windows in façade which are

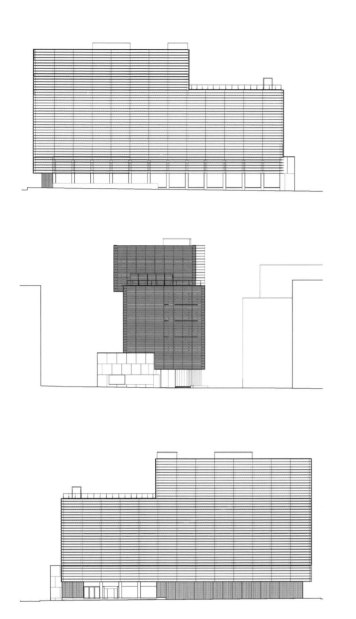

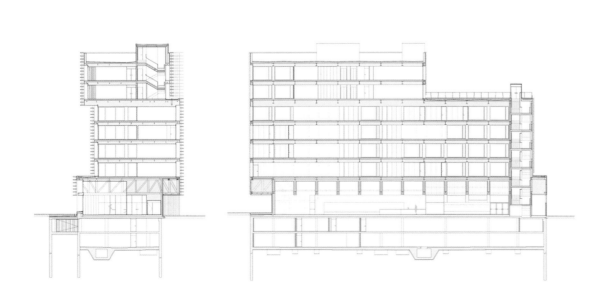

protected from the incidence of direct solar radiation through a continuous envelope of Aluminium slats. These slats give identity to the building façades: they are arranged horizontally along the main façades of the building (longitudinal façades) and in vertical on the beams.

The use of a framed structure with pillars in façade allows open floor plans that are organised from two vertical cores, leaving the rest of the surface free. The partitioning is done through lightweight partition wall. In this way, the building allows a versatility of occupation in time.

Façade detail
1. Rigid extruded polystyrene insulation, e = 50 mm
2. Sandwich panel cladding: Aluminium plate anodized e = 2 mm rock wool insulation e = 30 mm sheet steel Sendzimir e = 1 mm, fixed on battens
3. Aluminium plate e= 2 mm glued on fiberboard and cement viroc e = 2 cm
4. Window type Schüco AWS 75 bs.hi
5. Structure of galvanised steel tubular profiles 50x50 / 2 mm
6. Perlite projected e = 25 mm, for fire protection of structural elements
7. Steel tubular profile 50x20 / 2 mm
8. Galvanised steel plate e = 1.5 mm
9. Steel reinforced brickwall block 30 cm
10. Cf 110x60x20/3 mm

立面节点
1. 刚性挤塑聚苯乙烯绝缘层，e=50mm
2. 夹层板包层：阳极氧化铝板，e=2mm；石棉绝缘层，e=30mm；Sendzimir钢板，e=1mm，与条板固定
3. 铝板，e=2mm，粘在纤维板上；水泥板，e=2cm
4. Schüco AWS 75 bs.hi型窗
5. 镀锌钢管型材结构，50x50/2mm
6. 突出的珠岩，e=25mm，作为防火结构元件
7. 钢管型材，50x20/2mm
8. 镀锌钢板，e=1.5mm
9. 钢骨砖墙砌块，30cm
10. Cf 110x60x20/3mm

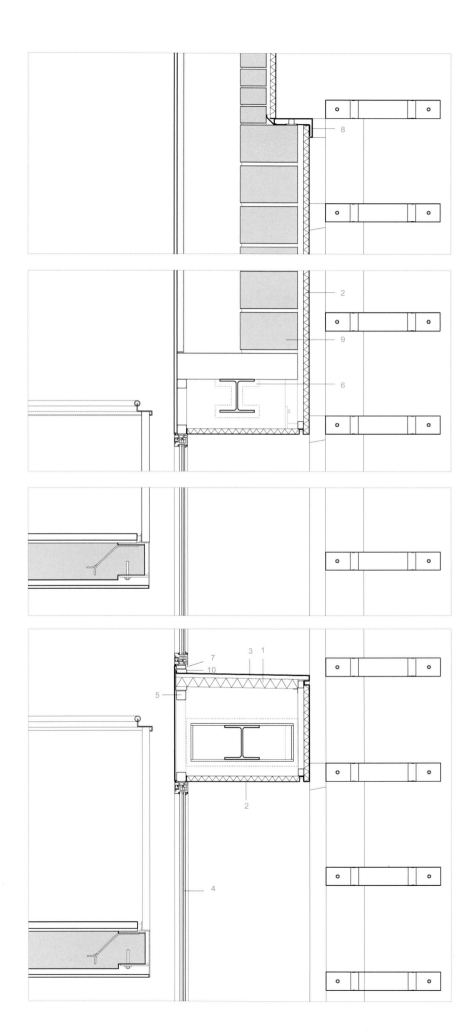

项目概况

建筑位于巴塞罗市中心的老城区，它的外观呈现为一个由半透明金属网所保护的玻璃盒子，金属网帮助遮挡了光线和视线。

建筑由三层结构组成，以不同的角度叠加起来：底层的透明结构与地面的几何造型相一致，与其所在公共空间建立起了连续的联系。2~5层的空间与罗姆街平行，向外突出，为建筑入口提供了一个巨大的门廊。这种变化让办公室拥有了规则的长方形楼面布局，有助于功能配置和灵活使用。与此同时，它还在侧面的广场营造出更多的空间。设置6~7层的第三层结构造型介于以下两层结构之间，比下层缺少的一块空间正好形成了一个可以从办公楼6层进入的露台。

细部与材料

作为一座办公楼，建筑需要自然采光。连续的铝板条结构避免了日光直射，配合宽大的玻璃窗为室内带来了舒适的自然光线。这些板条让建筑立面极富特色：它们沿着建筑的主立面横向排列，并且纵向形成了纵梁。

框架结构（在立面上设置支柱）的运用实现了开放的楼面布局。楼面围绕着两个垂直中心展开，其他空间都是自由的。设计以轻质隔断墙进行空间分区，保证了建筑的广泛用途。

Science Park

科学园

Location/地点: Linz, Germany/德国，林兹
Architect/建筑师: Caramel architekten ZT gmbh
Gross floor area/总楼面面积: 52,218m², phase 1: 16,509m²; phase 2-4: 39,796m² /一期工程16,509m²；二期工程39,796m²
Completion date/竣工时间: 2009

Key materials: Façade – Aluminium, glass
Structure – truss
主要材料： 立面——铝材、玻璃
结构——桁架

Overview
The program involved designing several individual buildings which would be interwoven as well as tied to the existing University of Linz campus. The plan was to take into consideration the neighboring residential buildings as well as the natural form of the slope and the katabic winds, which play an important role in keeping the city cool, and the poor condition of the building lot was not to be overlooked either.

Finally, the horizontal bending of the elongated blocks arose out of consideration to the existing structures. Moreover, the height of the building corresponds to the upper edge of the slope to the north and at the same time to the eaves of the residential buildings to the south. The subsequent bend in the buildings on the south side, however, results not only out of consideration to the neighbors but plays with the front edge, which, as seen along with all the building sections as a whole, ultimately modulates the overall form.

All in all, it was essential for the project to "crouch" into the landscape. Due to its enormous span and the deflection, the unit has been designed as a bridge structure, in which two massive cores support a 160-metre-long steel truss frame.

Detail and Materials
The design of the façade conforms to the structural system of the truss frame. The parapets are not arranged randomly but have been placed to coincide with the points of greatest deflection. In this way the outward impression is diversified while the interior is marked by greater individuality. A fascinating tension, the effect of which is intensified by the arrangement of the lamellae at different levels and intervals.

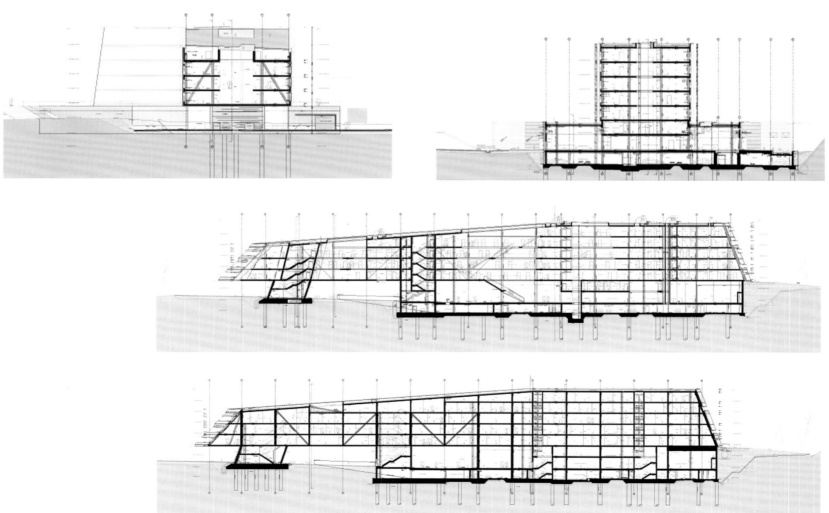

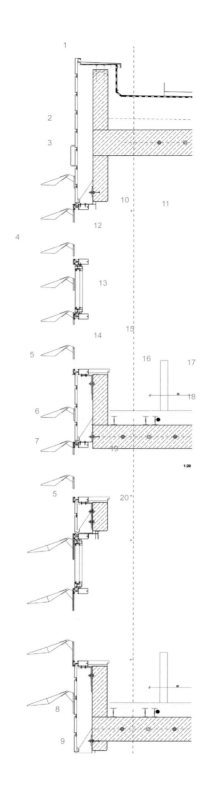

Standard façade section
1. Top of attic +23.45
2. Top of raw ceiling +22.66
3. Element lit from behind bottom of raw ceiling +22.38
4. Permanently sun protection extruded blades widths in 20cm, 30cm, or 45cm
5. Façade pressure released
6. Top of raw ceiling +19.42
7. Bottom of raw ceiling +19.26
8. Top of raw ceiling +16.20
9. Bottom of raw ceiling +15.94
10. Glare protection blind
11. Bottom of finished lintel 220.5cm above top of finished floor
12. Skylight
13. Tilt and turn window
14. Laminated safety glass
15. AW02
2.5cm Aluminium façade siding
2.5cm back ventilation
14cm hydrophobized thermal insulation
16cm reinforced concrete wall
16. Finished parapet height 45.5cm above top of finished floor
17. Feeder pillar h=550mm
18. Double floor can be opened
19. Protective edging ceiling stopped and smoothed
20. Pre fabricated concrete element parapets fastened to flat steel every 250cm

标准剖面节点
1. 阁楼顶部，+23.45
2. 原始天花板顶部，+22.66
3. 原始天花板底部，背光元件，+22.38
4. 永久性遮阳叶片，宽度为20cm，30cm，45cm
5. 释放立面压力
6. 原始天花板顶部，+19.42
7. 原始天花板底部，+19.26
8. 原始天花板顶部，+16.20
9. 原始天花板底部，+15.94
10. 眩光保护罩
11. 过梁底部，高于地面220.5cm
12. 天窗
13. 斜转窗
14. 夹层安全玻璃
15. AW02
2.5cm铝面护墙板
2.5cm后置通风
14cm疏水性隔热层
16cm钢筋混凝土墙
16. 护栏高度，高于地面45.5cm
17. 馈电柱h=550mm
18. 可打开的双层地面
19. 天花板防护边缘
20. 预制混凝土元件，护墙板固定在扁钢上，间隔250cm

项目概况

项目涉及多座独立建筑的设计，它们在林兹大学的校园内相互交织，形成了统一的整体。设计综合考虑了周边的住宅楼、天然的斜坡地形、下降风以及建筑场地的不良条件。其中，下降风对保持城市凉爽有着重要的作用。

最终建筑呈现为细长而弯折的水平结构。建筑的高度既考虑到了北面斜坡上边缘的高度，又考虑到了南面住宅楼屋檐的高度。建筑的弯折造型不仅体现了与周边环境的联系，还巧妙地处理了建筑的前端。从建筑剖面来看，建筑是一个完整的整体。

建筑俯卧在景观之上，形成了巨型桥梁结构，跨度极大，以两个厚重的核心支撑着160米长的钢桁架。

细部与材料

立面的设计与钢桁架结构系统相一致。护墙的排列并不是随机的，而是遵循了偏折点的角度。这既丰富了建筑外观的多样性，又保证了建筑内部各个空间的独立性。不同楼层和间隔区域之间的薄片突出了建筑布局的张力。

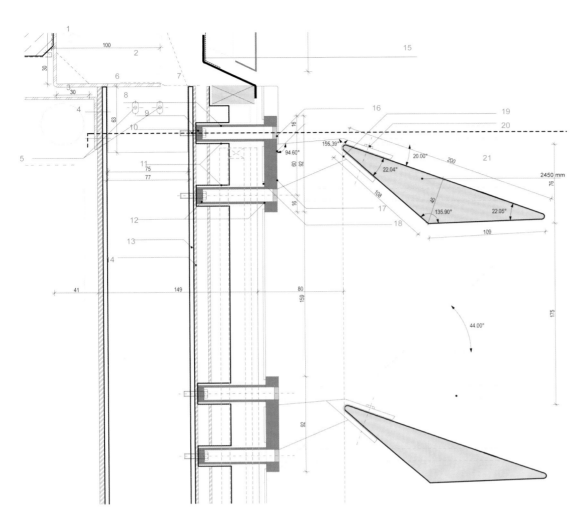

Façade section

1. Welded steel frame
2. Encircling steel frame loose connection intake 15 mm relative deformation
3. Aluminium angle
4. Blinds for glare protection(no cable guides, not electrically powered)
5. Stiff sheeting screwed to the support posts
6. Steel frame with slits
7. Vapor-tight membrane
8. SS pipe Ø max.16mm
9. Threaded screw
10. Insertion joint with pre-cut threads
11. Screw channel milled
12. Fixed joint welded
13. Steel C-profile insertion joint at every second post
14. Transom + mullion façade schüco system FW60+HI
15. Aluminium L-profile clamped to molding
16. Tight seal revisable
17. Fin bracket welded to end plate
18. Vertical molding passes through (flat Aluminium profile 60 mm)
19. Fins + bracket color in NCS chosen by client
20. Fins mounted according to manufacturer's specifications
21. Fin 200mm
    Rise 45mm
    Extruded Aluminium
    Element length 4900 mm
    Brackets mounted every 2450
22. End plate
    fyk = min. 190 N/mm²
23. End plate bracket welded to SS pipe
24. Pane embedded 13 mm
25. Flat pressure profile 60 mm
26. Horizontal molding runs continuously across vertical pressure profile

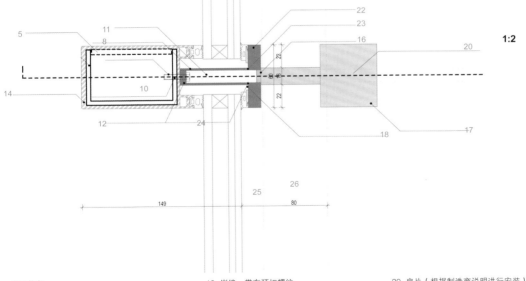

立面节点
1. 焊接钢框架
2. 环绕钢架，松动连接，通风口15mm相对变形
3. 铝角
4. 眩光保护罩（无缆线，非电动）
5. 刚性薄板，以螺丝固定在支柱上
6. 有缝钢框架
7. 气密薄膜
8. 不锈钢管，最大直径16mm
9. 双头螺栓
10. 嵌缝，带有预切螺纹
11. 螺纹槽
12. 固定接头，焊接
13. C形钢，每隔两根支柱有一个嵌缝
14. 气窗+窗框立面，schüco系统FW60+HI
15. L形铝材，与装饰线夹紧
16. 密封，可调整
17. 扇片支架，焊接在端板上
18. 垂直穿透式装饰线（平铝型材，60mm）
19. 扇片+支架，颜色由客户选择
20. 扇片（根据制造商说明进行安装）
21. 扇片，200mm
    上升，45mm
    挤制铝材元件，4900mm
    支架安装间隔，2450mm
22. 端板，fyk = min. 190 N/mm²
23. 端板支架，焊接在不锈钢管上
24. 嵌板，13mm
25. 平压线，60mm
26. 水平装饰线，沿着垂直压力线

Hotel Centar

辛塔尔酒店

Location/地点: Novi Sad, Serbia/塞尔维亚，诺维萨德
Architect/建筑师: MITarh team (Bransilav Mitrovic, team leader, Jelena Kuzmanovic, Nemanja Zimonjic, Igor Pantic, Ognjen Krašna, Siniša Tatalović)
Photos/摄影: MITarh and Rade Kovač
Built area/建筑面积: 4, 364m²

Key materials: Façade – Aluminium curtain
Structure – concrete, steel
主要材料: 立面——铝幕墙
结构——混凝土、钢材

Overview

Hotel "Centar" is segment of awarded (I prize) competition design for conceptual architectural and urban design of the IM "Matijevic" commercial building and public garage in Novi Sad. Competition task required with three different facilities: commercial building - later Hotel - for IM Matijevic, smaller business building and public multi-level garage. Idea was to design the fragments and organize the functions in one compact and harmonious unity. The basic principle in designig space in the central area of Novi Sad was to keep the unity of the entire building area, and to design structures perceived as a single mass, which will correspond to the surrounding buildings and area. Hotel is the first of these tree buildings, which is built. The public garage and smaller business buildnig are expected in the future.

Hotel is located across the beginning of the main pedestrian zone in Novi Sad and across of the Serbian National Theatre, as a dominant point in space. It was conceived as a cubic form, and follows years of research on the topic: "surface depth", "layers of membranes" and cubic design of Branislav Mitrović and MITarh team. Location of the hotel is very exposed and has dominate role in the city boulevard. The desire to make a distinctive, but unobtrusive and simple form in space, led us to balance the gap in the area by providing counterpart and counterweight to the power of geometry of the Serbian National Theatre.

Detail and Materials

Medium to achieve a unified form, or a single volume is the envelope - the building façade, both homogeneous and transparent. This ambivalence is achieved by fine-structuring of material, in this case the vertical metal strips with periodic interruptions, that let the light inside and filter the outside surroundings. This transparent and gossamer cubic form loses the prerogative of ordinary elements on the façade. Maximum transparency is achieved by the membrane, but also the strenght of the compact cube. The user who is located in the inner space has no sense of confinement or claustrophobia, as he essentially has a very clear view, while the passer from the outside perceives the object as a hermetic form, and only can peer into the privacy of the interior only in certain direct views. This design is a new purport of the relation between the inner and outer space. With its volume building shapes urban public space, and adjusts object to urban and investors' high dencity requirements.

Façade section

1. Terrace
   - Reinforced concrete slab
   - Vapour barrier
   - Thermal insulation – High density mineral wool
   - PVC foil
   - Perlite concrete reinforced with galvanised mesh from 140mm - 160mm thk.
   - Waterproofing layer 10mm thk.
   - Cement screed 30mm thk.
   - Ceramic tiles with adhesive 10mm thk.

2. Inclined Roof
   - Reinforced concrete slab
   - Vapour barrier
   - Thermal insulation - High density mineral wool (specific weight 400kg/m3)
   - PVC foil
   - Perlite concrete reinforced with galvanised mesh from 140mm - 160mm thk.
   - Waterproofing layer 10mm thk
   - Cement screed 30mm thk.
   - Ceramic tiles with adhesive 10mm thk.

3. Fire Protected Façade Perimetre
   - Reinforced concrete beam
   - Thermal insulation 60mm thk.
   - Fireproof gypsum board 15mm thk.
   - Airspace 60mm thk.
   - Tempered glass 8mm thk.

4. Fixed Sun Shade
   - Vertical sun shade 200mm width, step 300mm
   - Oval shape – type "Hunter Douglas"
   - Sun shade height 3370mm (from floor to floor)

5. Steel Beam – Sun Shade Holder
   - Sun shades are fixed to metal beam supported by reinforced concrete beam

立面节点

1. 平台
   钢筋混凝土板
   隔汽层
   隔热层——高密度矿物棉
   PVC箔
   珍珠岩混凝土，镀锌网加固，厚度为140mm~160mm
   防水层，10mm
   水泥浆找平，30mm
   瓷砖，带有黏合剂，10mm

2. 斜屋顶
   钢筋混凝土板
   隔汽层
   隔热层——高密度矿物棉（单位重量400kg/m³）
   PVC箔
   珍珠岩混凝土，镀锌网加固，厚度为140~160mm
   防水层，10mm
   水泥浆找平，30mm
   瓷砖，带有黏合剂，10mm

3. 防火立面
   钢筋混凝土梁
   隔热层，60mm
   PVC箔
   防火石膏板，15mm
   空气层，60mm
   钢化玻璃，8mm

4. 固定遮阳板
   垂直遮阳片，200mm宽
   椭圆形Hunter Douglas型
   遮阳板高度3370mm（楼层对楼层）

5. 钢梁——遮阳板底座
   遮阳板固定在金属梁上，整个结构由钢筋混凝土梁支撑

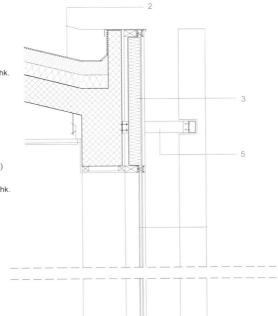
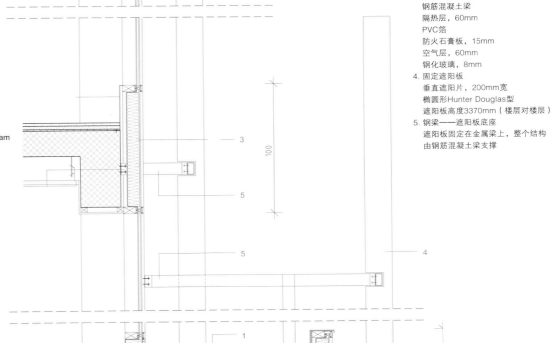

项目概况

辛塔尔酒店是诺维萨德的IM马蒂耶维奇商业楼和公共停车场综合项目设计竞赛作品的一部分。竞赛要求建造三个不同的设施：为IM马蒂耶维奇建造的商业楼（即后来的酒店）、小型商务楼和公共停车楼。设计方案对它们进行了分别的设计，并使其融为一个和谐的整体。设计的基本原则是保证建筑区域的统一性和独立结构的差异性，使其与周边建筑和区域联系起来。辛塔尔酒店是该项目的首个开发工程，公共停车楼和小型商务楼将在未来实现。

酒店位于诺维萨德一个主要步行区的起点，在塞尔维亚国家剧院对面。酒店呈立方体造型，实践了MIT建筑事务所团队的多年研究成果："表层深度"、"层次薄膜"和立方体设计。酒店的位置十分显眼，在城市的林荫道上扮演着重要的角色。建筑师以打造一座独特而简洁的建筑为目标，在塞尔维亚国家剧院对面建造了一座能与其旗鼓相当的美妙建筑。

细部与材料

建筑表皮帮助建筑实现了统一的造型和空间——建筑立面完整而通透。精致排列的金属条富有独特的韵律感，既能为室内引入光线，又能过滤外部的景象。这座带有半透明"薄纱"的立方体建筑独树一帜。薄膜和紧凑的立方体结构共同实现了建筑的通透感。内部空间的使用者不会有限制感或幽闭恐惧，因为他的视野十分清晰；外面的行人则会将酒店看成一座封闭的结构，只有从特定的角度才能窥到室内的一角。设计在室内外建立起了一种全新的联系，让建筑重塑了城市公共空间，同时又满足了投资者的高密度建设要求。

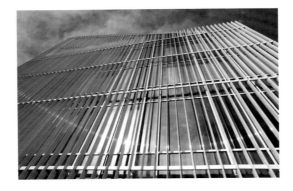

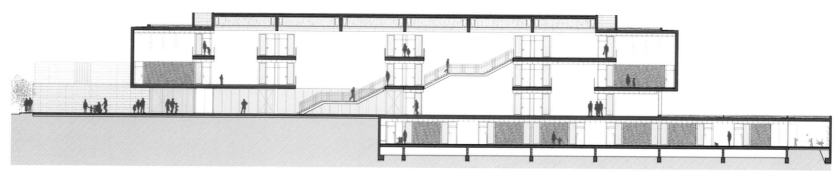

School Isabel Besora
伊莎贝尔贝索拉学校

Location/地点: Tarragona, Spain/西班牙，塔拉戈纳
Architect/建筑师: NAM ARQUITECTURA, Nacho Alvarez Martinez
Photos/摄影: José Hevia
Completion date/竣工时间: 2012
Built area/建筑面积: 4,024m²

Key materials: Façade – Aluminium Structure – concrete
主要材料：立面——铝
结构——混凝土

Overview

The building sits on a triangular plot of reduced dimensions and in an environment characterised by direct contact with the railroad on one side and a set of residential buildings of brickwork on the other. This explains the implementation of the building on the edge of the sun itself and also the density in the main body of four floors, a strange height in such buildings. This body, which basically contains all the classroom program and services, is attached to it forming an "L" on another volume in which the two largest program spaces are located, the kitchen and gym. Besides these, the architects also find in the rest of general services such as reception, concierge, secretarial, management, staff rooms, offices, library and services.

In each of these buildings, a central corridor separates the small service program aimed at the street's largest classroom program that spill into the sun. This major route also relates in a linear section through staircase and a skylight that illuminates and gaps related to various different areas of the centre.

Separating the building in respect of railways are the indoor games, sports court and wooded recreation area to fluff the possible visual and acoustic contact. During the day, a series of outdoor porches also allow covered movement of students to library, dining room, gym and locker rooms.

Detail and Materials

The whole building is constructed with concrete and formwork wooden tablet (vertical and horizontal). Once posed this simple and basic structure, the construction is completed by the use of glass as an enclosure in various sizes and finishes, and the vertical steel trusses of dark grey that allow entry and control of light in classrooms and other spaces. The upper part is enclosed with Aluminium panels.

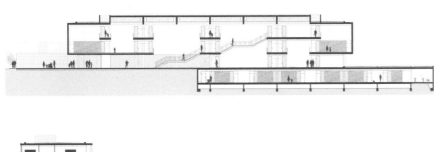
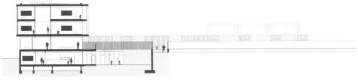
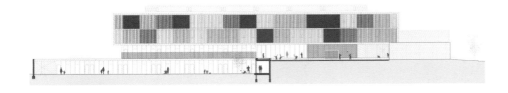

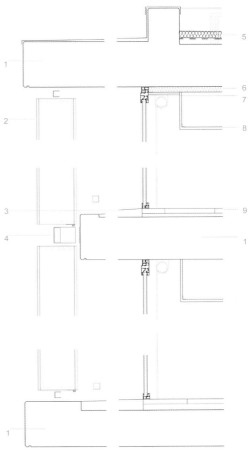

Façade detail
1. Concrete
2. Aluminum folding LAMA RAL 7004
3. Concrete slope
4. LAMAS mechanism
5. Insulation cover
6. Interior ceiling insulation
7. Aluminum joinery with thermal bridge, double glazing glass, colour RAL 7004
8. Ceiling plates of plaster micro perforated 60x60cm
9. Pavement terrazzo 40x40cm

立面节点
1. 混凝土
2. 折叠铝板LAMA RAL 7004
3. 混凝土坡
4. LAMAS机械系统
5. 隔热层
6. 内层天花板隔热
7. 铝窗框（断热），双层玻璃，RAL 7004色
8. 微孔石膏天花板60x60cm
9. 水磨石地面40x40cm

项目概况

建筑位于一块三角形的场地上，一面是铁路，另一面是砖砌结构的居民楼。这种地理位置决定了建筑的朝阳朝向和建筑主体不常见的四层楼高。建筑主体基本设置着所有教室和服务设施，与另一座结构共同形成了L造型。另一座结构内设置着两个最大的功能区——餐厅和体育馆。除此之外，建筑师还在建筑内添加了其他基础设计，包括前台接待、门卫、文秘、管理、员工休息室、办公室、图书馆等。

建筑内部的中央走廊将空间一分为二，一侧是小型服务区域，另一侧是朝阳的大型教室。走廊通过楼梯上下相连，屋顶的天窗为各个区域提供自然采光。

建筑在靠近铁路的一侧设置着室内游戏室、运动场和树林休闲区，弱化了铁路在视觉和听觉上对学校所造成的影响。一系列露天门廊引导着学生前往图书馆、餐厅、体育馆和更衣室。

细部与材料

整个建筑由混凝土和木板模架（分为水平和垂直两种）构造而成。在这个简单的基础框架下，建筑添加了各种尺寸的玻璃和深灰色钢条。钢条形成了入口，并控制了教室和其他空间的采光。建筑顶部则采用铝板包围起来。

CASP 74
卡斯普74办公楼

Location/地点: Barcelona, Spain/西班牙，巴塞罗那
Architect/建筑师: Jaume Bach, Eugeni Bach (Bach Arquitectes)
Photos/摄影: José Hevia
Construction area/建筑面积: 3,965.95m²

Key materials: Façade – sliding Aluminium blinds
主要材料： 立面——滑动铝遮阳帘

Overview
Designing an office building in the heart of the Example in Barcelona 150 years after the implementation of Cerdá Plan requires that a debate on how to conceive while on how they should be the façades of the building to suit the needs of the 21st century and so the same dialogue and respect of history. From this will, the architects have designed the building Casp 74, using a contemporary language recovers the traditional solutions of the Example prototypical façades but adapted to new needs and technology. The program is building 27 offices distributed in PB + 5, and two business premises with loft across the street giving Casp. In the basement there is a parking area for 34 vehicles.

Detail and Materials
The balcony, booklet blinds, cornices and colours are reinterpreted here to get a building not only integrated into the built landscape, but also does so with solutions energy savings both active and passive.

The street façade is subject to strict regulation of the rules of the Example in Barcelona, and solved by fixed panels formed by special stoneware pieces upright framed by a thin stainless steel frame and a sliding Aluminium blinds giving complete privacy and light control needs.

The stoneware pieces are Bach's own design arquitectes, made by Ceramic stoneware Cumella dyed extruded and cooked to a temperature of 1250 °. These parts section rectangular 120 x 80 mm clear enamel glaze randomly with bright, so that the changing effect of bright and spare parts mate gives a vibration to the façade which gives complexity.

The façade of the courtyard of the block is resolved with a grid of large steel framed windows of the houses. Sun protection and privacy are solved here with a blind lacquered Aluminium folding recreated a front moving and changing from abroad.

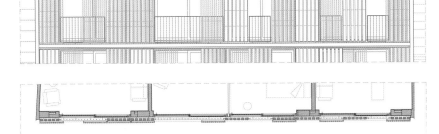

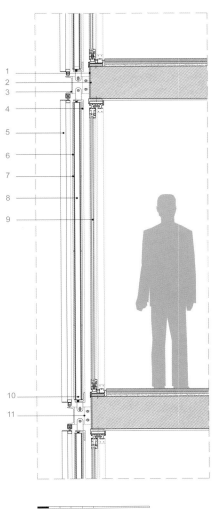

Façade detail
1. Continuous deck over forged e=10mm
2. Würth metalized fabric insulation
3. UPN 160. FERR 700 colour painted series of amount ferrum
4. Fixed frame formed by I90'1 or parts subject to special perimeter stone
5. Aluminium skid mark gray lacquer technal lead formed by rectangular vertical profiles
6. Special piece stoneware 120x80x680mm
7. Ceramic cumella. colour KT-7-1. long faces in some parts are enameled 3 mm neoprene between separation of the pieces of stoneware
8. Galvanised steel profile 50x30x3mm rigid and support for parts of sandstone
9. Sliding window brand model gti techno lacquer grey Aluminium lead
10. The last hole to allow side welding with internal support mat. ventilation should allow entry of water
11. Double crown set with high strength bolts. Holes for bolts withnotch for adjustment of the crane plumbing

立面节点
1. 连续的露台地面e=10mm
2. 金属涂层纤维绝缘
3. 涂色铁制结构UPN 160. FERR 700
4. 由I90'1零件组成的固定框架
5. 矩形垂直剖面组成的灰漆引线，铝制滑动记号
6. 特制石板120x80x680mm
7. 陶瓷板，色彩KT-7-1，部分结构涂有3mm氯丁橡胶，与石板隔开
8. 镀锌型钢50x30x3mm，支撑砂岩板
9. 滑动窗，配有深灰色铝引线
10. 单面焊接孔，带有内部支撑垫，通风孔
11. 双冠高强度螺栓，螺栓孔的凹槽可供调整

项目概况

卡斯普74办公楼选址在巴塞罗那塞尔达规划（Cerdá Plan）实施150年后的中心试验区内，因此其设计重点即为如何构思建筑的外观，一方面满足当代需求，另一方面尊重历史。建筑师从这两点出发，运用现代建筑语言诠释传统的外观样式，以适应现代需求和技术要求。建筑内部包括27间办公室以及两间带阁楼的商业事务所办公空间，地下停车场可容纳34辆车。

细部与材料

阳台、遮阳帘、檐口等结构以及色彩的巧妙诠释使得建筑与周围的景观完美融合，同时又实现了节能的目的。

临街外观的设计严格遵守赛达尔规划法则，特质的石器结构固定在不锈钢框架内，滑动铝遮阳帘阻挡强光照射，并保护了室内的隐私。石器结构由建筑师设计打造，并经染色和1,250度的高温烧结而成。这些结构被分割成规格为120x80毫米的条状，并饰以更加明亮的色彩，使得整体外观更加精美，更具多样性。

朝向庭院一侧的外观采用宽大的钢窗打造，遮阳帘结构的运用则起到保护隐私的作用，同时也形成了运动变化的立面效果。

Jinan Vanke Marketing Centre
济南万科营销中心

Location/地点: Jinan, China/中国，济南
Architect/建筑师: Tsushima Design Studio
Photos/摄影: Masao Nishikawa
Area/面积: 1,557m²
Completion date/竣工时间: 2013

Key materials: Façade – Aluminium alloy, glass
主要材料：立面——铝合金、玻璃

Overview
Located in the city Centre of Jinan (Capital of Shandong Province), the Jinan Marketing Centre is the highlight of the flagship mixed-development project for Vanke to venture into this region.

As part of the master planning for the project, the marketing Centre is the first area to be showcased to the public. The design concept is for the architecture to buffer the immediate urban city experience, serving as an important circulation transition as people leave behind the commercial space and enter into the quiet residential zone, which is sited next to a lush green hill.

Detail and Materials
The architect and landscape designer collaborated very closely to create a tranquil experience as people approach the project. The entire glass door system on the first floor can be fully opened towards the landscape to allow the spatial flow between indoor and outdoor space, while the façade on the second floor is designed with Aluminium louvers providing light and view while ensuring privacy.

A customised metal mesh (sandwiched between 2 panes of glass) is used in the entrance canopy and part of the front elevation. During the day, this creates a contrast between the semi-transparent materials with the dark bronze Aluminium. During the night, the glass-mesh hybrid material, combined with the integrated lighting fixtures, creates a shiny icon with a gold sheen, showcasing a very different character and atmosphere from the day.

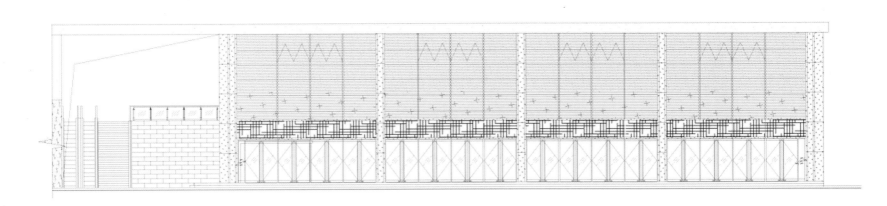

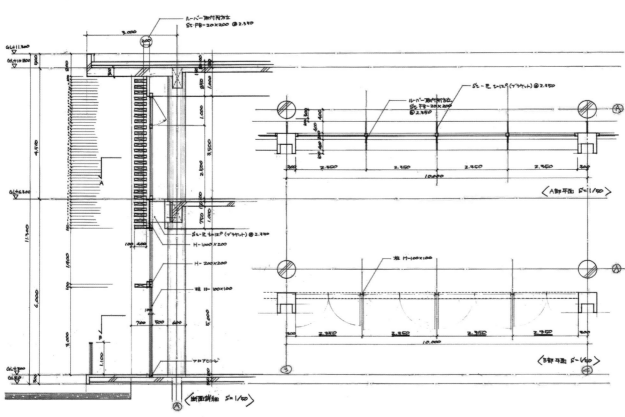

Façade detail
1. 100X100mm aluminium alloy column
2. 80X300mm aluminium alloy louvers
3. 2.5mm thick aluminium panel
4. Customized glass with wiremesh sandwich
5. 2.5mm thick aluminium panel (light)
6. LED lighting fixture
7. 2.5mm thick aluminium canopy
8. 15mm thick double glazed reinforced glass
9. 50mm thick insulation
10. 2.5mm thick aluminium panel
11. 6low-e double glazed reinforced glass (operable window)
12. Aluminium alloy mullion
13. 8low-E double glazed reinforced glass
14. Aluminium alloy horizontal mullion
15. Metal door handle

立面节点
1. 100X100mm铝合金柱
2. 80X300mm铝合金百叶
3. 2.5mm铝板
4. 定制玻璃金属丝网
5. 2.5mm铝板（轻）
6. LED灯
7. 2.5mm铝遮阳篷
8. 15mm双层钢化玻璃
9. 50mm绝缘层
10. 2.5mm铝板
11. 低辐射双层钢化玻璃（可控窗）
12. 铝合金竖框
13. 低辐射双层钢化玻璃
14. 铝合金横框
15. 金属门把手

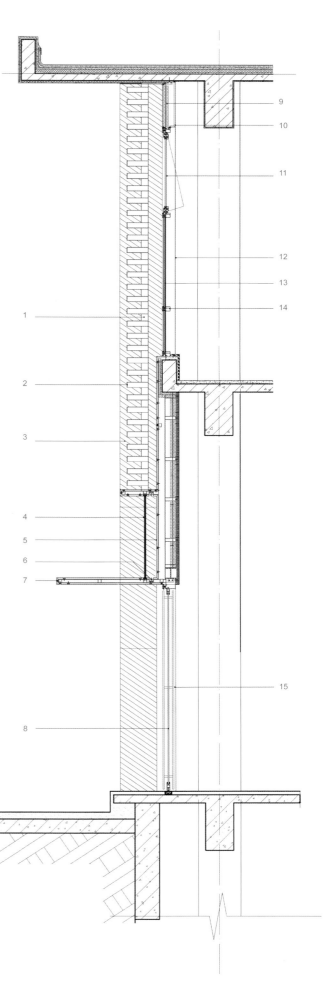

项目概况
济南万科营销中心位于山东省济南市市中心，其目的是宣传销售万科集团在该地区新开发的混合开发项目。
作为项目总体规划的一部分，营销中心是首个面向公众开放的设施。建筑的设计理念是缓冲直接的城市体验，在喧嚣的商业区和宁静的住宅区之间形成重要的过渡。住宅区就位于一座郁郁葱葱的小山旁边。

细部与材料
建筑师和景观设计师在项目开发中实现了紧密的合作，共同为人们营造了宁静的体验。一楼的整个玻璃门系统都可以打开，在室内外空间形成了流畅的连接。二楼的立面由铝合金百叶设计而成，既能透视光线和风景，又能保证隐私。

定制金属网（中间夹有两层玻璃板）被用于入口遮阳篷和建筑正面的一部分。白天，这种半透明的材料与深色的铝材形成了鲜明的对比。夜晚，玻璃金属网与灯具结合起来，形成了一层金色的光泽，让建筑呈现出与白天完全不同的效果。

A Change of Skin
维特鲁夫广场项目表皮翻新

Location/地点: Paris, France/法国，巴黎
Architect/建筑师: Atelier du Pont (Anne-Cécile Comar, Philippe Croisier, Stéphane Pertusier)
Photos/摄影: Luc Boegly
Completion date/竣工时间: 2013

Key materials: Façade – aluminium louvre
主要材料：立面——铝百叶

Overview

"Square Vitruve" is a real-estate project in co-ownership built on a concrete slab on the site of the former village of Charonne. It is virtually inaccessible, totally without vegetation and shapeless, and was built in the '70s and '80s without any consideration for the city and how people use it. Along with the construction of the tramway on the nearby boulevard, the ongoing Major Urban Renovation Project for the Saint-Blaise neighbourhood is opening up this forgotten corner of Paris.

The major rehabilitation of this seriously degraded building containing 56 social housing units was carried out while it was in use in a highly complex urban, technical and statutory environment and has played a part in preserving the social fabric of a neighbourhood undergoing massive change.

Detail and Materials

This intervention on the building's shell has been conducted in constant dialogue with the city's council and inhabitants in line with ambitious specifications. These specifications sought to confer a strong identity on the building, foster diversity, imagine new uses and improve the energy performance of a building whose occupants have described it as a "thermal sieve" in order to achieve Paris City Council's Climate Plan.

Basing its analysis on sun studies to assess the impact of the nearby high-rise buildings, Atelier du Pont suggested adding balconies wherever this made sense. This was a simple idea that was hard to put into practice, yet provided the apartments with outdoor areas and completely remodelled the architecture of the façades.

Designed as a kind of clever, oversized Meccano kit on a site that cannot be accessed by heavy plant, this new skin was installed without putting machinery on the concrete slab and using no cranes or pods. The balconies are suspended from the roof, and all the materials and technical solutions have been designed to avoid overloading the existing structure and disrupting residents' daily life.

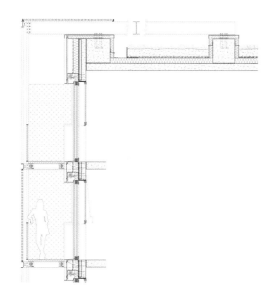

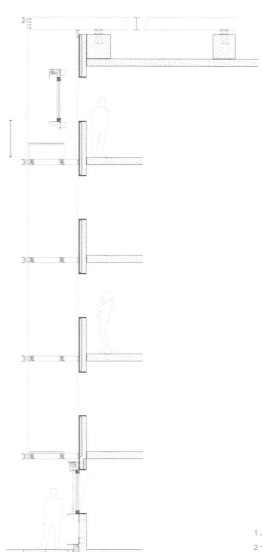

项目概况

维特鲁夫广场是一个混凝土板结构的地产开发项目，位于前夏洪尼村。建筑建于20世纪70、80年代，交通十分不便，完全没有考虑到城市规划和人们对建筑的使用方式。随着附近林荫道上有轨电车的开通，巴黎圣布莱斯城市改造项目重新想起了这个角落。

建筑的主要改造计划包括56间社会住宅单元的开发。项目位于一个高度复杂的环境之中，对保护经历着重大变化的社区的社会结构起到了重要作用。

细部与材料

建筑外壳的改造是建筑师与城市议会和居民不断协商讨论后得出的结果。根据他们的需求，建筑获得了强烈的存在感，拥有多样性，且具有良好的能源效率。居民将建筑称为"热筛子"，正在寻求成为"巴黎市议会气候计划"的一部分。

建筑师以对太阳的研究为基础，分析了附近高楼对项目的影响，为建筑添加了阳台。这个简单的想法很难付诸实践，却为公寓提供了额外的露天区域，并且能进一步完善建筑的立面。

建筑师充分发挥自己的想象力，将外立面拆分成巨大的钢铁拼装模型，用起重机安装在混凝土楼板上。整个阳台结构从屋顶悬下来，所有材料和技术方案都避免了原有结构的负载过重，也不会影响居民的正常生活。

Balcony façade
1. Balcony mounting bracket
2. Green roof
3. Aluminium & wood frames
4. Steel security railing
5. Fixed Aluminium louver
6. Vertical lacquered profiled steel sheets on top of exterior insulation
7. Roller blind box
8. Blinds with adjustable slats

阳台立面节点
1. 阳台安装架
2. 绿色屋顶
3. 铝木框
4. 钢制安全围栏
5. 固定铝百叶
6. 外层绝缘上的垂直涂漆型钢板
7. 卷帘遮阳盒
8. 带有可调节板条的遮阳帘

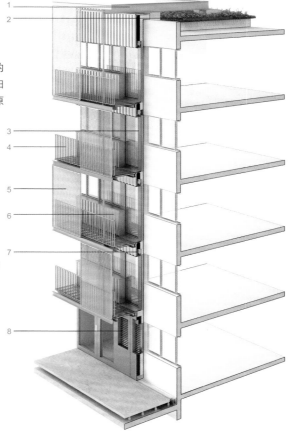

IMED Elche Hospital
埃尔切IMED医院

Location/地点: Elche, Spain/西班牙，埃尔切
Architect/建筑师: Jose Escobedo/ JAS ARQUITECTURA, Luis Rodrigo ARTLUX
Photos/摄影: Andrés Diaz
Site area/占地面积: 9,250m²
Gross floor area/总楼面面积: 40,000m²
Completion date/竣工时间: 2011

Key materials: Façade – aluminium cube, glass
Structure – concrete
主要材料: 立面——铝块、玻璃
结构——混凝土

Overview
The hospital company IMED had his own company architect to develop the hospital project. They contracted Jose Manuel Escobedo, the architect of the young studio JAS ARQUITECTURA to create the aesthetics, appearance and personality of the building. To develop the design of the building from floor plans that could not be modified required a mind openness in order to forget aesthetical unbreakable premises and to consider different design solutions and alternatives. The main target was to express that, despite the rigid functional needs, aesthetical and functionality can coexist.

Detail and Materials
The grey aluminium and dark glass curtain wall façade is wrapped with the white aluminium cubes up. The various white aluminium cube volumes have made possible to break the monotonous surface that usually these kinds of buildings have. These cubes have been displaced one from the other, generating a volumetric composition that allows including the indoor light in the global shadow- light effect. The light façade composition is therefore aesthetic, functional and energetically sustainable.

Elongated water sheets delimit the entrances to the building, which are located in three of its façades. These elements as well as the composition and location of the white aluminium cubes generate a very easy outer circulation. With this solution, the different ways of access to the building are easily understandable for the patients and visitors.

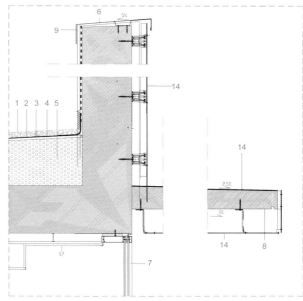

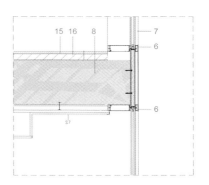

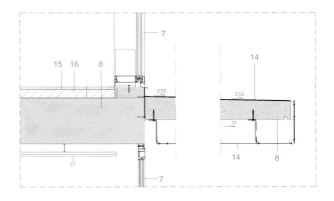

项目概况

IMED医疗集团有自己的建筑师团队对医院进行开发设计。他们委托JAS建筑工作室对建筑的外观、个性进行美化。建筑的楼面规划是固定的，JAS建筑工作室必须打破常规的审美，用独特的设计方案来进行设计。设计的目标是突出美观性和功能性的和谐统一。

细部与材料

由铝材和玻璃组成的灰色幕墙上部以铝块包裹。不同尺寸的白色铝块打破了单调的建筑表面。这些铝块相互作用，形成了独特组合，与玻璃幕墙共同形成了变化的光影效果。轻质立面组合兼具美观性和功能性，同时又生态环保。

建筑入口的长条水池划出了建筑的边界，分别位于建筑的三个立面之外。这些元素与白色铝块的组合形成了简单的外部交通流线。这样一来，患者和访客都能轻易地找到进入建筑的通道和入口。

Roof
1. Gravel
2. Geotextile Polypropylene. Polyethylene
3. Self protected waterproof membrane
4. Polystyrene 5cm
5. Arlit 15cm
6. Aluminium

Façade
7. Curtain wall 5/12/4+4
8. Reinforced concrete
9. PVC 1.5mm
10. High density polyethylene
11. Plaster panel 46mm+2×15mm
12. Colored vinyl
13. Stone wool 50mm 0.04W/mk
14. Alucobond white

Horizontal surfaces
15. Porcelain
16. Concrete
17. Granite

屋顶节点
1. 碎石铺面
2. 土工布 聚丙烯–聚乙烯
3. 自我保护的防水薄膜
4. 聚苯乙烯5cm
5. Arlit 15cm
6. 铝

立面节点
7. 幕墙 5/12/4+4
8. 钢筋混凝土
9. PVC 1.5mm
10. 高密度聚乙烯
11. 石膏板 46mm+2×15mm
12. 彩色乙烯树脂
13. 石棉50mm 0.04W/mk
14. 白色铝塑板

水平面节点
15. 瓷砖
16. 混凝土
17. 花岗岩

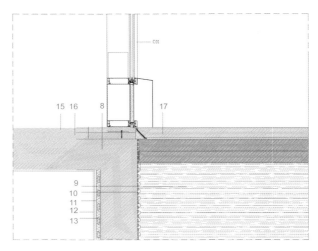

Policemen House

警察之家

Location/地点: Bilbao, Spain/西班牙，毕尔巴鄂
Architect/建筑师: COLL-BARREU ARQUITECTOS
Photos/摄影: COLL-BARREU ARQUITECTOS
Built area/建筑面积: 12,792.56m²
Completion date/竣工时间: 2012
Budget/预算: 16,215,310.26 €/16,215,310.26欧元

Key materials: Façade – aluminium sheet
主要材料： 立面——铝板

Façade material producer:
外墙立面材料生产商：
Curtain walls - Espalu
Folded Aluminium Panels – Imar (Poligono Industrial Granada, AB1 48530 Ortuella, Vizcaya)

Overview

The building accommodates the central Public Safety services for a one-million people urban area. The ground on which the building sits was unstable. It was uncertain. It is the result of having quickly filled an open mine, with pits up to 50m deep, which formed a cavernous place in its physical reality and its memory.

Upstream, other mine sites have been transformed, also visible from the site, which confirm the history of this exchange between matter and soil, the life of the city and that of men, the densification of the territory and the air.

Policemen House's concatenated planes are perceived from distant areas and set up a close relationship with the steep topography of the slope. Above this vacillating place, buildings do not have any mass.

Detail and Materials

Folded aluminium sheets acquire enough inertia to be supported in the air. The present construction is a system of layers. It is paper. The architects don't appreciate that the current architecture tries to simulate a massive construction, because the massive construction no longer exists. In this building corresponds, at the same time, to the mobile condition of the land.

It is appropriate that the policemen house has been made of paper, it is part of a democratic contract: the police are a piece of paper. The buildings are paper, as they are our security, our convictions, our society: uncertain, temporary, fleeting.

项目概况

建筑肩负着城区100万居民的公共安全服务。项目所在的场地地面并不稳定,因为场地是由露天矿场快速填充而成,矿坑的深度可达50米,曾经呈现出一个巨大的洞穴空间。

场地上游的其他矿场也经过了改造,从场地上都可以看到。它们反映了物质与土壤、城市生活与人类生活、地域与空气之间的微妙关系。

警察局的连锁式楼面即使在远处的区域也能感知到,与斜坡地势形成了紧密的联系。建筑没有任何体块建在这片不稳定的区域之上。

细部与材料

折叠铝板的惰性足以在空气中抵御侵蚀。目前的建筑结构是一个层叠式系统。建筑师并不赞同不断追求大体块的建筑,因为大体块建筑结构已经不再适用。这座建筑充分地应对了地块的不稳定条件。

建筑师让警察局看起来就像是纸质的一样,正如现代社会的民主合约一样:警察也是纸上(法律)所规定的一部分。建筑就像纸,因为它代表着我们的安全、信仰和社会,同时也充满了不确定性、临时性和短暂性。

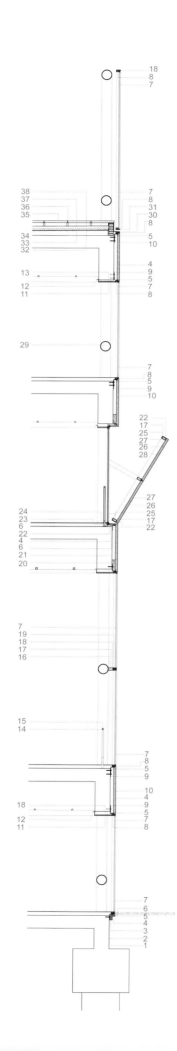

Façade detail

1. Delta drain
2. Waterproof insulation
3. Stainless folded sheet. e=3mm
4. Rock wool insulation
5. Hot galvanised steel angle. e=8mm
6. Galvanised folded sheet. e=3mm
7. Double glazing
8. Lacquered aluminium sheet Technal-cortine wall
9. Folded galvanised steel. e=1mm
10. Aluminium sandwich panel
11. DM board e=19mm. double coated aluminium lacquered aluminium sheet
12. Strip false ceiling
13. Pine wood batten
14. Lacquered steel tubular profile ø40.3
15. Lacquered steel tubular profile ø40.3
16. Hot galvanised steel tubular profile 60.5mm, c/60cm
17. Neoprene band
18. Lacquered aluminium sheet. e=3mm
19. Hot galvanised steel folded sheet. e=4mm
20. Hot galvanised steel tubular profile 40x5
21. Hot galvanised steel angle
22. Polished lacquered and transparent steel folded sheet
23. Hot galvanised steel tubular profile 50x20x2
24. Fixed window lacquered profile
25. Hot galvanised and lacquered tubular profile 120x60x8
26. Extruded polystyrene insulation
27. Lacquered aluminium sheet. e=1mm
28. Stamped polished lacquered and transparent aluminium sheet e=1mm
29. Lacquered steel tubular profile ø273mm
30. Lacquered aluminium double folded sheet. e=3mm
31. Lacquered aluminium angle. e=4mm
32. Extruded polystyrene thermal insulation Roofmate SLA. e=5cm
33. Extruded polystyrene thermal insulation Roofmate SLA. e=5cm
34. Slope formation lightweight concrete
35. Granite tile 60x60x5cm
36. P.V.C adjustable foot
37. Propylene fibre reinforced mortar layer
38. Geotextile felt

立面节点

1. 排水槽
2. 防水绝缘层
3. 不锈钢折叠板e=3mm
4. 石棉隔热层
5. 镀锌角钢e=8mm
6. 镀锌折叠板e=3mm
7. 双层玻璃
8. 涂漆铝板Technal-cortine墙面
9. 折叠镀锌板e=1mm
10. 夹心铝板
11. DM板e=19mm，双层涂漆铝板
12. 条状假吊顶
13. 松木板条
14. 涂漆钢管型材ø40.3
15. 涂漆钢管型材ø40.3
16. 镀锌钢管型材60.5mm, c/60cm
17. 氯丁橡胶带
18. 涂漆铝板e=3mm
19. 镀锌折叠钢板e=4mm
20. 镀锌钢管型材40x5
21. 镀锌角钢
22. 抛光涂漆透明折叠钢板
23. 镀锌钢管型材50x20x2
24. 固定窗涂漆型材
25. 热镀锌涂漆管120x60x8
26. 挤塑聚苯乙烯隔热
27. 涂漆铝板e=1mm
28. 印花抛光涂漆透明折叠铝板e=1mm
29. 涂漆钢管型材ø273mm
30. 涂漆双层折叠铝板e=3mm
31. 涂漆角铝e=4mm
32. 挤塑聚苯乙烯隔热Roofmate SLA.型e=5cm
33. 挤塑聚苯乙烯隔热Roofmate SLA.型e=5cm
34. 坡形轻质混凝土
35. 花岗岩砖60x60x5cm
36. PVC可调节脚线
37. 丙烯纤维加固砂浆层
38. 土工布毛毡

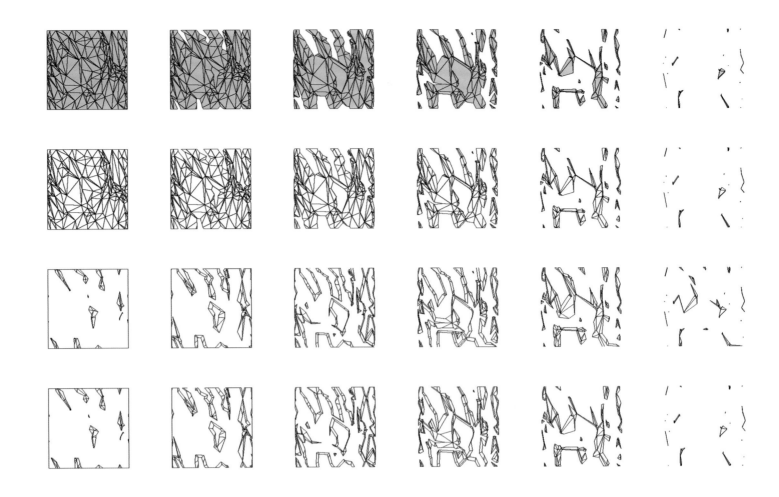

Supermarket in Athens
雅典超市

Location/地点: Athens, Greece/希腊，雅典
Architect/建筑师: KLab Architecture – Konstantinos Labrinopoulos
Design team/设计团队: Enrique Ramírez, Veronika Vasileiou, Elena Skorda, Haris Heizanoglou, Mark Chapman
Photos/摄影: Panos Kokkinias
Completion date/竣工时间: 2013

Key materials: Façade – perforated Aluminium, wood
主要材料：立面——穿孔铝板，木材

Overview
The store was fully operational during the construction period. Most of the works were happening during the weekends after the store was closing at 9.00pm at Saturday.

The task for KLab was very demanding. The supermarket building existed in 3 floors before the refurbishment and looked old and neglected. The store has altered through the years with additions of extra spaces that transform the interior to almost a labyrinth with very difficult navigation through the shelves and circulation between the floors. The exterior was also outdated. The quality of the products had nothing to do with the quality of the space.

The main idea was to change the notion of the supermarket as it exists today in Greece – indifferent big boxes with long alleys of products. The research started from recently designed supermarkets and continued to the open food markets. For as the supermarket had to be resolved as an interpretation of an urban market place where you go to different shops to find different goods and interfere with other people. Some of the designs as the galvanised stroll were used to show the aesthetics of farmers market, some like the marble counter, the old fish markets in the islands and the age treated wooden shelves, the old bakeries on a mountain village.

Detail and Materials
The supermarket lies on a corner of a piazza and to most of the people the piazza itself has the name of the supermarket. The store was there before anything else was built and the architects thought that they had to give back an aesthetic value to the town. They had never considered an approach designing the façade as per design a supermarket façade, but they believed

that they had to tell a story and design something appealing and enthusiastic. They chose to emphasize the corner that the two sides of the building develop and to create a sense of movement that will correspond to the perspective lines of the piazza and the movement of the cars along the two sides of the supermarket. In a way a 3D wavy Aluminium laser cut façade became a modern second skin protecting the supermarket from the intense sunlight. The design of the perforation has its origins from the Mediterranean culture in food that is being represented here by the olive tree leaf. The exterior walls were painted in olive gray. Wood cladding created the base of the building. Nature and technology coexist in a supermarket as in façade and interior design.

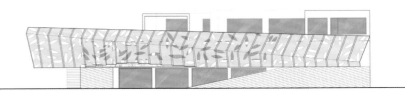

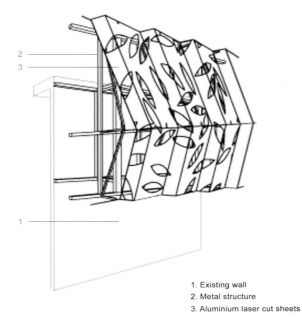 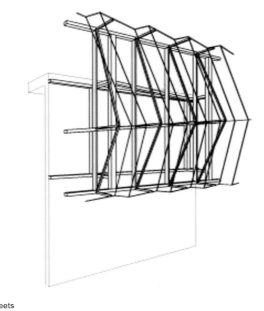

项目概况

项目的施工基本全在周六晚9点以后和周日完成,对建筑师KLab建筑事务所的要求十分严格。超市大楼在翻新前共有三层,十分破败。超市近年来一直在扩充内部空间,室内几乎变成了一个货架组成的迷宫;建筑外部也已经过时,完全无法和内部销售的商品相匹配。

设计的概念是让建筑打破希腊现有超市建筑的模式:巨大的盒式建筑,商品沿着狭长的走道排列。建筑师对最近新建的超市和露天市场进行了调查研究。对超市来说,最重要的问题就是如何在不同的店面与消费者之间建立起联系。设计采用了一些手段来向传统市场致敬,例如,镀锌板对应农贸市场,大理石柜台对应水产市场,做旧的木架对应老面包店。

细部与材料

超市位于广场的一角,大多数人都以广场的名字来称呼超市。超市已经建成了很长时间,建筑师认为它应该给城市带来独特的美学价值。建筑师试图打造一个与众不同的建筑立面,他们认为立面必须富有吸引力而又热情友好。他们选择了突出建筑的转角位置,形成于广场透视线以及车流相配合的运动感。一个三维波浪形激光切割镂空铝板立面成为了超市的第二层表皮,保护其不受强烈的阳光照射。穿孔板的设计源于地中海美食文化中的橄榄树叶。整个外墙被漆成了橄榄灰色。建筑底部采用了木板包层。建筑的立面和室内设计体现了自然与技术的完美结合。

1. Existing wall
2. Metal structure
3. Aluminium laser cut sheets

1. 原有墙壁
2. 金属结构
3. 激光切割铝板

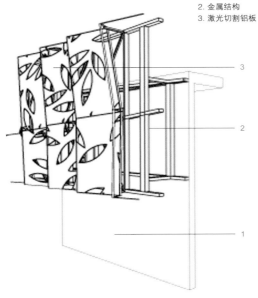 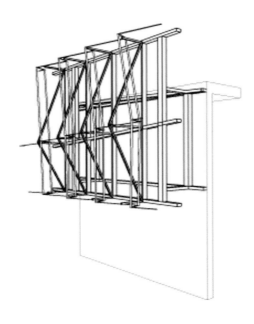

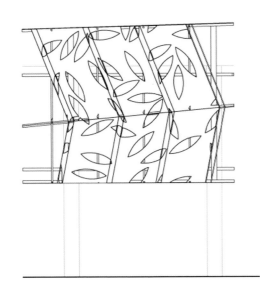 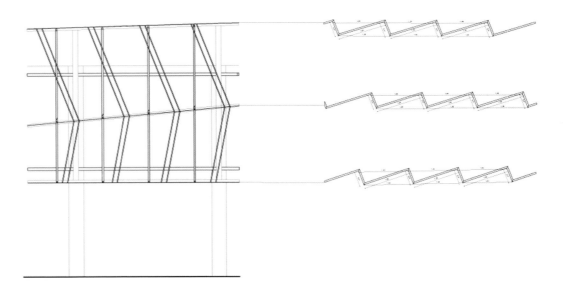

Innsbruck Furniture Store

因斯布鲁克家具店

Location/地点: Innsbruck, Austria/奥地利,因斯布鲁克
Architect/建筑师: Zechner & Zechner ZT GmbH, Vienna
Photos/摄影: Thilo Härdtlein, Munich
Site area/占地面积: 6,900m²
Construction area/建筑面积: 29,550m² (17,100m² above ground, 12,450m² underground) (地上面积17,100m2, 地下面积12,450m2)
Key materials: Façade – perforated Aluminium sheet, glass
Structure - steel
主要材料: 立面——穿孔铝板、玻璃
结构——钢材

Overview

The building consists of four above-ground floors and two below-ground floors. On the ground floor, near the furniture store, there is a branch of the grocery chain MPreis, with a connecting bistro and terrace. The first and second floors are taken up by the furniture retailer's display space, and on the top floor there is a small restaurant with terrace, from where you can enjoy the view of the surrounding mountains. Storage and admin areas are located on the third floor.

A spiral ramp inside the building connects the display floors, allowing customers to get an overview of the store's range by just strolling through the building.

The two below-ground floors accommodate customer parking as well as various building services areas. The ground floor is slightly raised, by about 1m, from the surrounding ground level, in order to provide natural illumination and ventilation to the underground parking. Daylight enters via the gap this creates, which helps prevent the impression of a closed-off and dark garage.

To allow the building to be appreciated from all angles, the loading bays for all incoming and outgoing goods have been integrated deep within the building, and as far as technically possible for access, covered by façades. This gives the building the lack of a front or a back, allowing it to be appreciated from all sides.

With an investment of around 30M euros for the new Leiner site in Innsbruck, special attention has also been given to energy efficient construction techniques – going well beyond legal requirements. Using innovative building services technology, resources are conserved and energy use and CO_2 emissions reduced. As a comparison: the reductions in CO_2 emissions made by the project using innovative building services technology is equivalent to that released by 100

0 10 20 40m

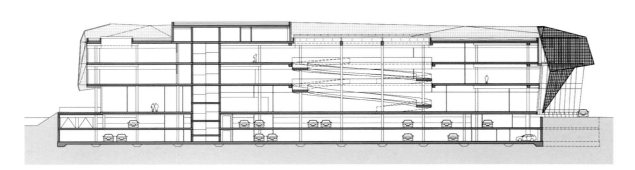

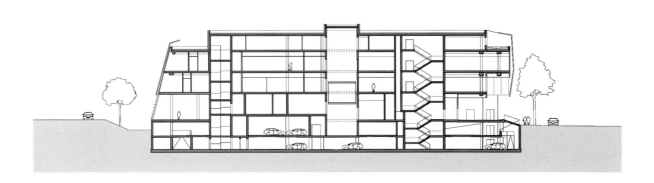

people travelling by car 30,000km a year each. This is the equivalent to 80 average, 150 m2, detached homes with oil heating.

Detail and Materials
A Green Tablecloth with Swan Design
The thermally active façade is predominantly formed of sandwich panels. Display windows are used only in limited areas, including opening the building's closed skin, at a busy corner on a roundabout, with a two-story glass façade that gives a view of the furniture store's range. The building will receive another opening in the entrance area and along the side of the grocery market.

The external shape of the building is, however, primarily created by a suspended projecting façade, which lies like an angular cloth above the geometrically simple structure below. This awakes associations of a tablecloth, a cloak, and of folded packaging. This external skin is 'pulled up' in particular areas, such as the glass façades at the roundabout, the MPreis façade and at the entrance area. The prismatic façade of perforated aluminium sheets produces different reflections, depending on façade angle, reduces the building's mass and gives the building a very striking exterior.

The furniture retailer's logo is three swans. From this logo, an abstract perforation pattern was developed for the outer envelope. The perforations are larger where the rooms behind are illuminated by windows, and smaller in other areas to prevent climbing. In particular areas the perforations also trace out the three-swan logo.

The façade cavity is illuminated in the evening, continually altering the appearance of the building as the lighting changes. The outer envelope then appears lighter than and as transparent as a curtain, and the building's volumes start to light up.

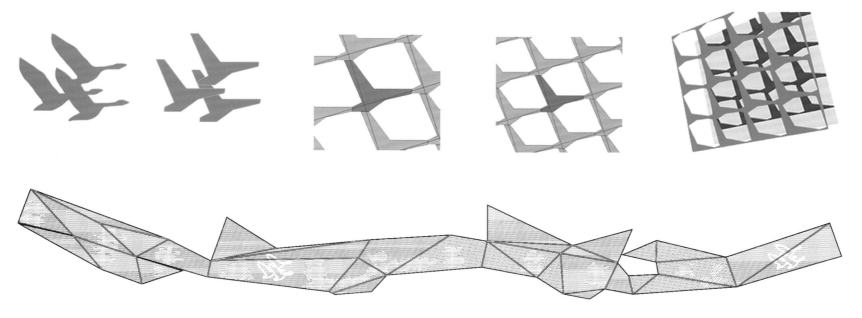

项目概况

建筑分为地上四层和地下两层。一楼家具店的旁边是一家连锁便利店，与小酒馆和露台相连。二、三楼是家具商的展示区，顶楼是一家带有露台的餐厅，在那里，你可以享受群山的美景。仓库和行政区都位于四楼。建筑内部，一条螺旋形坡道将展示楼层连接起来，消费者可以边在楼内漫步，边纵览店铺商品。

地下两层是顾客停车场和各种建筑服务区。一楼被抬高了1米，保证了地下停车场的自然照明和通风。日光通过缝隙进入地下，让停车场变得开放而明亮。

为了让建筑从各个角度都十分美观，装货平台被嵌入了建筑内部，被外立面包裹起来。这让建筑没有前后之分，从各个角度看都十分完美。

项目耗资约3,000万欧元，十分注重节能技术，远超过奥地利的国家标准。创新建筑服务技术的运用实现了资源和能源的节约，减少了二氧化碳排放量。项目通过创新建筑服务技术所减少的二氧化碳排放量相当于100个人一年行车30,000千米所排放的二氧化碳量的总和，也相当于80个采用燃油供暖的150平方米的独立住宅。

细部与材料
带有天鹅图案的绿色桌布
建筑的热力作用立面主要由夹层板组成，仅在有限的区域使用了展示橱窗，在建筑表皮上的开口、建筑转角采用了双层高店铺展示橱窗。建筑的另一个开口是零售店旁的入口区域。

建筑的外部造型主要由悬浮的立面构成，立面像一块棱角分明的布料覆盖在下方简单的几何结构上。这一设计令人联想起了桌布、斗篷和包装纸。这层表皮在特殊的区域会被"揭开"，例如转角的玻璃橱窗展示和入口的便利店。穿孔铝板的棱形立面在不同的角度上呈现出不同的反射，赋予了建筑多变的外观。

家具商的标志是三只天鹅，建筑师以此抽象出了穿孔铝板上的图案。镂空穿孔在采用窗户采光的房间外就较大，在禁止攀爬的区域就较小。在某些特定的区域，穿孔图案组成了天鹅标志。

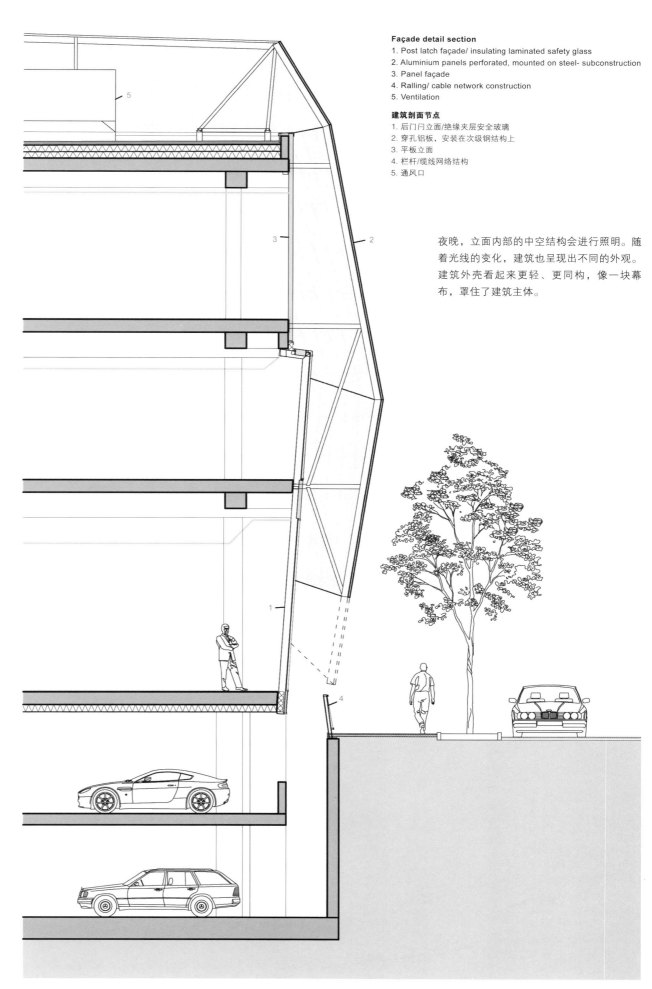

Façade detail section
1. Post latch façade/ insulating laminated safety glass
2. Aluminium panels perforated, mounted on steel- subconstruction
3. Panel façade
4. Ralling/ cable network construction
5. Ventilation

建筑剖面节点
1. 后门闩立面/绝缘夹层安全玻璃
2. 穿孔铝板，安装在次级钢结构上
3. 平板立面
4. 栏杆/缆线网络结构
5. 通风口

夜晚，立面内部的中空结构会进行照明。随着光线的变化，建筑也呈现出不同的外观。建筑外壳看起来更轻、更同构，像一块幕布，罩住了建筑主体。

Knox Innovation, Opportunity and Sustainability Centre (KIOSC)

诺克斯创新机遇可持续中心

Location/地点: Victoria, Australia/澳大利亚，维多利亚
Architect/建筑师: Woods Bagot – Bruno Mendes
Photos/摄影: Peter Bennets
Built area/建筑面积: 1,750m²
Completion date/竣工时间: 2012

Key materials: Façade – double glazing, alumimium fins
主要材料： 立面——双层玻璃、铝翅片

Overview

The Knox Innovation, Opportunity and Sustainability Centre (KIOSC) will promote career pathways in traditional and emerging industry trades as they relate to the application of manufacturing, engineering, electro-technology and renewable energy solutions.

The design comprises a two story building with the main functional areas grouped and separated on each floor. The ground floor accommodates five VET/TAFE labs for use by both the Schools and Swinburne University, whilst the first floor (which is the main entry point to the building) accommodates the Discovery Centre.

Detail and Materials

The building floor plates have been developed to maximise the potential of the site and to provide optimum levels of natural light. The floor plates have been designed for long term flexibility by minimising the number of internal columns and load bearing walls.

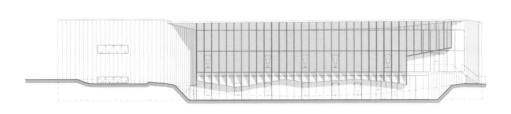

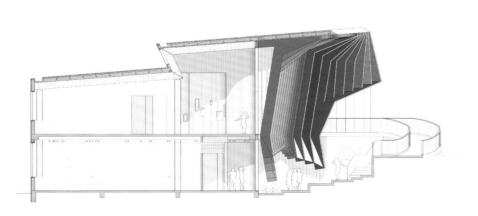

Glazing has been optimised to the building's north and south elevations to maximise the potential for natural lighting and to minimise the energy required for artificial lighting. Solar shading has been provided to glazed walls to minimise direct solar loads and reduce glare.

The Building's distinctive façade is part of the sustainable imperative of the design. The large eve acts as a canopy, while the blades serve as a screen, positioned in response to the angle of the sun.

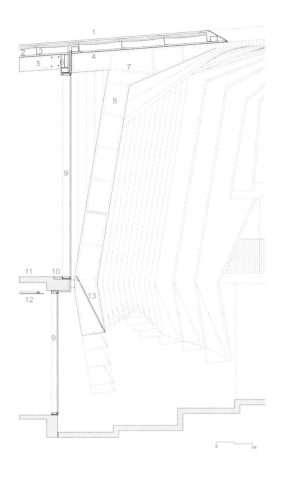

Section detail	剖面节点
1. Roof sheets	1. 屋顶板
2. Insulation	2. 隔热层
3. Roof purlins	3. 屋顶桁条
4. EXOTEC cement sheeting	4. EXOTEC水泥板
5. 2x466mm laminated timber beams	5. 2x466mm夹层木衡量
6. 400 UB steel membrane	6. 400 UB 钢膜
7. Internal steel frame to baldes	7. 翅片内置钢架
8. 3mm thick painted Aluminium sheets	8. 3mm厚涂漆铝板
9. Full height performance double glazing	9. 全高高效双层玻璃
10. Recessed hydronic heating system	10. 内置式热水功能系统
11. Polished concrete slab	11. 抛光混凝土板
12. Suspended plasterboard ceiling	12. 石膏板吊顶
13. 3mm thick perforated Aluminium screens	13. 3mm厚穿孔铝板

项目概况

诺克斯创新机遇可持续中心为传统和新兴行业的从业人员提供职业发展路线，涉及制造业、工程业、电子科技和可持续能源等多个方面。

中心有两层楼高，分布着主要功能区。一楼是5个供学校和斯温伯恩大学所使用的职业技术与继续教育实验室；二楼（即建筑的主入口）内设置着发现中心。

细部与材料

建筑楼面的开发最大限度地利用了场地的潜力，提供了充足的自然采光。楼面尽量减少了内部立柱和承重墙的数量，保证了长期的灵活性。

建筑的南北立面都采用了大面积的玻璃幕墙，最大限度地利用了自然采光，有效减少了人工照明所需的能源。玻璃幕墙上装配着遮阳板，以减少日光直射和眩光。

建筑独特的立面是可持续设计的体现。巨大的屋檐形成了天篷，而翅片结构则像是屏风，随着太阳的角度而设置布局。

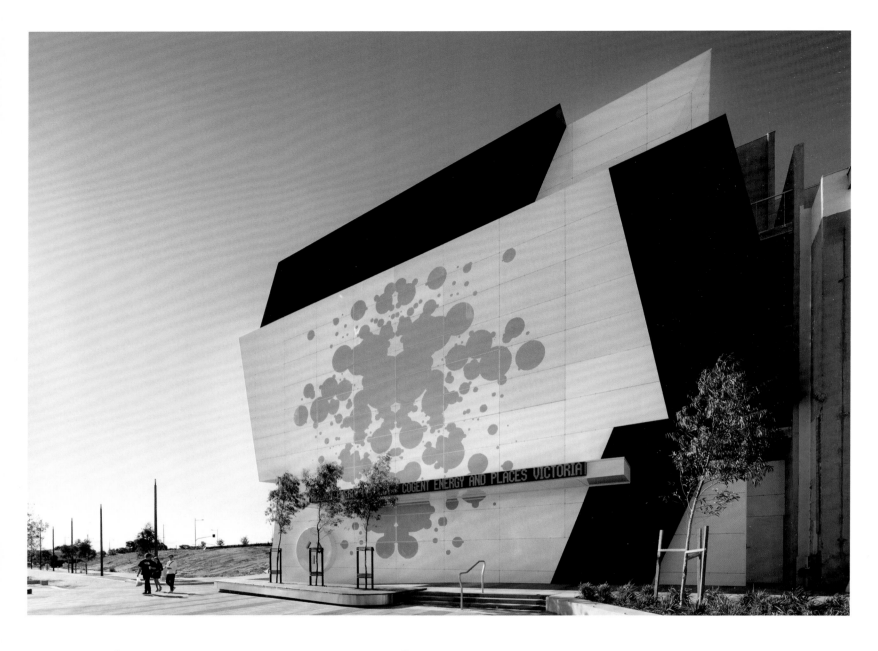

Precinct Energy Project
区域能源项目

Location/地点: Victoria, Australia/澳大利亚，维多利亚
Architect/建筑师: PETER HOGG + TOBY REED ARCHITECTS
Photos/摄影: John Gollings

Key materials: Façade – powder coated Aluminium, Mastermesh perforated metal Structure – steel, concrete
主要材料： 立面——涂粉铝板、Mastermesh穿孔金属板 结构——钢材、混凝土

Overview

The introduction of Cogeneration into the urban fabric has resulted in a new type of public building. To effectively utilize waste heat generated by the cogeneration electricity production process, and to minimise the extent of the underground pipe network, the site must be centrally located. The P.E.P. is situated on the corner of Station North Plaza and the City Street Mall (soon to house Lyons new Municipal Centre). On a pivotal and highly constrained site, the architects have endeavoured to give the building a highly sculptural form and to activate the surrounding area, while allowing for the stringent requirements of the "green machine" inside.

The P.E.P. aims to provoke discussion about the environment, society's power consumption and our future in a warming world in a fun way, without preaching. The big power-points, the big switch, the "circuit diagram" lighting display and the giant "cogeneration diagram" on the north elevation are all intended to encourage further investigation. The frieze on the rear elevation (to a future laneway) utilises engineers' notations for the machinery inside and aims to educate children and adults about the production of the energy we all use. The moving dot matrix display on the front canopy displays information about power production, consumption and green house gas savings from the building.

Detail and Materials

The Precinct Energy Project (P.E.P.) is a perforated metal façade on a steel frame attached to concrete tilt panel walls, forming a large cavity wall containing ducts, vents and lighting.

The architects consulted intensely with the builder and the Nepean Group regarding the method of construction and how to achieve the look and detail they were after. They wrote this in the specification, as well as providing a file of the 3D model of the project for accurate dimensions. Although the geometry looks quite simple, it is in fact quite complex as (particularly on the west elevation) there are many intersecting angles which cause every panel to be a parallelogram. Therefore they used the 3D model to output dimensioned "true" elevations of each section so that each panel could be made and assembled correctly.

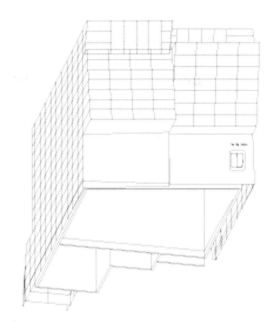
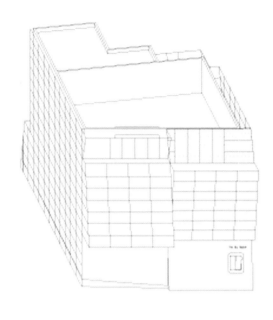
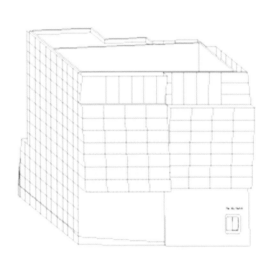
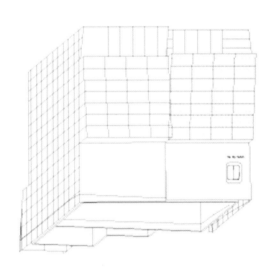
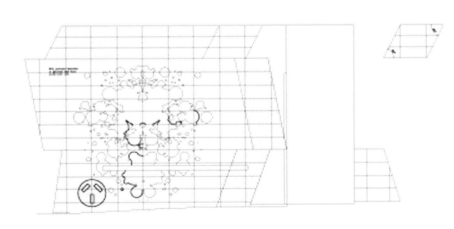
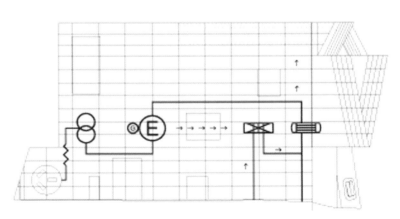

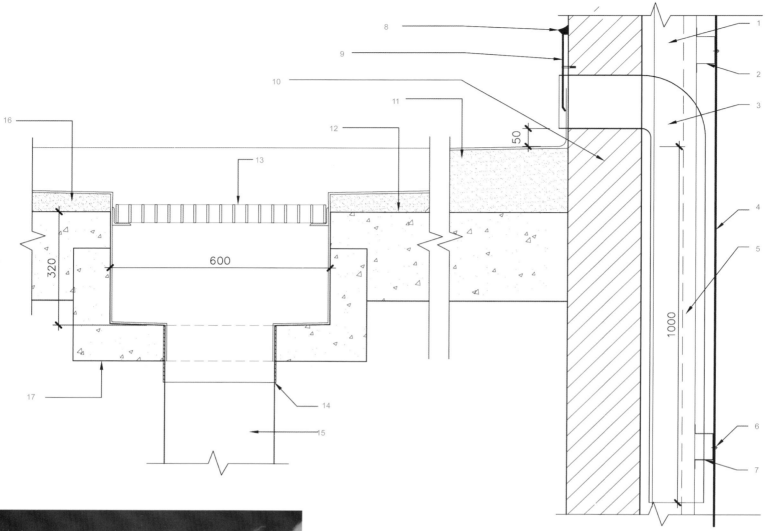

Façade section
1. Vertical Z purlin secured to pre-cast wall
2. Top hat batten fixed to vertical Z purlin
3. 150mm diam. colorbond overflow pipe near base of parapet wall
4. Perforated metal façade panels
5. Discharge to overflow pipe nom. 1000mm below roof level
6. Tek screw to match perforated screen colour
7. Cut top hat batten ground downpipes as required
8. Silicone capling seal to flashing
9. Colorbond over flashing to precast panel min. 50mm overlap
10. Precast concrete wall panel to engineers design
11. 3 layers Tremco water proofing membrane
12. Mortar bed fall to sump at min 1:50
13. Galvanised steel grate
14. Galvanised steel collar
15. Stormwater waste pipe from roof sump, connect to DP to south boundary
16. 3 layers Tremco water proofing membrane
17. Precast sump cast into roof slab

立面剖面节点
1. 垂直Z形桁条，固定在预制墙壁上
2. 大帽板条，固定在Z形桁条上
3. 直径150mm镀铝锌钢溢流管，靠近防护墙底部
4. 穿孔金属墙板
5. 溢流管排放，距离屋顶下沿1000mm
6. Tek螺栓，与穿孔网板同色
7. 环绕落水管的切帽板条
8. 硅树脂封层
9. 镀铝锌钢防水板，至少有50mm重叠
10. 预制混凝土墙板
11. 三层Tremco防水薄膜
12. 污水沟灰浆层，最小比例1:50
13. 镀锌钢格栅
14. 镀锌钢肩
15. 屋顶排水管，与南侧的落水管相连
16. 三层Tremco防水薄膜
17. 预制污水沟，铸入屋顶板

项目概况

废热发电在城市中的应用促生出一种全新的公共建筑类型。为了有效地利用废热发电过程中所产生的废热和最小化地下管道系统的范围，发电场所必须设在中央。区域能源项目坐落在站北广场和城市商业街的一角。由于建筑位于重要的枢纽地带，建筑师赋予了建筑雕刻般的造型，既活跃了周边的区域，又保证了内部"绿色机器"的运转。

区域能源项目旨在以一种有趣的方式掀起人们对环境、社会电力消耗和全球变暖的讨论。巨大的电源插座、开关、电路图照明展示和建筑北立面上巨大的"废热发电图解"都鼓励人们进行进一步的探索。建筑后面的绿绒面墙面彰显了内部机械结构的存在感，旨在教育儿童和成年人我们所使用的能源的生产过程。活动的显示屏展示了有关电力生产、消耗以及建筑如何减少温室气体的信息。

细部与材料

区域能源项目的立面由固定在钢框上的穿孔金属板构成，钢框附着在混凝土斜板墙壁上，在中空结构里添加了管道、通风口和照明。

建筑师与建筑工人和Nepean集团就建造方法进行了详细的讨论，以确保实现期待的外观和细部设计。他们将其写入了设计说明中，并提供了尺寸精确的三维立体模型。建筑造型看起来十分简单，其实非常复杂（尤其是西立面），由许多不同的角度构成，导致了每块金属板都呈现为平行四边形。因此，建筑师利用三维立体模型来展示每个剖面的精确尺寸，保证各个板块正确的制作和安装。

Brno Observatory and Planetarium
布尔诺天文馆

Location/地点: Brno, Czech Republic/捷克，布尔诺
Architect/建筑师: Rudiš-Rudiš architekti s.r.o
Photos/摄影: Filip Šlapal

Key materials: Façade – perforated and sheer aluminium sheets with silver coating
Structure – reinforced concrete
主要材料： 立面——银色涂层穿孔铝板
结构——钢筋混凝土

Overview

The building represents principles and objectives of Science Exploratorium in a form of a multivisual Centre: to explore, to explain and to encourage protection of the natural world as for the field of astronomy. The task was to manage the expected considerations of the use and, at the same time, to seek for an impressive architectural concept with the aesthetics portraying the exactness and perfection of the "space" technology.

The design intention was to come up with more than just a house with interesting expositions and programs. The house itself was to be conceived as an object demonstrating an interest in nature and science and inspiring visitors. We are surrounded by nature – i.e. actually the park, and the building illuminates the natural history. Thus the logical design consequence was to interconnect the building with the surrounding park so that a formal impression of transparency and visual connection could be set up. Large glazing allows for an interesting view out and through the building into the extensive green. The decision to demolish the most problematic part of the original building formed an overall architectural concept. By doing so, a central section was established and a new reasonable layout introduced an entrance hall, Exploratorium and an access to the terrace with the panoramic view over the city. On the north side of the original building another 2-storey addition was built to house the institution's offices; this is connected to the technical and technological backstage of the large planetarium.

Detail and Materials
The building has been assembled from a reconstructed 50's segment, reconstructed 90's segment, a new additions both on the north and in the vacant (after demolition) central site. The segmented character of the building's layout and its varying height obliged us to unify the historically developed parts into an integral complex design. It seamed to be important to outfit the façade so that the material and technical performance could communicate the building's idea and content. So the elementary design feature is a suspended perforated metal cladding. The transparent "cloth", interrupted just by the large glazing has proved to do so – it appears both impressive light and unambiguous. The skin is subtle and somewhat translucent as the day changes light and weather conditions. The gentle texture imparts a fragility and lightweighted aesthetics. The surface's structure is formed by both perforated and sheer aluminium sheets treated by a special silver coating. The formal composition idea evolved just by a coincidence during a digital processing of a photography of south bohemian woods. The posterization process created a vertically stripped structure out of the wood trunks image. The black&white image was then radically transposed into a memorable artistic form.

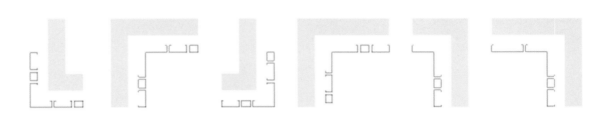

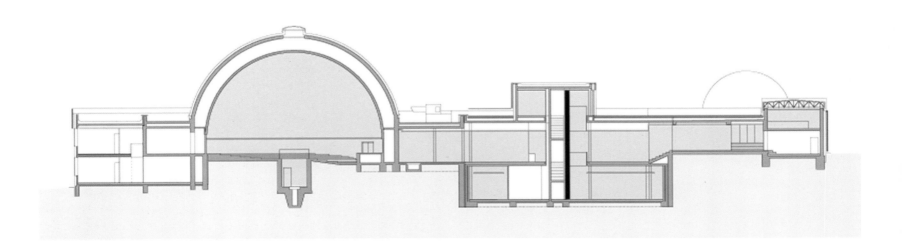

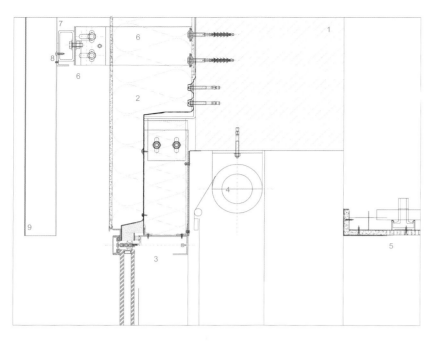
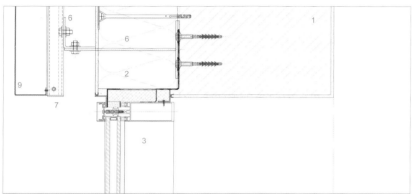

Façade detail
1. Structure of reinforced concrete
2. External thermal insulation system
   10mm fine rendering (dark grey); 20mm mineral wool board
3. Fixed glazed in Aluminium profiles – schueco FW50
4. Shielding blind of fabric
5. Acoustic ceiling
6. Steel clamping elements (galvanised, visible parts: dark grey paint)
7. Horizontal profile – steel 40/80 – 3mm(galvanizing +dark grey paint)
8. L profile connecting sheet elements into panels (Al 40/60 -3mm, dark grey coating)
9. Vertical sheet elements (Al – 2mm, 3mm; special silver coating)
   -Perforated sheet – width of elements: 200mm, 400mm
   -solid sheet –width of elements: 100mm

立面节点
1. 钢筋混凝土结构
2. 外墙隔热系统
   10mm精细打底（深灰色）；20mm矿物棉板
3. 固定在铝型材上的玻璃装配，schueco FW50型
4. 遮阳屏
5. 吸音天花板
6. 钢钳元件（镀锌，可见部位带有深灰色涂漆）
7. 水平型钢40/80 – 3mm（镀锌+深灰色涂漆）
8. 铝板L形连接片元件（铝40/60 –3mm，深灰色涂层）
9. 垂直片元件（Al – 2mm, 3mm；特殊银色涂层）
   – 穿孔铝板，宽度200mm, 400mm
   – 实心板，宽度100mm

项目概况

建筑体现了科学探索馆的新模式，为人们提供了探索和了解天文领域的自然世界，鼓励人们保护自然。项目的任务是实现建筑的预期使用价值，同时寻找到与"太空"技术相匹配的美观建筑形式。

设计的目标不仅是打造一座充满趣味展览和功能的建筑，这座建筑本身还必须能够体现自然科学的趣味性，从而吸引游客。游客被大自然所环绕——建筑外围的小公园和建筑本身都体现了自然的历史。设计让建筑与环绕的公园相互联系起来，形成了半透明的视觉体验。大面积的玻璃装配让建筑内部的有趣景象渗透到了绿地之中。设计拆除了原有建筑问题最大的部分，形成了完整的建筑概念。这样一来，建筑中央形成了入口大厅、探索馆和眺望平台的入口。项目在原有建筑北面新建了一座两层高的附属楼，用来设置办公设施，与天文馆的技术后台连接起来。

细部与材料

建筑由一座建于20世纪50年代的重建结构、一座建于20世纪90年代的重建结构、北面的新建结构和中央场地的空白组成。建筑布局的分段式特征和错落的高度要求建筑师将这些碎片统一成一个综合的整体。建筑立面的统一处理以材料和技术表现了建筑的理念和内涵，主要体现为悬挂的穿孔金属包层。这层半透明的表皮仅在玻璃幕墙处偶有中断，呈现出出色的轻盈感和通透感。金属表皮十分精妙，随着日光和气候的变化会呈现出透明的感觉。温和的纹理传递出脆弱而轻盈的美感。

立面结构由经过特殊银色涂层处理的穿孔薄铝板组成。铝板上的镂空图案模仿了南面波西米亚树林的形象。在照片处理的过程中，树干变成了垂直的板条结构。建筑师顺势就将这一黑白照片的形象转移到了铝板设计中。

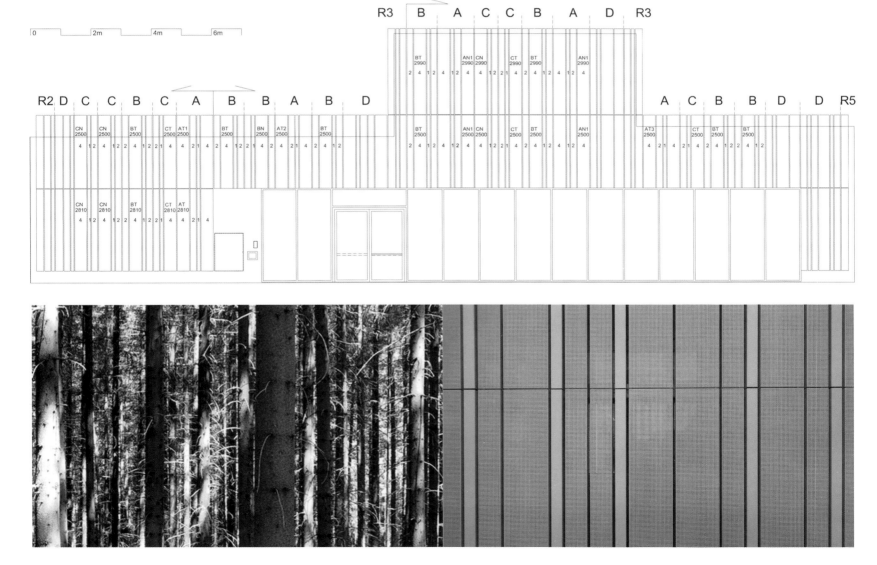

Head Offices of the Telecommunications Market Commission, CMT

CMT总部办公楼

Location/地点: Barcelona, Spain/西班牙，巴塞罗那
Architect/建筑师: BATLLE I ROIG ARQUITECTES
Photos/摄影: Enric Batlle, Joan Roig, architects
Gross floor area/总楼面面积: 12,000m²
Completion date/竣工时间: 2010

Key materials: Façade – aluminium slat
主要材料：立面——铝板条

Overview

The building for the Telecommunications Market Commission (CMT) situated in the 22@ district, stands on a long, narrow site that presents its main façade to Carrer Bolivia and is delimited to either side by a passage. One of the old Can Tiana factory buildings stands at the centre of the site, and the project sets out to recover and incorporate it into the CMT's functional programme. The main volume comprises three basement floors for car parking and eleven floors above grade with offices and services. The ground floor, providing the function of access and entrance, connects with the old mill building, the original structure of which is conserved as an auditorium with capacity for 330 persons, a large meeting room and services for CMT employees. The roof of the mill was adapted for use and connects with the first floor.

Detail and Materials

The decision to bring a unitary treatment to the building's outer appearance led architects to protect its façade using a horizontal slat system throughout its volume that continues over the old factory, connecting the two. The slats serve to cover the upper terraces and installations, and form an awning at the ground floor entrance.

This formal freedom produces a singular form, faceting the faces of the building and moulding it as a unique, recognisable piece that finds its reason for being in an innovative relation between exterior and interior. The variation and superposition of exterior spaces and workspaces serve to direct the volume towards the old factory and establish a subtle, utilitarian relation. The distant presence of the sea and a south-facing orientation determine the correct position of the terraces.

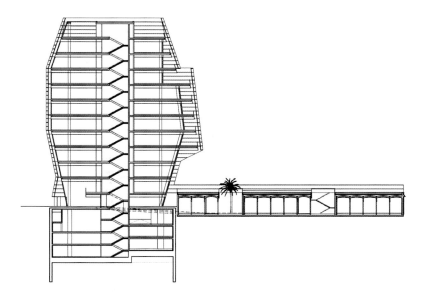

项目概况

CMT总部办公楼位于一块狭长的场地上,正面面对玻利维亚卡雷拉公司,两侧各有一条通道。建筑场地中央是一座旧厂房,项目将其融入了CMT办公楼的设计之中。建筑的主体结构由3层地下停车场和11层办公服务楼组成。一楼设置着建筑入口,与旧厂房相连。旧厂房被改造成了一个可容纳330人的礼堂,可供CMT员工举办会议。厂房的屋顶通过改造与办公楼的二楼相通。

细部与材料

建筑师对建筑外观进行了统一处理,利用水平的板条系统包围了整个结构,并将其延伸到了旧厂房楼顶,使其与办公楼连接起来。外部空间和工作空间的多变性和叠加感在办公楼和厂房之间形成了微妙的关系。屋顶平台远眺大海,朝向南方。

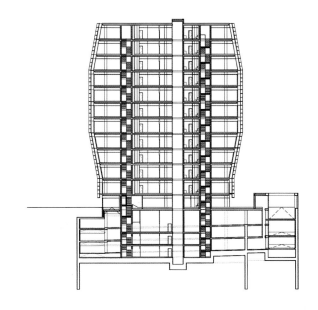

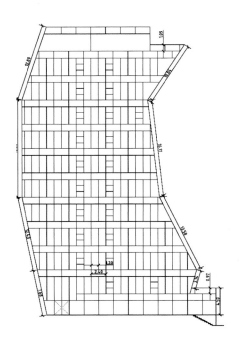

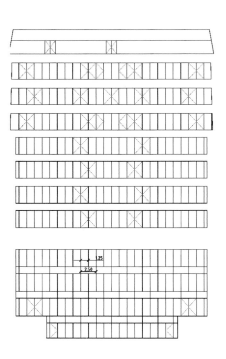

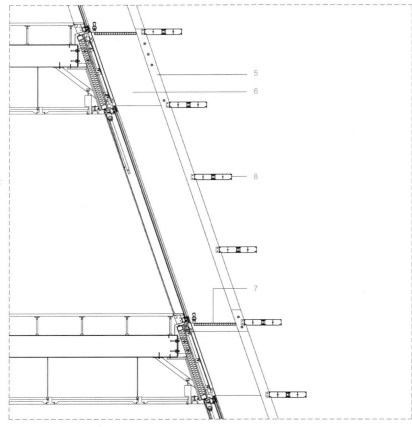

Façade detail 立面节点

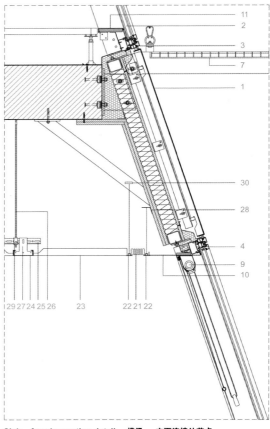

Slab – façade meeting detail 横梁 – 立面连接处节点

Façade detail
1. Hanging tray made of lacquered composite panels. colour defined by the architect
2. Double glazing 8T/158/6+4 with a brown hue selective layer on side 2
3. Flat Aluminium timber (13mm)
4. Aluminium framework (65mm)
5. Painted and galvanised structural steel flat bars (12mm Thk.)
6. laser-cut galvanised steel sheet (8mm Thk.)
7. Galvanised steel grating 30*30mm
8. special extrusion Aluminium profile with a dovetail joint
9. Rolling curtain, features defined by the architect
10. Lacquered Aluminium sheet molding (2mm Thk.)
11. Laminated glass 8T+8T
12. Pavement made of granulated prefabricated pieces 30*60*5cm
13. Adjustable galvanised steel supports
14. Concrete compression layer with 150*150*5mm steel mesh (5cm Thk.)
15. Geotextile
16. Thermal insulation made of extruded polystyrene boards 4cm Thk.
17. Asphalt waterproofing membrane
18. Regularization mortar layer (2cm Thk.)
19. Foamed concrete sloping layer
20. Vapor barrier
21. Ventilation grille made of extruded Aluminium anodized finishing frameless with fixed horizontal blades
22. Metallic guide suspended from the slab width 32mm
23. False ceiling micro perforated metallic boards with an acoustic barroer
24. "C" shaped metallic profile 100*30*4000mm
25. "U" shaped support and bracing profile (1.5Mm thk.)
26. Threaded rib
27. Hanger for c shaped profile (1.5mm Thk.)
28. "L" shaped profile 25*25*1.2mm
29. Metallic fastener
30. Metallic crossbeam between the façade braces

立面节点
1. 涂漆复合板制成的悬挂托盘
2. 双层玻璃装配8T/158/6+4，第二面涂有棕色薄膜
3. 平铝板（13mm）
4. 铝框架（65mm）
5. 涂漆镀锌结构扁钢条（12mm厚）

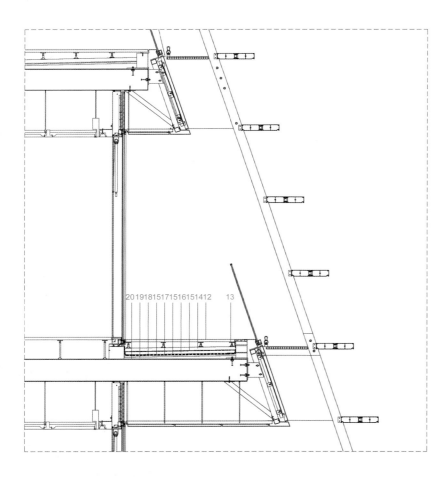

Façade detail balcony 立面阳台节点

6. 激光切割镀锌钢板（8mm厚）
7. 镀锌钢格栅30x30mm
8. 挤制铝型材，带有鸠尾接合
9. 卷帘
10. 涂漆铝板嵌线（2mm厚）
11. 夹层玻璃8T+8T
12. 预制颗粒材料铺装30x60x5cm
13. 可调节镀锌钢支架
14. 混凝土压缩层，带有150x150x5mm钢网（5cm厚）
15. 土工布
16. 挤塑聚苯乙烯板制成的隔热层（4cm厚）
17. 沥青防水层
18. 规格化砂浆层（2cm厚）
19. 泡沫混凝土找坡层
20. 隔汽层
21. 挤制阳极饰面无框铝和固定水平扇片所制成的通风格栅
22. 由楼板上悬垂下来的金属导向，32mm宽
23. 穿孔金属板假吊顶，带有隔音层
24. C形金属型材100x30x4000mm
25. U形支架型材（1.5mm厚）
26. 螺纹挡边
27. C形型材悬轴
28. L形型材25x25x1.2mm
29. 金属紧固件
30. 立面支架间的金属横梁

Nursery in Zarautz

萨劳特斯托儿所

Location/地点: Zarautz, Spain/西班牙，萨劳特斯
Architect/建筑师: Ignacio Quemada Arquitectos Slp
Photos/摄影: Alejo Bagué
Built area/建筑面积: 692m²
Completion date/竣工时间: 2011
Cost/造价: 765,794 €/765,794欧元

Key materials: Façade – colour aluminium slat, aluminium plate
Structure – concrete
主要材料：立面——彩色铝条、铝板
结构——混凝土

Overview

The plot is a right-angled triangle, practically isosceles, with one side in line with the street, another with a future 8 storey block and the hypotenuse facing towards the Abendaño canal but separated from it by a green strip (albeit not maintained).

As if it were a children's game, the architects provided a long room along the side that faces the forthcoming new block, bringing all the service spaces together, and then the corridor and the five classrooms. To make the layout fit the plot, they rotated one of the 1-2 years classrooms, placing it on the hypotenuse and we moved the other one forwards, leaving a multi-use space, while the three 0-1 years classrooms were left adjoining the centre. The remainder of the triangle was left as an outside garden and play area.

During the course of the project, the Local Authority decided to split the multi-use space and join up the two 1-2 years classrooms, which meant that the location of the entrance had to be moved.

The simplistic geometrical design of the floor plan has a volumetric impact. The classrooms are taller rooms with single-pitched roofs opening towards the play area, to look like boxes on top of the rest of the building which is covered with a flat roof.

As the existing nursery is grey with a few touches of colour, they have taken those colours as the "material" from which to build the elevations of the new building.

Without being sure exactly how they were going to build it, for the purposes of the tender they planned to use children's coloured pencils as a theme for the elevations. The short timescale and budget meant that this idea translated into a standardised construction system. Finally, they designed and built it from LUXALON 84 R enamelled aluminium slats, selecting six colours from the reduced exterior colour chart.

Detail and Materials

The vertically placed slats enforced a 100mm spacing which extended to all elevations, resulting in precise adjustments to the locations of all the windows.

The colours are deployed in accordance with a repeated sequence whereby white slats alternate with coloured slats, such that in front of the spaces in the east and south elevations the pattern of colour is transformed into a vertical lattice in which the coloured slats cross over the windows while the white ones are interrupted.

The corners are covered with folded aluminium plates, which have the same visible thickness in each elevation as the slats themselves. With this uniform approach to the elevations, the space, while being made up of the collection of the flat roof and the boxes with pitched roofs, is not decomposed into separate pieces, rather can be understood as a whole, a "sculpture" composed of a mass of strips of different colours.

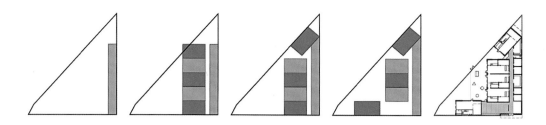

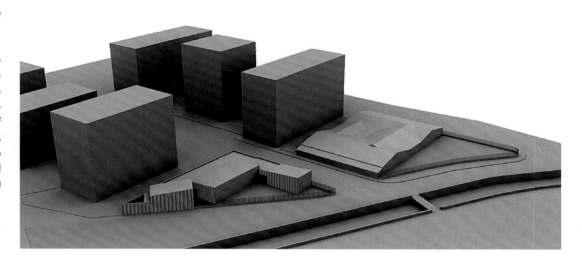

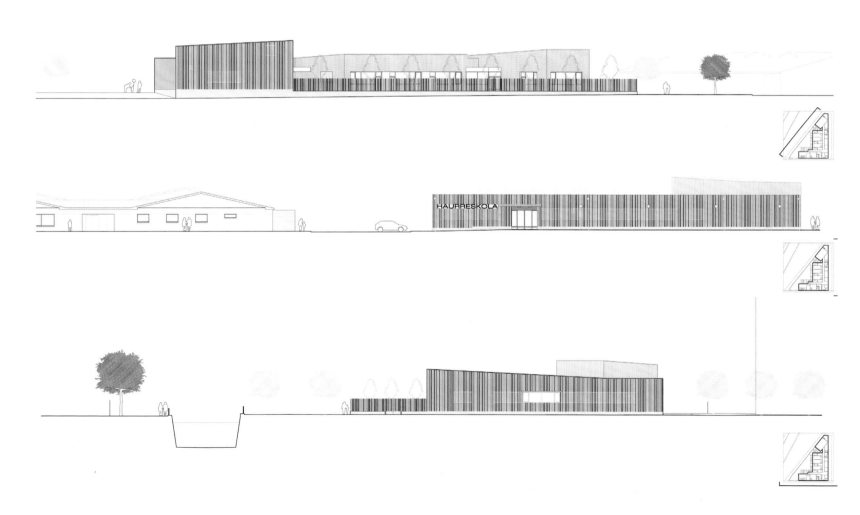

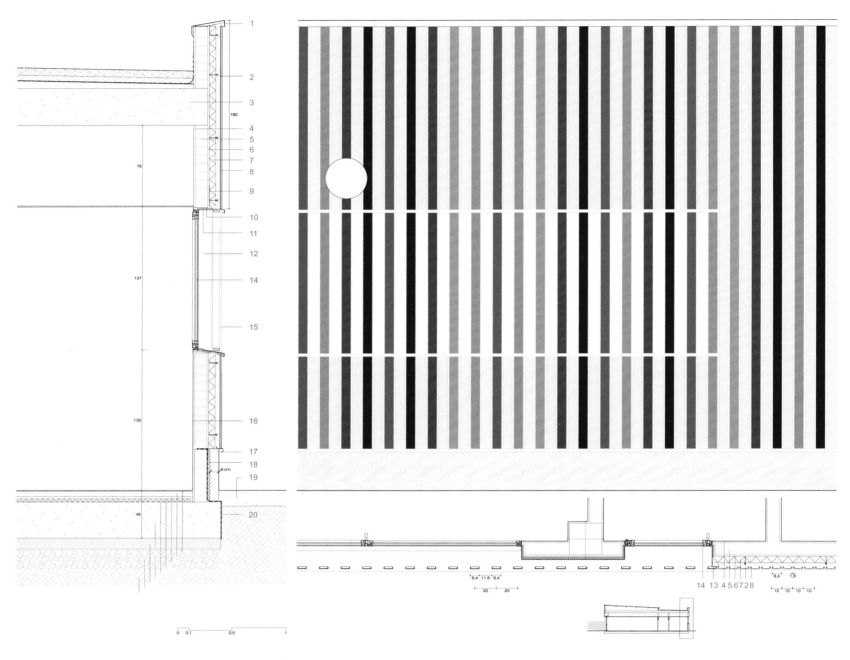

Construction detail
1. Aluminium sheet coping
2. LUXALON profile
3. One-way slab of joists and concrete joist-filler blocks, 350 mm
4. Plaster, 15 mm
5. 1/2 foot of concrete perforated brick
6. Waterproof cement rendering, 15 mm
7. Mineral wool insulation, 60 mm
8. Lacquered aluminium LUXALON profile, 84R type recessed join profile V5 system
9. Self-adhesive SIKA MULTISEAL strip
10. 4 mm folded steel sheet lintel
11. Lacquered aluminium upper moulding sheet, 2 mm
12. Lacquered aluminium sheet side cover, 2 mm
13. Lacquered aluminium sheet, 2 mm
14. Aluminium frame with thermal bridge breaking
15. Lacquered aluminium LUXALON profile lattice/louver. Passing one out of two, fixed on a extruded aluminium tube.
16. PVC coated skirting board
17. In-situ concrete skirting board with vertical plank formwork
18. Polydros, 2 cm
19. Concrete base with 15.15.6 welded wired mesh and printed finish
20. Perimetral waterproofing

结构节点图
1. 铝板顶盖
2. LUXALON型材
3. 托梁和混凝土填块单向板，35mm
4. 石膏板，15mm
5. 1/2'混凝土多孔砖
6. 防水水泥抹面，15mm
7. 矿物棉绝热层，60mm
8. 涂漆铝LUXALON型材，84R型嵌入式结合型材V5系统
9. SIKA MULTISEAL自粘条
10. 4mm折叠钢板过梁
11. 涂漆上层模制铝板，2mm
12. 涂漆铝板侧盖，2mm
13. 涂漆铝板，2mm
14. 防热桥效应铝框
15. 涂漆铝LUXALON型材格架/百叶，固定在挤制铝管上
16. PVC涂层壁脚板
17. 现场浇筑混凝土壁脚板，带有垂直厚木板模壳
18. Polydros, 2cm
19. 混凝土底座，带有15x15x6焊接钢丝网和印制涂层
20. 边缘防水

项目概况

项目位于一块直角三角形的场地。两条直角边，一边与街道平行，另一边的对面将建造一座8层的高楼。三角形的斜边朝向阿本达诺运河，中间隔了一条绿化带。

建筑师将所有服务空间、走廊和5间教室都设置在朝向未来新建大楼一侧的长条空间内。为了让布局与场地相称，他们旋转了1~2岁幼儿的教室，将其设置在斜边一侧，0~1岁幼儿的教室则被设置在了场地中央。三角形的其他部分留下成为了露天花园和游乐区。整个过程就像孩子的拼图游戏一样。在项目实施期间，当地有关部门决定将多功能空间分开，将1~2岁幼儿的教室合并，这使得托儿所入口的位置也改变了。简洁的楼面布局拥有独特的空间效果。教室的空间略高，采用单坡屋顶，朝向游乐区；其他空间则采用平屋顶。

由于托儿所原来的外观是灰色的，带有少量色彩，建筑师决定将这些色彩移植到新建筑的外立面上。他们决定在外立面上以儿童彩绘铅笔为主题。紧迫的时间和预算最终导致这一主题转化成了标准的结构系统。最终，他们利用LUXALON 84 R铝条和6种色彩构成了建筑立面。

细部与材料

垂直的彩色板条统一采用100毫米的间距，被应用在了各个立面上，保证了所有窗口的精确定位。

白色板条与色彩板条形成了重复的序列。在建筑的东、南两个立面，色彩形成了垂直格架，彩色板条穿越了窗口，白色板条则被拆除。

建筑的四角覆盖着折叠铝板，其厚度与板条一样。建筑立面的统一感让平屋顶和斜屋顶结合起来，形成了一个整体。整个建筑就像一座由彩条组成的"雕塑"。

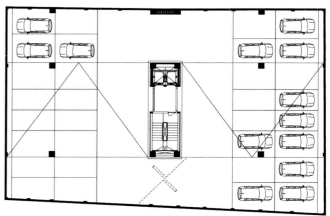

LoMa Chapalita
洛马大厦

Location/地点: Guadalajara, México/墨西哥，瓜达拉哈拉
Architect/建筑师: ELÍAS RIZO ARQUITECTOS
Photos/摄影: Marcos García
Area/面积: 7,279.90m²
Completion date/竣工时间: 2012

Key materials: Façade – perforated aluminium
主要材料：立面——穿孔铝材

Façade material producer:
外墙立面材料生产商：
Hunter Douglas exterior blinds

Overview

On an emblematic avenue in Guadalajara, a short distance from a city landmark in the heart of a cosmopolitan area, this building rises in an attempt to become a reference in Guadalajara's cityscape. The project aims to create an imposing building that is kind both to its surroundings and the environment. The project consists of 6 commercially available office levels plus commercial spaces. The office levels have the possibility of splitting into 4 private stalls giving the building a dynamic and variable trait with up to 24 distinct office spaces.

Detail and Materials

The design features a building with clean straight lines that offers a continuously changing façade. The office building will appear as a translucent block that appears to float over a glazed base since the ground floor is covered in floor-to-ceiling glass. The office levels are wrapped in a double skin that works for sunlight control and natural ventilation. The passive systems aim to take advantage of Mexico's temperate weather. The inner layer in this double skin is made of floor-to-ceiling sliding glass panels. The outer layer is made up of vertical mesh louvers made of perforated Aluminium. The translucent panels can be freely adjusted to aid in controlling sunlight incidence on the office space. The movement of the exterior louvers will grant the building dynamic component and a constantly changing façade that answers to the specific needs of temperature and ventilation control.

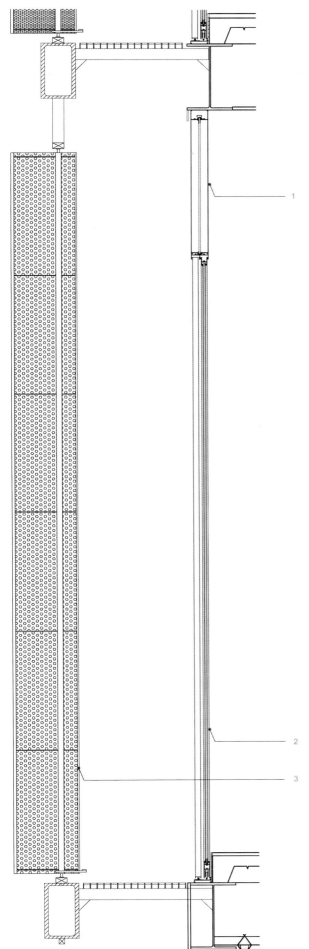

Façade detail (Window with 12mm glass, pivot hinges in ironwork)
1. Fixed window
2. Sliding window
3. Aluminium screen

立面节点（窗户由12mm玻璃组成，采用铁质枢轴铰链）
1. 固定窗
2. 滑动窗
3. 铝幕墙

项目概况

项目位于瓜达拉哈拉的标志性街道上，距离市中心的城市地标很近，希望成为城市街景的一部分。项目旨在打造一座宏伟的建筑，同时对周边环境呈现出友好的姿态。项目由6层办公楼和商业空间组成。办公楼层分成4个小块，赋予了建筑运动而多变的特征。建筑最多可设置24个独立办公空间。

细部与材料

建筑设计线条简洁直接，拥有不断变化的立面。办公楼看起来像是悬浮在玻璃底座上的半透明方块，整个建筑的一楼都被落地玻璃所包围。办公楼层拥有两层表皮，起到了遮阳、通风的作用。被动式系统充分利用了墨西哥温和的气候。内层表皮由落地滑动玻璃板构成。外层表皮由穿孔铝材制成的垂直网状百叶构成。半透明铝板可以根据日照方位和办公室需求任意调节。外层百叶的运动保证了建筑动感的造型，不断变化的立面满足了室内对温度和通风的需求。

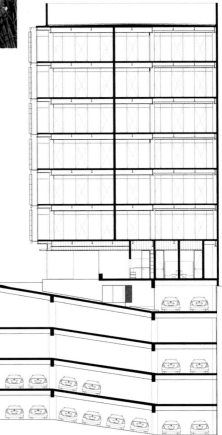

Nursery in the Jardines De Malaga in Barcelona

巴塞罗那马拉加花园托儿所

Location/地点: Barcelona, Spain/西班牙，巴塞罗那
Architect/建筑师: Enric Batlle, Joan Roig architects
Gross floor area/总楼面面积: 745m²
Completion date/竣工时间: 2010

Key materials: Façade – perforated and folded steel, aluminium slat blind
Structure – concrete
主要材料: 立面——穿孔折叠钢材、铝条百叶
结构——混凝土

Overview

The redesign of the Malaga gardens, located in the block interior bounded by Nicaragua, Berlin, Numancia and Avenida Josep Tarradelas streets, in Barcelona, promoted the allocation of a small site for the construction of an education facility: a nursery. The rectangular shaped site is attached to a dividing wall one of its long sides closing the block interior, and faces the gardens on the other.

The small area of the site prevented the program to be developed in a single story, as would have been desirable. It was decided to organize it in two floors, which allowed all the classrooms to face south. The need for playgrounds and porches on both levels was solved with a structural offset in section. The classroom access is done, in both levels, trough a wide corridor illuminated by a patio attached to the neighbor division wall. The entrance, located at one end of the volume, was protected by another offset of the superior volume that created a cantilever over the access.

Detail and Materials
The play of volumes and geometries imposed to the building by the small area of the site was nuanced by some moderation on the election of materials, choosing concrete for the structural elements for the less sunny side Façades, and metal, both opaque and perforated, on the sunniest ones.

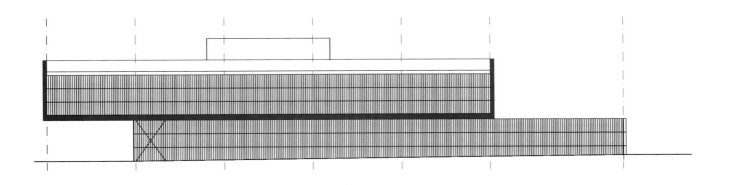

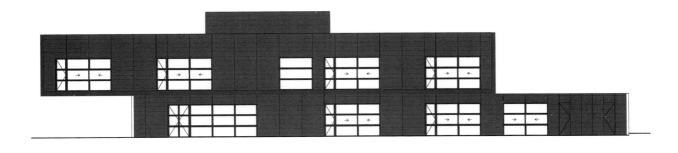

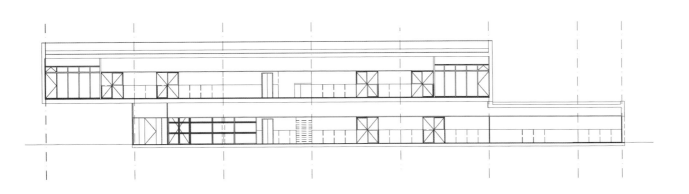

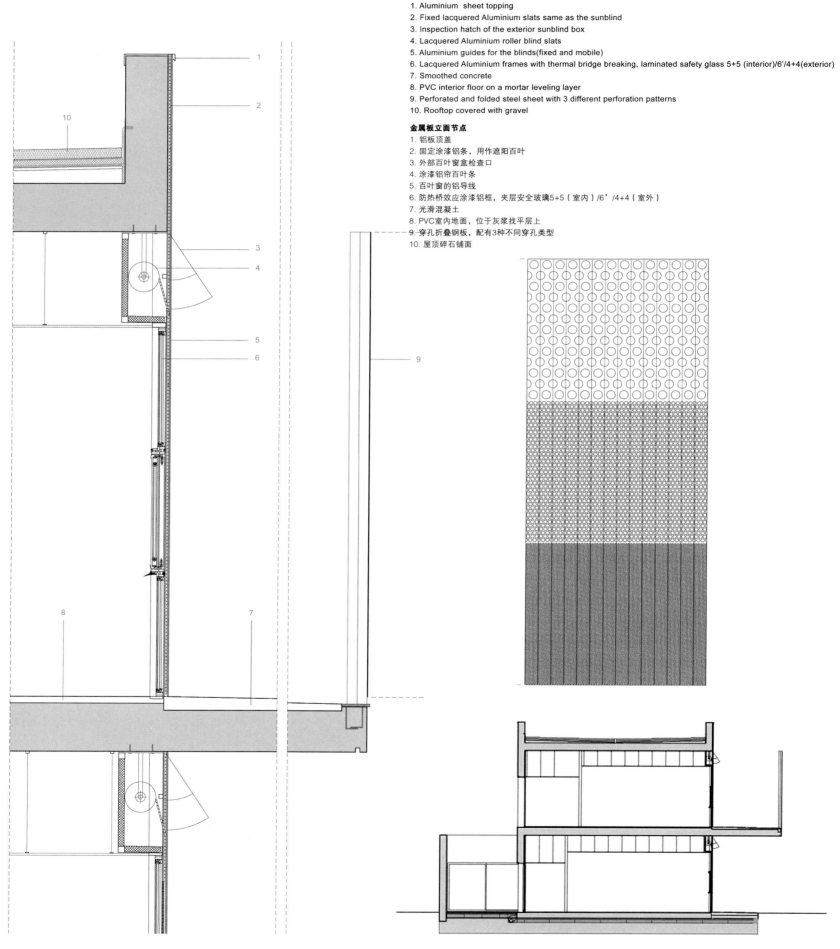

Steel sheet façade detail
1. Aluminium sheet topping
2. Fixed lacquered Aluminium slats same as the sunblind
3. Inspection hatch of the exterior sunblind box
4. Lacquered Aluminium roller blind slats
5. Aluminium guides for the blinds(fixed and mobile)
6. Lacquered Aluminium frames with thermal bridge breaking, laminated safety glass 5+5 (interior)/6'/4+4(exterior)
7. Smoothed concrete
8. PVC interior floor on a mortar leveling layer
9. Perforated and folded steel sheet with 3 different perforation patterns
10. Rooftop covered with gravel

金属板立面节点
1. 铝板顶盖
2. 固定涂漆铝条，用作遮阳百叶
3. 外部百叶窗盒检查口
4. 涂漆铝帘百叶条
5. 百叶窗的铝导线
6. 防热桥效应涂漆铝框，夹层安全玻璃5+5（室内）/6'/4+4（室外）
7. 光滑混凝土
8. PVC室内地面，位于灰浆找平层上
9. 穿孔折叠钢板，配有3种不同穿孔类型
10. 屋顶碎石铺面

项目概况

巴塞罗那的马拉加花园街区由尼加拉瓜街、柏林街、努曼西亚街和何塞普大道环绕。街区在改造项目划分出一小块场地建造托儿所。长方形的建筑场地朝向花园，其中一个长边是分隔墙。

场地的局限性倾向于打造单层建筑结构。但是，最终建筑师决定打造两层楼，让所有的教室都朝南。他们利用平移保证了两层楼都拥有运动场和门廊。人们从宽阔的走廊进入教室，走廊通过与分隔墙相连的天井获得采光。托儿所的入口位于建筑的一端，平移的上层结构形成了入口上方的保护顶盖。

细部与材料

建筑材料的选择进一步突出了空间与几何造型的巧妙组合。建筑师在背光面选用了混凝土和结构元件，在朝阳面则选用了不透明金属板和穿孔金属板。

ZAP' ADOS
活力滑板场

Location/地点: Calais, France/法国，加来
Architect/建筑师: Bang Architectes (Nicolas Gaudard and Nicolas Hugoo)
Photos/摄影: Julien Lanoo (www.ju-la.be)
Gross floor area/总楼面面积: 2,760m²
Completion date/竣工时间: 2011

Key materials: Façade – perforated aluminium mesh
Structure – concrete
主要材料： 立面——穿孔铝网
结构——混凝土

Overview

The operation takes place along a canal in St. Pierre, which is the former industrial district of Calais. It continues the urban renewal initiated by La Cité de la Dentelle (by Moatti & Rivière Architects) located a hundred meters downstream.

The existing building is a common industrial hall with no outstanding features, consisting of a concrete structure filled with precast concrete panels and a roof of cement sheets. The hall was once a roasted peanut factory, followed by various other incarnations (including a go-kart track) before being abandoned for several years. Prior to handing and processing it had been dilapidated, vandalised and had become structurally unsafe. The first task was to open the dark hall before curettage and structural recovery. This was achieved by removing precast concrete panels on the eastern and western façades to release through-views and bring natural light into the heart of the building.

Detail and Materials

The architectural expression is unified by a common envelope made of expanded metal, which turns the silhouette from a hanger into a prism protruding from a singular hybrid form.

The metal mesh allows spectators to watch activities inside and is gradually perforated from top to bottom. The mesh acts like a shutter, controlling direct sunlight and the colour is striking; it is deliberately conspicuous. This colourful mesh protects the equipment as the expanded metal is very resistant and anti-graffiti. It is doubled with a curtain wall to protect users from prevailing winds and reduce any noise nuisance to nearby houses. Outside the building the front square is treated using an orange frame to draw parking spaces, which overlap the textures of the existing floor coatings.

Construction System

Structure of the gables and volumes emerging: structural steel

West façade: the existing concrete columns and structural steel are "sandwiched" by painted gradually expanded Aluminium on the outside wall and the curtain wall noise attenuation inside

Acoustic wall on the indoor street: textile glass fiber coated with PVC stretched over two layers of cotton batting and put on a sheet of extruded PVC formed

Ceiling: industrial acoustic suspended baffles made out of melamine
Skate joinery: wood frame and covering in birch plywood from Finland coated by a clear glaze
External joinery: Aluminium with double glazing
Roofing of emerging volumes: self-protected bitumen

项目概况
项目位于圣皮埃尔区的一条运河旁，属于加来以前的工业区。原有建筑是一座毫不起眼的普通工业建筑，由预制混凝土板和水泥板屋顶组成。建筑曾是一家花生烘焙厂，随后又被许多其他用途（其中还包括卡丁车赛车场），目前已经被废弃多年。在项目开始前，建筑结构已经被破坏，变得十分不安全。设计的首要任务是改善昏暗的内部环境，拆除了东、西两面的预制混凝土板，让自然风景和日光进入了建筑内部。

细部与材料
建筑被统一的金属网表皮包裹起来，使其看起来像一个变化多端的混合结构。

金属网让行人可以看到建筑内部的景象，由上至下，网眼逐渐变大。金属网起到了百叶窗的作用，控制了太阳直射。金属网的色彩十分醒目，引人注意。彩色金属网的金属十分耐久，还可以防止涂鸦。内层幕墙保护建筑的使用者不受盛行风影响，也减少了对周边住宅的噪声污染。建筑门前的广场被改造成停车场，用橙色油漆画出了停车位。

结构系统
山形墙和新建结构：结构钢材。
西立面：原有混凝土柱和结构钢材之间添加了外墙涂漆铝网和内层减噪幕墙。
室内隔音墙：纺织玻璃纤维，带有PVC涂层，下方是两层棉絮，最底层是挤制PVC板。
天花板：由三聚氰胺制成的工业隔音挡板。
滑板设施细木结构：木框和芬兰产桦木胶合板，表面上釉。
外立面结构：铝板和双层玻璃。
新建结构屋顶：自护沥青。

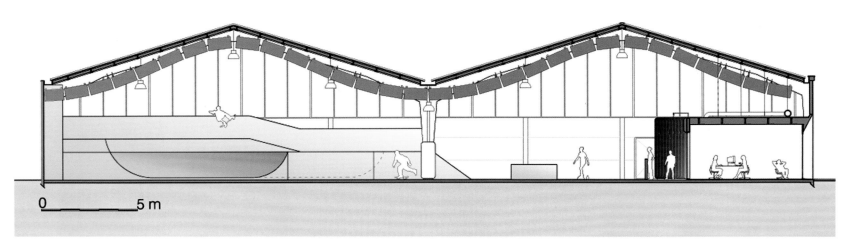

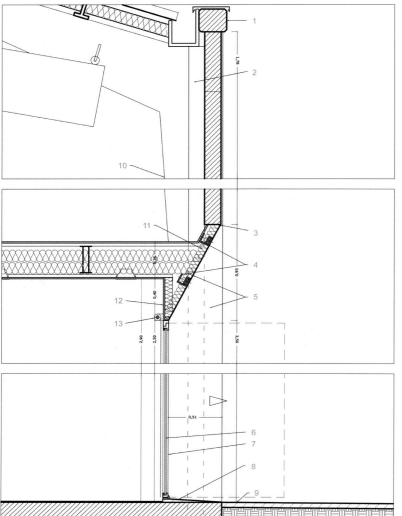
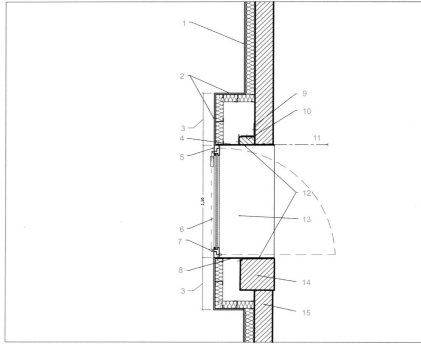

Horizontal opening cut on south façade
1. Thermal and acoustic lining: wool high density glass Glued on 80mm plasterboard 10mm
2. Gypsum board on metal framing + mineral wool 80mm
3. variable dimension
4. Fixed on masonry subframe
5. Carpenter (door) anodize Aluminium + firefighter Cremona
6. Double glazing
7. Fixed on behind the scenes to store opening
8. Fixed on masonry subframe
9. Fixing clip
10. Post created on metal stud foundation created in fixed head on existing lintel
11. Sawing panels
12. Table sheet metal Aluminium 30/10 anodized to joinery (no visible fixing)
13. Form threshold concrete + anodize Aluminium sheet metal
14. Existing concrete
15. Existing concrete panel

Vertical opening cut on south façade
1. Existing concrete beam
2. Post created on metal stud foundation created in fixed head on existing lintel
3. Drop of water creating
4. Secondary Aluminium frame
5. Arches and tole paintings in Aluminium 30/10 Anodized to fixed carpentry clipping (no visible fixing)
6. Double glazing
7. Carpenter (door) anodize Aluminium
8. Form threshold concrete + anodized Aluminium sheet metal (h = 2cm max)
9. Existing coated
10. Existing concrete
11. Mineral wool 100mm returned to play side
12. Gypsum board on metal framing + mineral wool 80mm
13. Store screen with scenes

南立面水平开口节点
1. 隔热吸音衬板：10mm高密度棉，由玻璃胶黏合在80mm石膏板上
2. 金属框架上石膏板+矿物棉，80mm
3. 可变尺寸
4. 固定在砌石副框架上
5. 木门阳极氧化铝+防火Cremona尼纶
6. 双层玻璃
7. 固定在店面后方
8. 固定在砌石副框架上
9. 固定夹
10. 金属立杆基架，一头固定在原有的过梁上
11. 锯切板
12. 阳极氧化铝板30/10（隐形固定）
13. 入口，混凝土+阳极氧化铝板
14. 原有的混凝土
15. 原有的混凝土板

南立面垂直开口节点
1. 原有的混凝土梁
2. 金属立杆基架，一头固定在原有的过梁上
3. 形成滴水
4. 次级铝框
5. 阳极氧化铝板上的拱门和涂漆，30/10，与固定木夹连接（隐形固定）
6. 双层玻璃
7. 阳极氧化铝板30/10（隐形固定）
8. 入口，混凝土+阳极氧化铝板
9. 原有的混凝土
10. 原有的混凝土板
11. 矿物棉100mm，滑板场一侧
12. 金属框架上石膏板+矿物棉，80mm
13. 风景屏

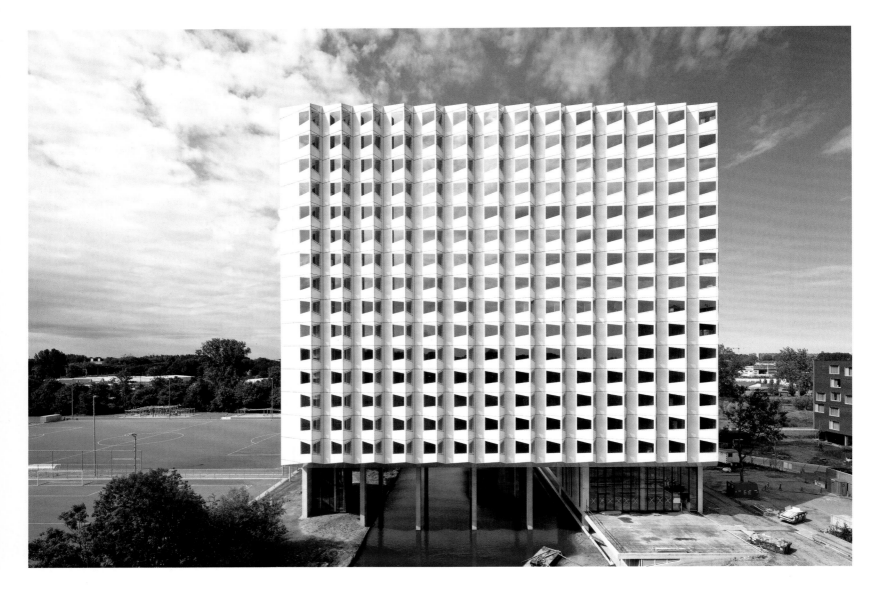

Student Housing in Delft
代尔夫特学生宿舍

Location/地点: Delt, The Netherlands/荷兰，代尔夫特
Architect/建筑师: De Zwarte Hond
Photos/摄影: Scagliola&Brakkee
Completion date/竣工时间: 2011

Key materials: Façade – white coated aluminium
Structure – concrete
主要材料：立面——白色涂层铝材
结构——混凝土

Overview
On the Balthasar Van der Polweg in Delft, De Zwarte Hond designed a student housing complex with a sunken bicycle storage facility. The building, comprising 15 floors, provides accommodation to over 400 residents. The conscientious integration of the bicycle shed into the landscape and the building's extraordinary architectural design are an enrichment for the TU campus.

The area is home to buildings of varying dimensions that are, however, in rhythmic proportion to each other. The housing complex is not of average height for the campus, but it fits into the surroundings because its location and height contribute towards the rhythm of the rest of the high-rises.

Detail and Materials
The most remarkable aspect of the complex are the side elevations. Their design was mainly influenced by one single environmental factor: noise. Because the building is located within the noise contour of the Kruithuisweg, every apartment has a bay window with one 'deaf' wall and one that can be opened. The wall cladding is white coated aluminium that is carried through to the upper and bottom sides of the tower. This hides the technical installations from view and provides the façades with a fitting finish. The end walls are made of concrete and glass in varying transparencies and hues of green. As a whole, it is an expressive building which, despite its contemporary style of architecture, ties in with the Brutalist architecture of the existing faculty buildings.

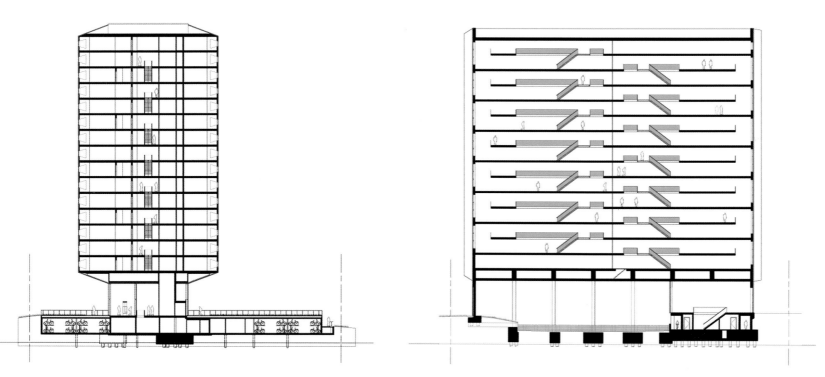

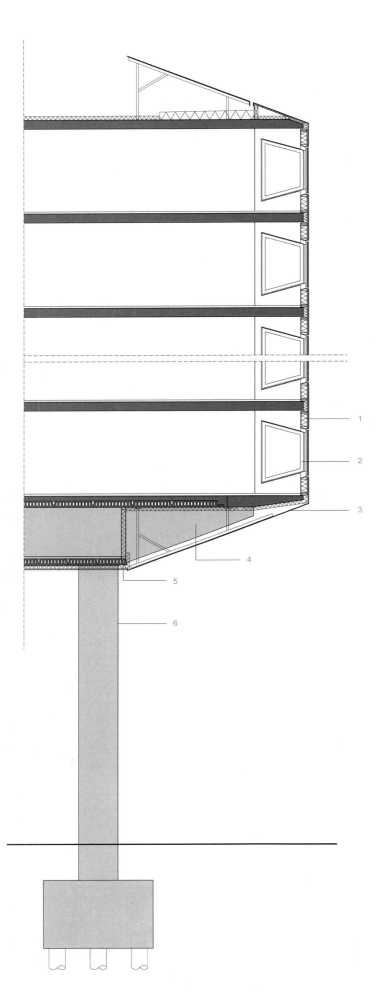

Detail
1. Wooden frame construction+Alipolic façade cladding
2. Alu frame
3. Alpolic façade cladding
4. Prestressed precast beam
5. Hollow core slab
6. Concrete column

节点
1. 木框结构+ Alipolic立面包层
2. 铝框
3. Alpolic立面包层
4. 预应力混凝土梁
5. 空心板
6. 混凝土柱

项目概况

建筑师在代尔夫特设计了一座学生宿舍，宿舍带有下沉式的自行车棚。建筑高15层，可为400多名学生提供住宿。与景观融为一体的自行车棚和建筑非凡的设计为大学校园添加了丰富的色彩。

校园内的建筑形式和规模各不相同。宿舍楼并不属于平均的高度，但是它以和谐的设计融入了环境之中，与其他高层建筑遥相呼应。

细部与材料

建筑最引人注目的设计就是建筑侧面，影响设计的主要环境因素就是噪声。由于建筑位于道路噪声圈之内，所有公寓都配备了带有隔音墙的可开式凸窗。白色涂层铝板包层将墙面从上至下包裹起来，既隐藏了技术装置，又提供了合适的饰面。端墙由混凝土和玻璃建成，玻璃拥有不同的透明度和绿色色调。尽管建筑的风格十分现代，但与校园内的野兽派建筑融合得十分彻底。

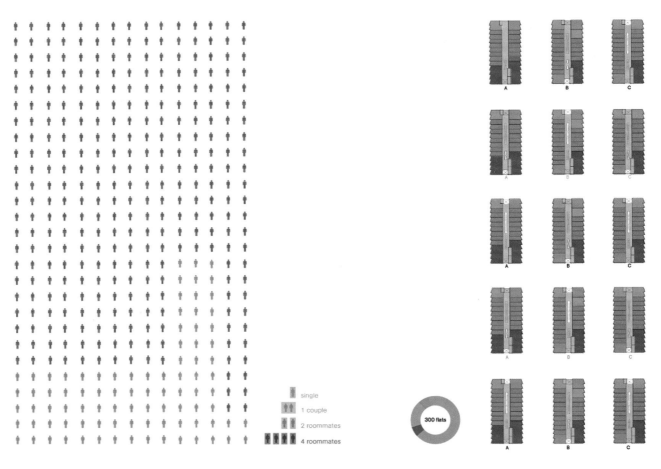

single
1 couple
2 roommates
4 roommates

300 flats

Library of South University of Science and Technology of China
南方科技大学图书馆

Location/地点: Shenzhen, China/中国，深圳
Architect/建筑设计: Urbanus(www.urbanus.com.cn)
Photos/摄影: Alex Chan (http://www.reappaer.com)
Site area/占地面积: 8,627.9 m²
Built area/建筑面积: 10,727.8m²
Building height/建筑高度: 16.5m
Design/Completion date/设计时间: 2010-2012
Completion date/竣工时间: 2013

Key materials: Façade – aluminium module
主要材质: 立面——铝模块

Overview

Nowadays, as books cease to be the primary carrier for the transfer of knowledge, the meaning of "The Library" is also shifting. Recognizing this, the design not only incorporates the conventional programs of libraries, but also attempts to excavate what is embedded in libraries--the public nature that is closely associated with today's society.

Located at core of the campus, the library features a slightly curved façade, maintaining a humble stance towards its surroundings. Students and teaching staff who commute between the academic area and the living area on a daily basis will always pass by the library, regardless of which direction they are heading. Such circulation gives rise to a corridor system that crisscrosses throughout the building, referencing the traditional Cantonese commercial arcade adapted to hot and rainy climates, attracting the public to walk to the inside. Along the north-south corridor are the public programs, including the main entrance lobby, academic auditorium, association activity room, book bar, and so forth. The corridor on the second floor extends from the west to the east – it passes by the book bar, skylight, multi-function hall, bamboo garden, reading area, semi-outdoor platform, and then finally ends at the Baishu Garden on the east. Overlapping circulation allows people to meet and communicate with each other; they can stop by to read or participate in academic activities, which will naturally become a part of their daily life. These create opportunities for a physical library to become more lively and inviting than a virtual library. On the top floor, 3,800 square meters of clear space serves an open-shelf reading area. In order to facilitate the role exchange between the modularized book collection area and the reading area, the structural load of this floor is customized for book collection areas; the column span is unified as 8,400*10,800mm.

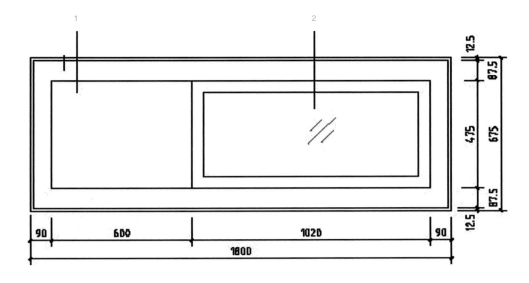
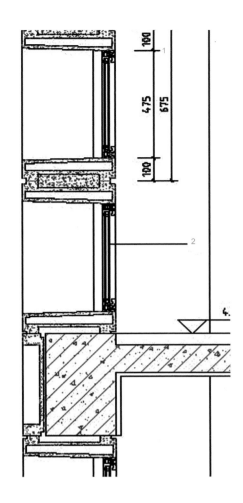

Typical GRC precast block
1. GRC precast block
2. IGU window

标准GRC预制块外墙详图
1. 玻璃纤维强化水泥预制块
2. 低辐射中空玻璃窗

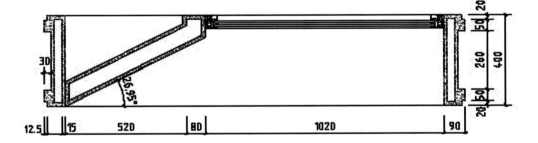

Detail and Materials

GRC (Glass Fiber Reinforced Concrete) was originally specified for the library façade. With the consideration of façade scale, structural load, sizes of fabricated structural components, local shading requirement, and other factors, the GRC unitised component was designed into a light and high-strength hollow module that had a dimension of 1,800*675*400mm; its hollow core was then filled with insulating materials, with mold release and curing treatment. Due to various reasons, the client requested the material to be changed before construction. Eventually, silver grey, semi-unitised Aluminium modules were assembled together in a staggered fashion. The Aluminium modules were both water-proof and sunlight resistant, and were assembled following the original approach. For the crisscrossing corridors, orange high-strength fiber cement boards were applied to the ceiling and the wall. The orange color extends from the outdoor public space to the indoor public area, inviting people at the corridors to enter the building.

项目概述

当书不再是唯一知识传播载体的情况下，图书馆的意义也在发生改变。我们在满足图书馆的传统功能要求的同时力图挖掘图书馆与当代社会特征紧密关联的公共性，图书馆位于校区中心，略微内凹的弧形轮廓，对环境形成谦逊的姿态。师生每日往返于教学区与生活区时，会从不同方向途经此地。顺应这种动线，生成了穿越建筑的十字形游廊系统。以期像传统的岭南骑楼一样，既能适应深圳的炎热多雨，又能吸引人走进去参与空间活动。主入口门厅、学术报告厅、社团活动室和书吧等公共功能被有意安排在南北向通廊的两侧。二层游廊自西向东途经书吧、天井、多功能厅、竹园、阅览区、半室外台地，最终到达东面的百树园。流线交叉给人们的日常穿越带来相遇和交流，停留、阅览和参与学术生活自然成为生活的一部分。使实体图书馆有机会比虚拟阅读更鲜活有趣。顶层是供开架阅览使用的近3800平方米的开敞式大空间。为便于模数化的藏书区和阅览区日后互换，整层结构板均按藏书区荷载来设计，柱跨统一为8400×10800（mm）。

细部与材料

图书馆外墙意图使用GRC（玻璃纤维增强混凝土），在综合考虑了立面尺度、结构承载力、可加工的构造尺寸、当地遮阳需求等因素后，GRC单元格被设计为尺寸1800×675×400（mm）的轻质高强的空心模块，中间填充保温隔热材料，经脱模养护而成。种种原因，甲方在施工前要求更换为传统材料。最后实施的是银灰色半单元式铝制模块错缝拼装。铝板模块集防水保温自遮阳于一体，延续了原尺寸和拼装方式。与外墙不同，十字形游廊选用了橘色高强度水泥纤维板作为天花和墙面装饰材料。橘色主题从室外公共空间延续至室内的公共区，将人们自然地从游廊引入到建筑中来。

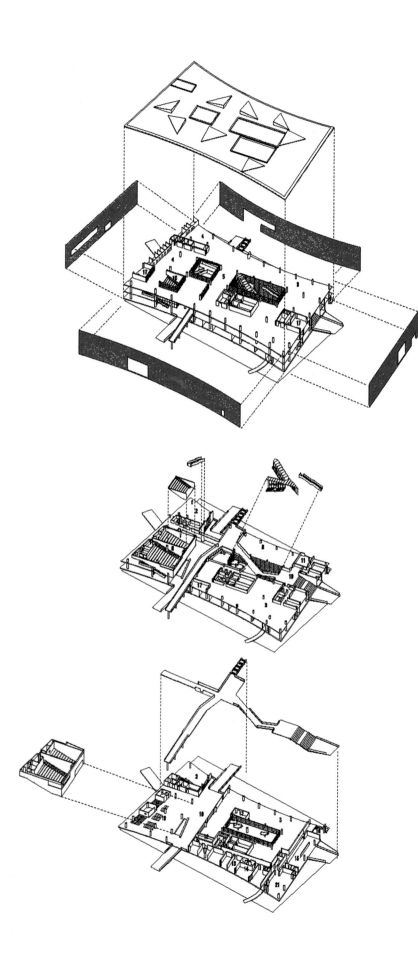

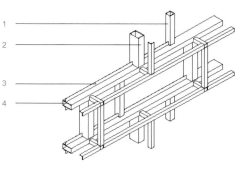
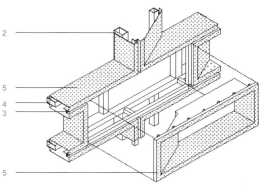
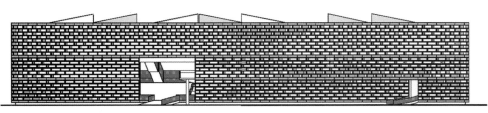
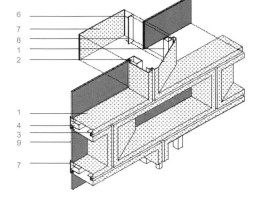
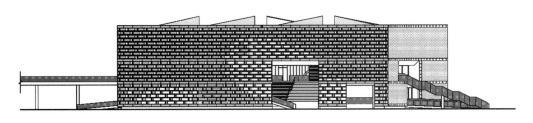

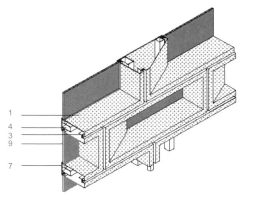

Assembly diagram

1. 70x5 steel square tube
2. 150x100x6 steel tube
3. L50x4 steel angle
4. 150x75x6 steel tube
5. 2.5mm single-layer aluminium sheet
6. Phillips panhead tapping screw ST4. 8x19
7. 2mm aluminium sheet
8. 4mm steel sheet
9. 6+12A+6 low-E hollow glass

铝板整体成型安装过程示意图

1. 70x5钢方通
2. 150x100x6钢通
3. L50x4角钢
4. 150x75x6钢通
5. 2.5mm单层铝板
6. 十字槽盘头自攻螺钉ST4. 8x19
7. 2mm铝板
8. 4mm钢板
9. 6+12A+6 low-E中空玻璃

Extension of Two Elementary Schools

两所小学的扩建

Location/地点: Courbevoie, France/法国，库布瓦尔
Architect/建筑师: BP Architectures - Mandataire
Gross floor area/总楼面面积: 2,200m²

Key materials: Façade – anodised aluminium
Structure – galvanised steel
主要材料: 立面——阳极氧化铝
结构——镀锌钢

Overview

Two schools that share the same court, two extensions binoculars by the similar treatment of their coating of Aluminium siding and even their front door overhang bring consistency and rhythm to a diverse neighbourhood.

Schools that we enjoy every day and find that strengthens the desire to learn by spaces qualitative balcony overlooking the city.

The purpose of the contest was to perform, on a plot residual, shared the extension of two separate elementary schools, met on the same block in a residential district of Courbevoie. The architect proposed to make each extension in continuity of existing sets and weextract the master plan designed by the City, or to realize two buildings instead of one.

Detail and Materials

The volumes consist of a mineral base (SCC stained) forming the ground floor and the wall surrounding the courtyard. Of vertical openings and a regular calpinage vertical concrete, punctuate and animate the façades. These bases are metal brackets volumes upstairs whose façades are clad in street of a frame made of smooth anodized Aluminium (three different colors) arranged vertically in an irregular and random acting as sunscreens.

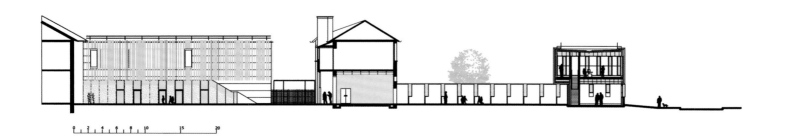

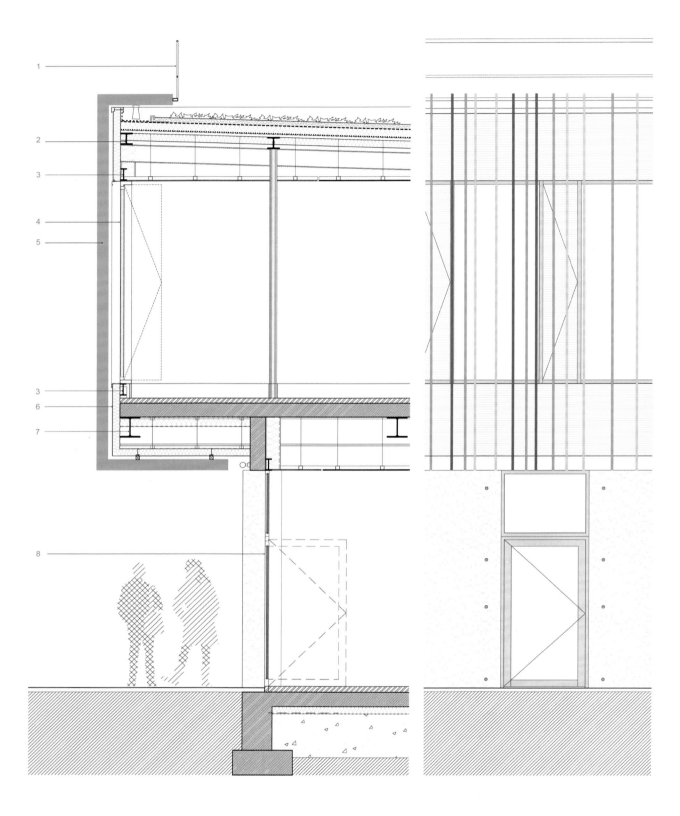

Façade detail
1. Security galvanised steel body
2. Edge beam steel HEB160
3. High IPE160 smooth steel
4. Joiner panel lacquered finish tole opening
5. Anodized Aluminium solar shading 150x30mm
6. Sandwich panel horizontal micro-rib e = 100mm h = 1000mm Type Promline 2000HB
7. Edge beam steel HEB280
8. Glass door opening

立面节点
1. 安全镀锌钢主体
2. 钢边梁HEB160
3. 光面钢IPE160
4. 漆面接合板，技术薄片开口
5. 阳极氧化铝遮阳板150x30mm
6. 夹层板，水平微挡边，e = 100mm，h = 1000mm，Promline 2000HB型
7. 钢边梁HEB280
8. 玻璃门开口

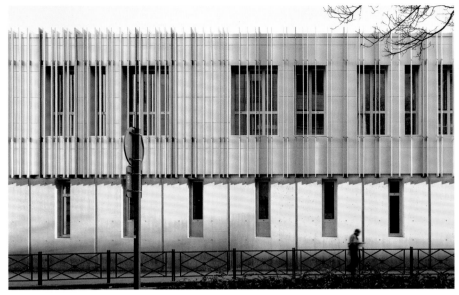

项目概况

两所学校共享一个庭院,它们的铝墙板十分类似,甚至前门的屋檐都是一样的。这些都为多样化的社区带来了统一感和韵律感。

项目设计的目标是对两所独立的小学扩建,使它们融合库尔布瓦这个住宅街区之中。建筑师的方案让扩建工程与原有布置和谐统一;他们根据城市总体规划,分别对两所学校进行了改造。

细部与材料

项目由一楼的岩石底座和环绕庭院的围墙组成。垂直开口和规则的垂直混凝土条让立面活跃起来。底座为上方结构提供了金属支架,上方结构被包裹在光滑的阳极氧化铝框(有三种色彩)中。铝框随机排列,起到了遮阳板的作用。

Hotel Well
维尔酒店

Location/地点: Terme Tuhelj, Croatia/克罗地亚，泰尔梅图尔吉
Architect/建筑师: MVA/Marin Mikelić, Tomislav Vreš
Photos/摄影: Ivan Dorotić
Built area/建筑面积: 12,250m²
Completion date/竣工时间: 2012

Key materials: Façade – aluminium composite panels, precast concrete
Structure – reinforced concrete
主要材料：立面——复合铝板、预制混凝土
结构——钢筋混凝土

Overview
The addition to an existing congress/wellness hotel is located in the thermal complex "Terme Tuhelj", Croatia, next to the protected historical park and baroque Curia "Mihanović". The starting point of the project was the idea of connecting all the existing and new facilities and integrating them into a new and meaningful ensemble.

With respect to the existing pavilion-like organisation, the architects developed a concept of a "detached" hotel, providing the users a connection with nature from all the inside spaces.

The program of the new part of a hotel is clearly vertically separated (public + accommodation). 126 rooms are divided into three smaller pavilions (along to the three old ones) and the "public" facilities are organised into elongated ground base. This base assimilates some of the existing "public" spaces and connects all old and new accommodation pavilions.

This organisation provided natural light even in congress halls with beautiful views to the Zagorje landscape. Dividingthe accommodation facilities into three pavilions provided all of the rooms with nice views to the scenery.

Terrain configuration is used to form an amphitheater in the Centre that provides a visual and spatial communication of the congress Centre and hotel lobby with the baroque Curia in the background. This space, besides the added value for the hotel and convention Centre, can adopt a whole range of different activities and could contribute to the social life of local community.

Detail and Materials
The selection of materials for the façade envelope marks the theme of duality (light / heavy). Side façades are "wrapped" with the envelope of lightweight Aluminium composite panels coloured in gold whose perforations allow views from the rooms. The rest of the façades and roof surfaces are covered with precast concrete panels, partly opening as perforated 'eyelids' in front of the windows.

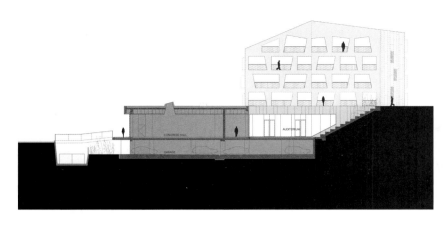

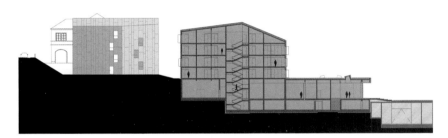

项目概况

这座会议休闲酒店位于克罗地亚的泰尔梅图尔吉地区，紧邻受保护的历史公园和巴洛克式的米哈诺维克城堡。项目的出发点是将原有设施与新建设施连接起来，将它们融合为一个整体。建筑师尊重了原有的亭馆式组织结构，形成了"独立式"酒店的概念，为使用者提供了与自然亲密接触的机会。

酒店新建部分的功能规划采用垂直分割的方式（公共区域+住宿区）。126间客房被分配在3个小楼之中（与原有结构相邻），而公共设施则设置在细长的基座之上。这个基座吸收了一些原有的公共空间，并且将新旧住宿空间连接在一起。这种组织结构保证了会议厅享有充足的自然采光和美妙的风景。住宿空间的零散分布让每间客房都能够享有自然美景。

项目利用地势在中央形成了一个露天剧场，从视觉和空间上将会议中心、酒店大堂和远处的城堡联系起来。除了为酒店和会议中心增添价值之外，这一空间还能举办各种类型的活动，对本地社区的社交生活十分有益。

细部与材料

外立面的材料选择体现了设计的双重性——轻盈和厚重。侧墙由金色轻质复合铝板覆盖，上面的孔眼保证了各个房间的视野。其他立面和屋顶表面都覆盖着预制混凝土板，并且在开窗处添加了一部分穿孔铝板。

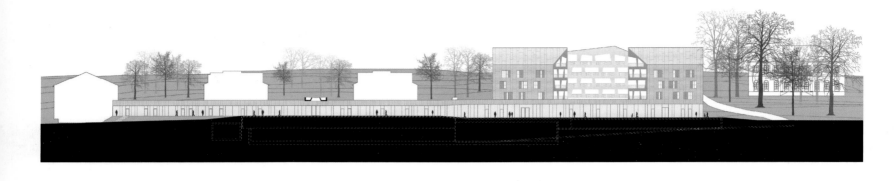

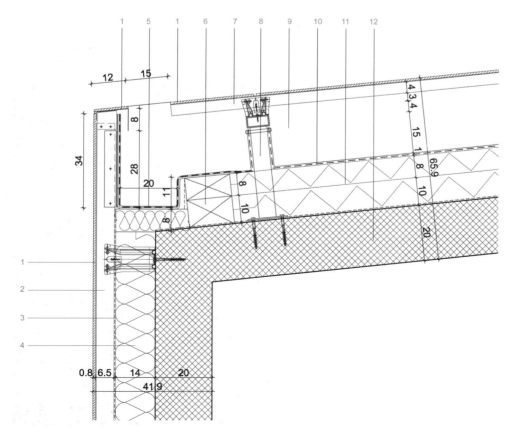

Roof detail

1. Façade/roof – riveted fiber-concrete panel 8mm
2. Ventilated air gap/vertical aluminium bearing structure 62mm
3. Rain dam
4. Mineral wool insulation 140mm
5. Rain gutter 120x180mm – galvanised metal sheet
6. Wooden beams – 100/160mm + 80/160mm
7. Horizontal aluminium bearing structure 40mm
8. Steel bearing structure – vertical profile D=60mm, horizontal profiles 40/80mm
9. Ventilated air gap
10. Waterproofing
11. Extruded polystyrene thermal insulation 180mm
12. Reinforced concrete 200mm

屋顶节点

1. 立面/屋顶——铆接纤维混凝土板8mm
2. 通风气口/垂直铝支承结构62mm
3. 雨挡
4. 矿棉隔热层140mm
5. 雨水槽120x180mm——镀锌金属板
6. 木梁——100/160mm+80/160mm
7. 水平铝支承结构40mm
8. 钢支承结构——垂直型材D=60mm，水平型材40/80mm
9. 通风气口
10. 防水层
11. 挤塑聚苯乙烯隔热层180mm
12. 钢筋混凝土200mm

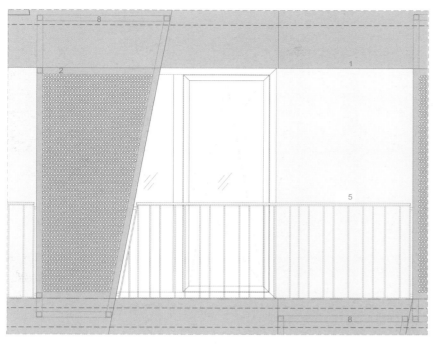

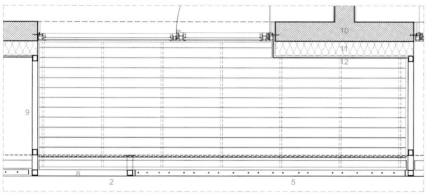

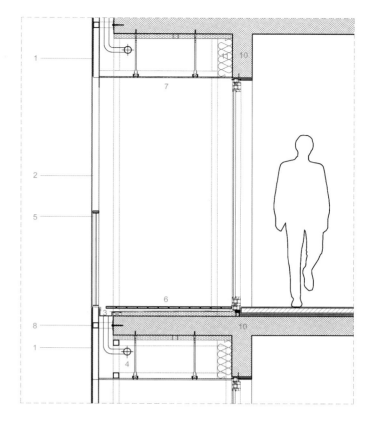

Façade detail
1. Façade – aluminium composite panel 3mm
2. Semi-transparent screen- perforated aluminium composite panel 3mm
3. Gutter
4. Drainage pipe
5. Railing – steel frame 60/10mm, vertical steel profiles 10/10mm and wooden handrail 60/20mm
6. Wooden planking – IPE LAPACHO 100/20mm
7. Outdoor cement board 12.5mm
8. Steel frame 60/60mm
9. Participation – steel frame 50/60mm with aluminium composite panel 3mm on both sides
10. Concrete wall 200mm
11. Extruded polystyrene insulation 150mm
12. Mineral render 20mm

立面节点
1. 立面——复合铝板3mm
2. 半透明遮阳板——预制复合铝板3mm
3. 排水槽
4. 排水管
5. 栏杆——钢架60/10mm，垂直钢型材10/10mm和木扶手60/20mm
6. 木板铺装——IPE LAPACHO 100/20mm
7. 户外水泥板12.5mm
8. 钢架60/60mm
9. 隔断——钢架50/60mm，两侧是复合铝板3mm
10. 混凝土墙200mm
11. 挤塑聚苯乙烯隔热层150mm
12. 矿物打底层20mm

Reconversion Post Site

邮局大楼改造

Location/地点: Aalst, Belgium/比利时，阿尔斯特
Architect/建筑师: Vereniging van Studiebureaus 'ABSCIS – PROVOOST – TECHNUM'
Built area/建筑面积: 31,000m²
Completion date/竣工时间: 2012

Key materials: Façade – lacquered aluminium, glass
主要材料: 立面——涂漆铝材、玻璃

Overview

The former post office site in Aalst was a mixture of industrial buildings of the former 'Filature du Canal'. The "Filatures& Urban Fabric(s): Masterplanstationsomgeving Aalst" masterplan, designed by the Christian Kieckens Architects, started from a site conversion, whilst keeping the main building of the former mill in the Manchester building along the Vaartstraat. The modifications to this building have been kept to a minimum because of organisational and programmatic interpretation, which corresponds with the scale of the building. The façades have been cleaned and re-pointed and equipped with new outside joinery work after its historical model. The building has been renovated on the inside also, in order for it to comply with current regulations, without however interfering with the warehouse's authenticity.

The new office blocks are strip shaped units, inserted in a north-south direction, guaranteeing the best east-west direction for the offices. The buildings are built perpendicularly onto the linear infrastructure elements (such as the river Dender, the railway line and the roads), strengthening its kinetic perception and resulting in a dynamic image of new buildings in an attractive, modern architecture against a backdrop of old factory units. The building is cut in two using a public internal road, which creates a shortcut to the train and bus station. It is on this public internal road that the main entrances to the buildings are located.

Detail and Materials

Whereas the existing Manchester building is characterised by its weight and solidity, the new office strips are light, full of dynamism and transparency. They were interpreted as mostly glass units, equipped with blinds with coloured slats. These slats are kind of a second façade around the layer of buildings, patches of coloured textile as it were, referencing at the same time to the industrial past of this site as a mill. The Werfplein façade is the only façade which can be seen from the front. Here the coloured slats are intensified and only sporadically interrupted by big windows with a view on the Werfplein.

Detail 1
1. Curtainwall in Aluminium
2. Invisible fixing with brackets, supporting structure in C-pillars
3. 18 mm plywood coverd with Aluminium 0,8 mm powder coated
4. Curtain wall: Aluminium sandwich panelpowder coated
5. Rock wool 6cm with moisture barrier
6. Insulation in mineralwool
7. Casingbead
8. Stucco
9. Casingbead
10. Connection drywall on concrete column, with casing bead
11. Stucco
12. Casingbead
13. Casingbead
14. Acrylic joint
15. Flashover façade 1m

节点1
1. 铝幕墙
2. 隐形支架安装，采用C形柱支承结构
3. 18mm胶合板，上覆0.8mm涂层铝
4. 幕墙：涂层夹心铝板
5. 6cm石棉，配有防潮层
6. 矿物棉隔热层
7. 边龙骨
8. 灰泥
9. 边龙骨
10. 混凝土柱连接干墙，配有边龙骨
11. 灰泥
12. 边龙骨
13. 边龙骨
14. 丙烯酸接头
15. 1m对齐立面

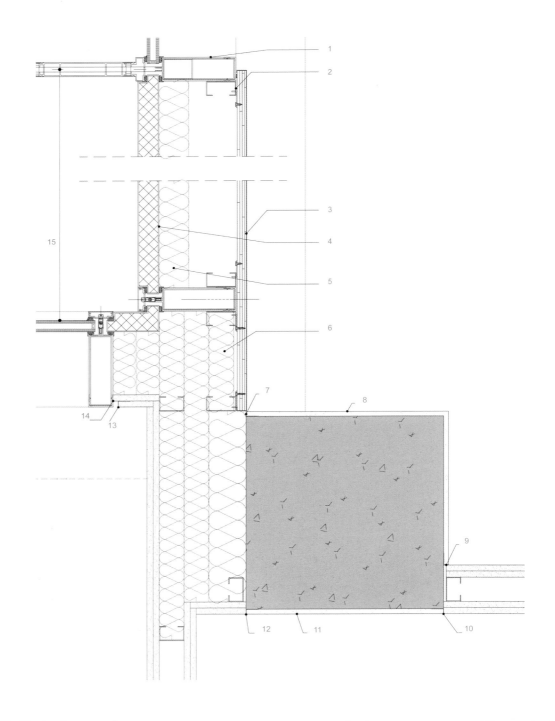

项目概况

阿尔斯特的邮局原址是"运河剿丝厂"的厂房。"剿丝厂城市规划"项目由Christian Kieckens建筑事务所设计，对场地进行了改造，同时保留了位于瓦尔特街上的前工厂大楼——曼彻斯特楼。建筑的改造程度被降到了最低，因为它的组织结构和功能设置都与建筑的比例相对应。建筑立面进行了清洗，重新配备了外部细木工装饰。建筑内部同样进行了翻新，既符合现有法律法规，又保留了厂房的历史特征。新的办公楼由南向北呈条形排列，保证了办公室的东西朝向。建筑垂直于其他直线形的基础设施（例如单德河、铁路和道路），突出了自身的动感，在一片旧厂房中呈现出独具一格的活力形象。建筑被一条内部公路一分为二，从而形成了通往火车站和汽车站的捷径。建筑的主入口正好位于这条内部公路上。

细部与材料

原有的曼彻斯特楼以厚重感著称，而新建的办公楼则轻盈而充满了动感。办公楼几乎全部采用玻璃幕墙，配有彩色板条百叶窗。这些板条类似于建筑的第二层立面；同时彩色条纹也令人想起了场地的工业背景。朝向工人广场的一侧立面的彩条更加密集，仅在几个巨大的开窗处有所缺失。

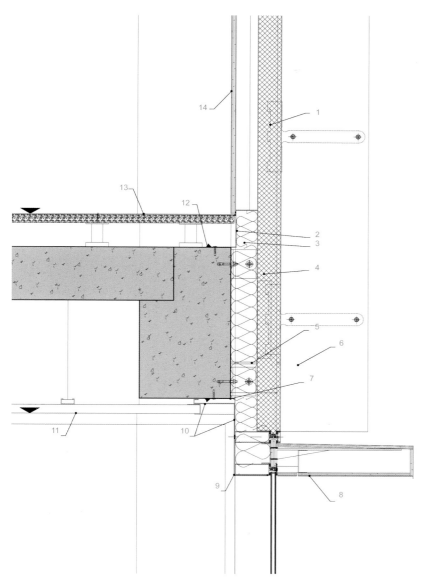

Detail 2
1. Anchored façadestrip
2. Sheet metal, steel ei60
3. Insulation in mineralwool
4. Curtain wall: sandwich panelAluminium powder coated
5. Thermal expansion joint in vertical profiles
6. Curtain wall: strip in Aluminium
7. Sheet metal, steel ei60
8. Framework covered in Aluminium, concealed fixing
9. Sheet metal, Aluminium 2mm
10. Sheet metal, Aluminium2 mm powder coated, concealed fixing
11. Climate ceiling
12. Rf-joint
13. Raised removable floor
14. 18mm plywood covered in Aluminium 0.8mm powder coated, concealed fixing with panel hooks

节点2
1. 固定立面板条
2. 金属板，EI60型钢
3. 矿物棉隔热层
4. 幕墙：涂层夹心铝板
5. 垂直剖面热膨胀伸缩缝
6. 幕墙：铝条
7. 金属板，EI60型钢
8. 铝框，隐形安装
9. 金属板，2mm铝
10. 金属板，2mm涂层铝，隐形安装
11. 气候天花板
12. Rf接缝
13. 抬高可移动地板
14. 18mm胶合板，富有0.8mm涂层铝，利用板挂钩实现隐形安装

Detail 3
1. Sheet metal, steel EI60
2. Insulation in mineral
3. Curtain wall: sandwich panel in (coated) Aluminium
4. Interruption vertical profiles in function of thermal expansion
5. Sheet metal, steel EI60
6. Curtain wall: Aluminium
7. Waterproof plywood
8. Rain barrier
9. framework in Aluminium, covered waterproof carrier, concealed fixing
10. curtainwall: Aluminium
11. safetyglass, insulating
12. sheet metal, Aluminium 2 mm, concealed fixing
13. mineral wool
14. moisture barrier
15. Rf- joint
16. raised removable floor

节点3
1. 金属板，EI60型钢
2. 矿物棉隔热层
3. 幕墙：涂层铝夹心板
4. 热膨胀垂直立缝
5. 金属板，EI60型钢
6. 幕墙：铝
7. 防水胶合板
8. 防雨板
9. 铝框，配有防水层，隐形安装
10. 幕墙：铝
11. 安全玻璃，隔热绝缘
12. 金属板，2mm铝，隐形安装
13. 矿物棉
14. 防潮层
15. Rf接缝
16. 抬高可移动地板

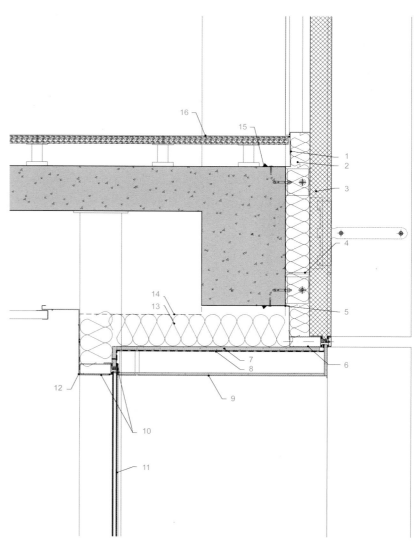

Parking Garage 'de Cope' at Papendorp

帕潘多普德科普停车楼

Location/地点: Utrecht, The Netherlands/荷兰，乌特勒支
Architect/建筑师: JHK Architecten
photos/摄影: Palladium Photodesign
Built area/建筑面积: 3,000m² office and ca. 460 parking places/3,000m2办公空间，约460个停车位

Key materials: Façade – perforated Platibond composite panels
主要材料: 立面——Platibond穿孔复合板

Façade material producer:
外墙立面材料生产商：
PLATINOX

Overview

Parking garage 'de Cope' at Papendorp in Utrecht proves once and for all that a parking accommodation need no longer be a grey and negative mass. With its clear design and transparent parking structure, this garage is truly an added value to its environment and makes parking a positive experience.

Temporary parking solutions very much mark the streets. In the meantime, parking regulations have been amended and new developments set into motion. In the context of this development, Kroon Group and JHK Architecten realised a parking garage combined with an office/business premises in Papendorp Zuid, with a total of some 500 parking places and 3x1000m² worth of office space. The majority of the parking spaces is intended for use by the neighbouring offices.

The design consists of two abstract units, connected by means of a crossway. Two spirals open up the various parking decks within the building. The parking garage functions separately from the three office storeys on the three top floors. The entrance to the offices is an entrance hall at ground level. Orientation within the garage is optimal both for motorists and pedestrians. You need not spend time looking for a spot or the exit: the design quite simply leads you along.

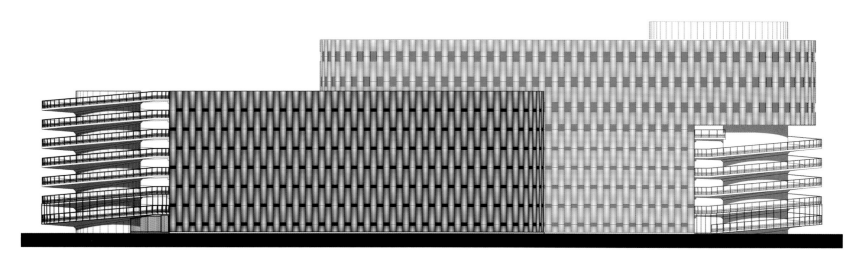
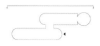

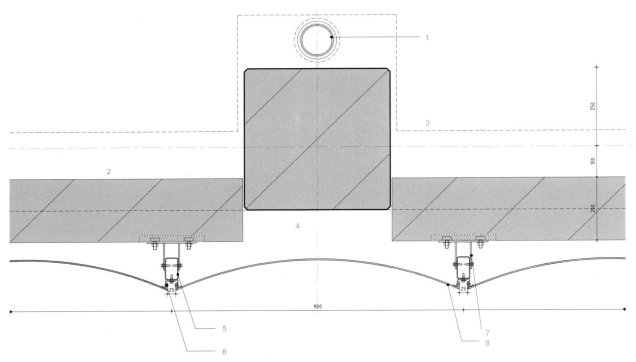

Detail 1
1. Drainpipe
2. Prefabricated concrete plinth
3. Contour prefabricated concrete floor
4. Prefabricated concrete column 450x450
5. Galvanised steel RHS k70/40/2
6. Aluminium Z-profile extrusion, glued and mechanically mounted with blind revits
7. Galvanised steel mounting
8. Platibond 4mm, gold colored arched with gradient perforation

节点1
1. 排水管
2. 预制混凝土基座
3. 外围预制混凝土楼面
4. 预制混凝土柱450x450
5. 镀锌钢RHS k70/40/2
6. Z形挤塑铝材，黏合安装在百叶螺丝上
7. 镀锌钢安装
8. Platibond复合板4mm，拱形，配有金色涂层和倾斜穿孔

Detail and Materials

The building's façade is the result of the desire to create a building that is not the same old parking garage with its typical construction of bar, balustrade and column. The level of ambition is high from an urban planning perspective, as the design must be able to compete with an originally planned office unit.

By building the façade around both the parking decks and the office storey's, there is no way to visually distinguish between the different functions. The façade consists of inward curved, perforated 'Platibond composite panels' with a golden coating, which give the Aluminium surface coat its golden metallic look. The panels were specially developed for this project in collaboration with the façade supplier. The gradual changes in the perforations and the staggered panels create the impression of a woven structure. The perforations guarantee the natural ventilation of the parking garage.

The construction of the panels also allows for even distribution along the façade in places where the façade curves. Form, color and detail give the building a different look depending on the weather and the light. Between the many office and industrial buildings, the garage is a welcome and pleasant change.

项目概况

乌特勒支的帕潘多普德科普停车楼向全世界证明了停车场不再是灰暗、消极的沉闷空间。简洁的设计和通透的停车结构让停车楼在环境中脱颖而出，提供了积极的停车体验。

临时停车位大多占用了公共道路。同时，停车规定也在不断改进，要求新建项目提供充足的停车位。在这种环境下，Kroon集团和JHK建筑事务所共同开发了一个与办公/商业空间相结合的停车楼。项目共有约460个车位和3层分别为1,000平方米的办公空间。大多数停车位都供办公空间所使用。

设计由两个抽象的结构组成，二者由通道相连。两个螺旋打开了建筑内部的停车平台。停车场与上面3层的办公楼分隔开。办公楼的入口设在一楼大厅。停车场的导向标识对机动车和行人都十分清晰，人们无需耗时间寻找车位和出口，设计能清晰地标识出来。

细部与材料

建筑立面摒弃了旧式停车楼的设计，没有典型的围栏、栏杆和立柱。为了与办公楼相匹配，设计采用了较高的楼高。

从停车场和办公楼层的立面上来看，无法区分二者的不同功能。立面由向内卷曲的Platibond穿孔复合板组成，上面涂有金色涂层，赋予了铝材表面闪亮的金属感。复合板为立面供货商特别为该项目所设计。渐变的穿孔和跳跃的复合板形成了编织的感觉。穿孔保证了停车楼的自然通风。

复合板的建造还保证了立面弯曲处的均匀分布。造型、色彩和细节让建筑在不同气候和光线下呈现出不同的外观。在众多的办公楼和工业楼之间，停车场显得独树一帜。

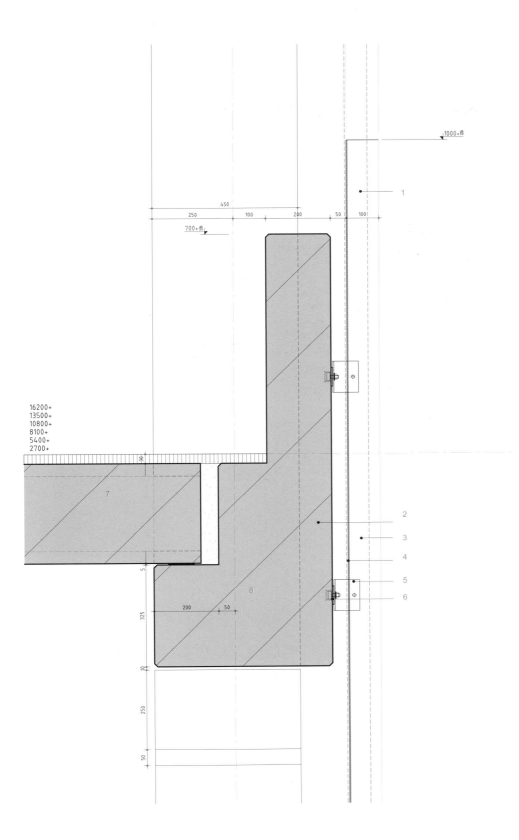

Detail 2

1. Façade panel as fall protection
2. Prefabricated concrete parapet
3. Galvanised steel box section k70/40/2
4. Platibond 4mm, gold colored arched with gradient perforation
5. Galvanised steel mounting
6. Cast-in anchor
7. Prefabricated concrete floor
8. Prefabricated concrete beam

节点2

1. 防跌落面板
2. 预制混凝土护墙
3. 镀锌方形钢管k70/40/2
4. Platibond复合板4mm，拱形，配有金色涂层和倾斜穿孔
5. 镀锌钢安装
6. 浇铸式锚点
7. 预制混凝土
8. 预制混凝土梁

Statoil Regional and International Offices
挪威国家石油公司办公楼

Location/地点: Fornebu, Norway/挪威，福内布
Architect and interior designer/建筑师与室内设计师: a-lab & Momentum Arkitekter (NO)
Landscape architect/景观设计师: Rambøll
Construction area/建筑面积: 117,000m²
Completion date/竣工时间: 2012

Key materials: Façade – prefabricated aluminium composite cassettes
Structure – steel and concrete (prefab)
主要材料：立面——预制复合铝板
结构——钢材、预制混凝土

Façade material producer:
外墙立面材料生产商：
Profile system: Schuco
Cladding: Etalbond by Elval Colour

Overview

After 20 months of construction, the Statoil regional and international offices at Fornebu outside Oslo, Norway is now complete. A-lab received the commission after winning first prize in an open competition in 2008. Statoil is a Norwegian energy producer, the 57th largest company in the world by revenue, with about 30 000 employees in 37 countries. From October 15th, 2500 of these employees will work in the new building - including Statoil's international department.

The new office building stands on the site of the old airport's multi-storey car park, most of the site being given over to the development of a new 5 hectare public park. One of the main goals of the Mikado inspired design has been to minimize the footprint of the building in the park, whilst creating a flexible and stimulating workspace- offering views over the fjord from most parts of the building and a spectacular view of the park, artworks and the fjord from the communal atrium space.

The project consists of five similar office modules, each 3 stories high, approx 140 meters long and 23 meters wide. Stacked on top of each other they form a climatized atrium and a multipurpose space on the ground floor. Due to the very short construction period, most of the building, including steel and concrete superstructure, façades and glazed structures, was prefabricated off-site. This leads to a high degree of precision despite the extremely fast on-site production. The steel superstructure enables the different office modules to cantilever up to 30 metres. Northern Europe's biggest mobile crane was used for the assembly of the steel trusses.

Detail and Materials

The buildings skin is integral to the building's energy efficiency. The façade has low U-values: 0,18 W/m2K for the solid panels (including profiles) and 0,6 W/m²K for the triple-glazed panels (excluding profiles). In combination with the exterior mounted solar shading from Vental, this creates an extremely energy efficient skin. Window area is kept to an optimal net 65% glass

area, towards the interior, thus ensuring an optimal balance between the daylight levels in the interior and potential solar-gain and the higher U-values associated with glass. G-values in the glazed panels varies within the interval 0,24-0,36, depending on the position in the building. Despite being a prefabricated façade composed of many separate elements the façade is extremely airtight, with a diffusion of (n50): ≤1,0h-1. The façade also allows for a very low normalised cold-bridge value of 0,02 W/m^{2}K.

White Long façades

The external profile that was developed especially for the project is the key to the elegant and seamless façade expression. This complex design and technical device also allows the achievement of the architects goal: no visible fixings in the entire prefabricated façade. This profile includes the track for the external solar shading, in the form of "traditional" blinds, as well as providing fixing for the rain-screen cladding cassettes. These are made of powder coated aluminium sandwich panels folded to create 220mm deep cassettes giving the façades a deep relief and also shedding precipitation effectively. The same prefabricated façade system is used on the interior where the office lamells overlook the climatized atrium. The same construction principles apply to the façade cassettes of the interior with the only difference being that they are finely perforated (Ø5mm, 25 %) in order to deal with the acoustic challenges of the interior atrium

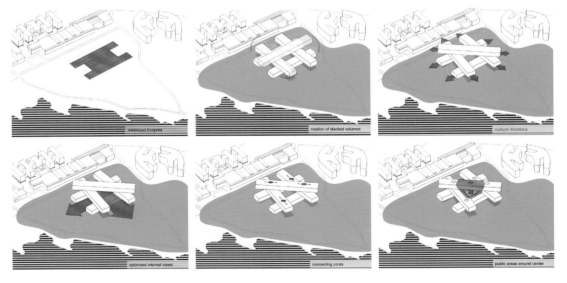

space: ensuring a low reverberation time in the atrium, despite the hard flooring.

Black Gable Ends

The solar shading in the gables was also developed specifically for the project, based on a previous system by Flex Fasader. Each gable façade is composed of 180 4,2m high and 400mm wide laminated glass lamellas which overlap slightly. Each lamella consists of two sheets of 10mm Parsol Grey glass laminated together with a dark laminate core, thus giving a the appearance of solid black. Whilst the solar shading in the long façades is fitted with manual override switches within the office space, the black glass fins in each gable all move together on a pre-programmed pattern angled individually to the orientation of each gable.. The solar shading is surprisingly "invisible" when seen for the office interior, which is fully glazed in the gables.

Façade detail
1. Prefabricated aluminium composite cassettes
2. Insulation
3. Solar shading
4. Sealed double glazed unit
5. Sedum roof
6. HSQ
7. Suspended gypsum ceiling
8. Poder coated aluminiums profile
9. Heating convector
10. Prefabricated concrete floor elements
11. Suspended prefabricated aluminium composite cassettes

立面节点
1. 预制复合铝板
2. 隔热层
3. 遮阳板
4. 双层密封玻璃窗
5. 植物屋顶
6. HSQ
7. 石膏吊顶
8. 涂层铝型材
9. 热空气对流器
10. 预制混凝土楼板
11. 悬挂的预制复合铝盒

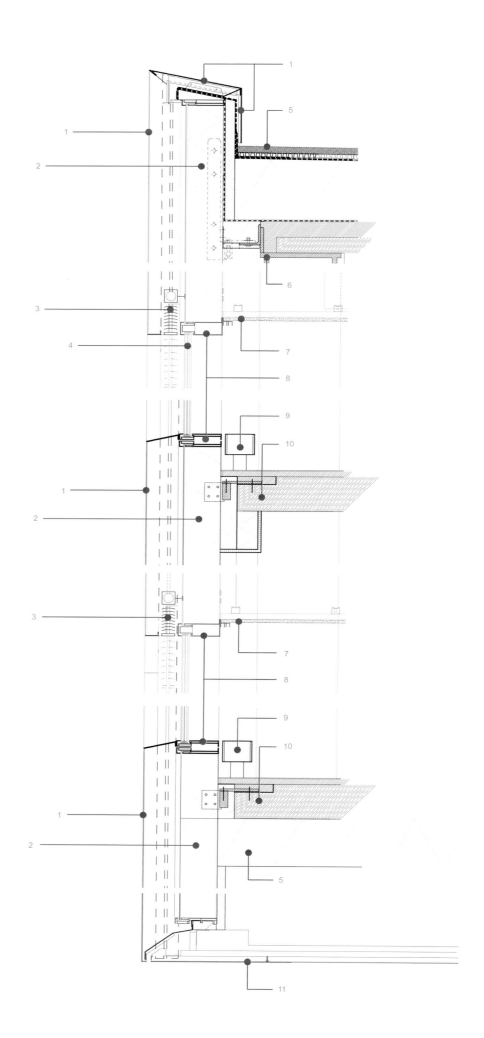

项目概况

新办公楼坐落在一个旧机场的多层停车场上方,大部分场地空间都被划为占地5公顷的公园。设计的主要目标之一是缩小建筑在公园中的占地面积,打造灵活的办公空间,让建筑内的大多数能享有峡湾的景色,为中庭空间提供公园、艺术品和峡湾的三重美景。

项目由5个相似的办公模块组成,每个模块3层高,约140米长,23米宽。它们叠加起来在一楼共同形成了一个中庭和多功能空间。由于工期极短,建筑的大部分结构(包括钢材混凝土上层结构、立面和玻璃结构)都是在场地外预制而成的。这保证了建筑的高效和精准性。钢结构保证了各个办公模块的悬臂可达30米。建筑的钢桁架装配动用了北欧最大的移动式起重机。

细部与材料

建筑表皮对建筑的能源效率至关重要。建筑的立面传热系数很低:实现板墙(包括剖面)部分为0.18 W/m²K,三层玻璃板(不包括剖面)部分为0.6 W/m²K。它们与外置Vental遮阳板结合起来,形成了极为高效的表皮。窗口面积保持在最佳比例:65%的净面积为玻璃区域,保证了室内日光照明和潜在热增量之间的平衡。玻璃板的太阳得热系数在0.24~0.36之间,根据其所在的位置略有不同。虽然建筑立面由许多独立的组件组合而成,它的密封十分良好,扩散值(n50)≤1,0h-1。此外,立面设计还保证了标准冷桥值低至0.02 W/m²K。

白色长边立面

项目特别开发的外轮廓是优雅的立面表达的关键。复杂的设计和技术装备还实现了建筑师的另一个目标:在整个预制立面上没有可见的固定设施。这个轮廓包括外部遮阳板(以"传统"百叶窗的形式出现)的轨道以及防雨覆盖层的固定设施。涂层夹心铝板折叠成220mm深的铝盒,赋予了立面深层浮雕的感觉,并且有效地抵抗了灰尘堆积。同样的预制立面系统也被运用在中庭的室内设计中,唯一的区别在于室内墙面是穿孔板(Φ5mm,25%),具有隔音效果,缩短了中庭的混响时间。

黑色山墙立面

山墙的遮阳板也是特别为项目所开发的。每个山墙立面都与180个4.2米高,400mm宽的夹层玻璃板组成。每个玻璃板由两层10mm Parsol Grey玻璃和一层深色夹心组成,看起来像是纯黑色一样。长立面的遮阳板在办公室内配有手动控制开关,而山墙的黑色玻璃翅片则根据每面山墙的朝向按照预定程序整体移动。从办公室内部看,遮阳板是看不见的,因为山墙全部采用玻璃幕墙装配。

YJP Administrative Centre
于家堡工程指挥中心

Location/地点: Tianjin, China/中国，天津
Architect/建筑师: HHD_FUN
Photos/摄影: Gang Wei, Zhenfei Wang
Site Area/占地面积: 18,000m²

Key materials: Façade – prefabricated aluminium
Structure – steel reinforced concrete, steel
主要材料： 立面——预制铝材
结构——钢筋混凝土、钢材

Overview
YJP Administrative Centre is a temporary building in Tianjin Binhai CBD. Enclosed verandahs are arranged around the building to afford visual access to the CBD area from within the building. This allows the occupants to survey the surrounding construction site.

Detail and Materials
The size of openings within the façade relates directly to the lighting requirements for particular activities within different areas of the building. The porosity of the façade is designed to produce the required conditions for these activities. The integration of the density of the patterned façade with the various inner functions forms a key focus of the project.

The façade apertures serve as view frames. Aperture size and orientation is varied in a continuous manner introducing topological difference across the façade. The whole façade is constructed from six forms, reflected to give twelve types of identical components, making the building process highly efficient. This meant that the building is to be constructed in less than seven months.

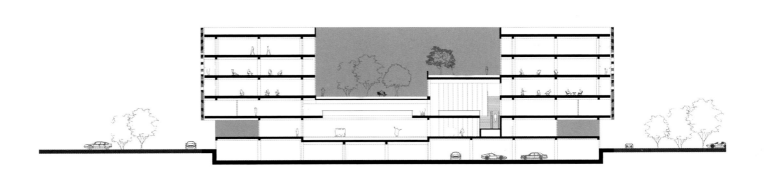
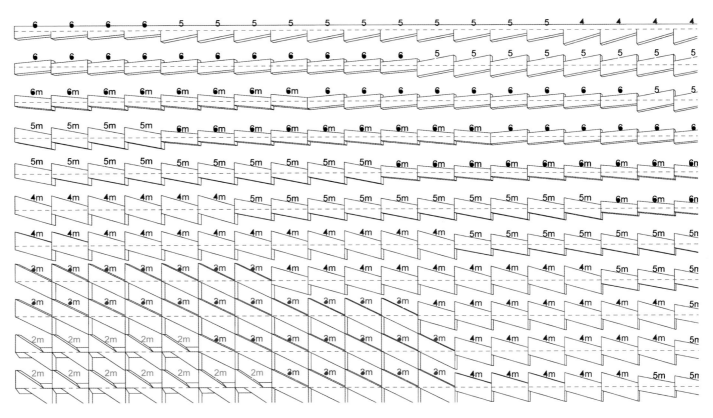

255

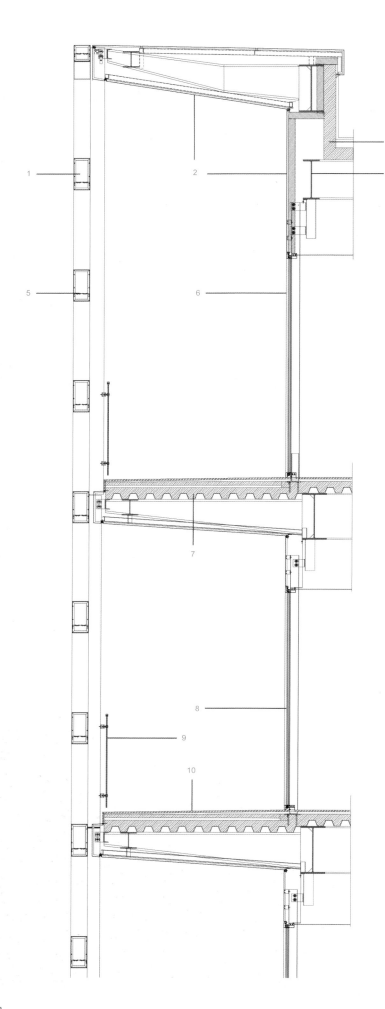

Façade detail
1. Prefabricated aluminium façade component
2. Aluminium panel
3. 150mm XPS insulation
4. Steel beam
5. Façade components bolted together at joint
6. Aluminium frame glass curtain wall
7. Steel reinforced concrete slab
8. Double-glazing
9. Tempered safety glass
10. Stone floor finish

立面节点
1. 预制铝立面
2. 铝板
3. 150厚发泡绝缘材料
4. 钢梁
5. 立面接合处
6. 铝框玻璃幕墙
7. 钢筋混凝土板
8. 双层玻璃窗
9. 钢化安全玻璃窗
10. 石质地面

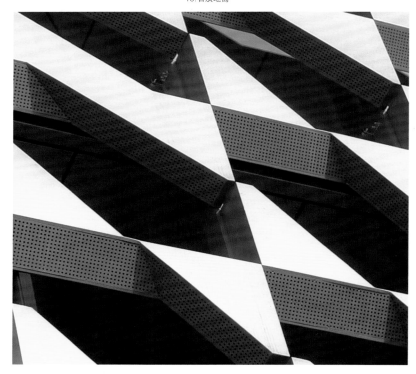

项目概况

于家堡工程指挥中心是位于天津滨海工业区内的一幢临时建筑，围合在一个院子之内，从建筑内部可以看见商业区内的景象，从而便于工作人员勘察周围的建筑选址。

细部结构与材料

外观上的开口设计考虑到建筑内部用于举办不同活动的相关区域的光线需求，而设计的关键因素就是将前面各个开口孔洞的孔隙度与建筑室内环境的不同功能完好的融合在一起。

立面上的开口构成景窗，其在规格和朝向上连续变化，从而突出了立面的多样性。立面最后限制在6种不同的模块内，这6种连续变化的模块通过光线的反射即可形成12种造型，大大地缩短了工期。这就意味着建筑可在7个月内完工。

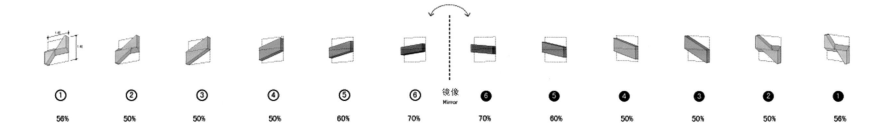

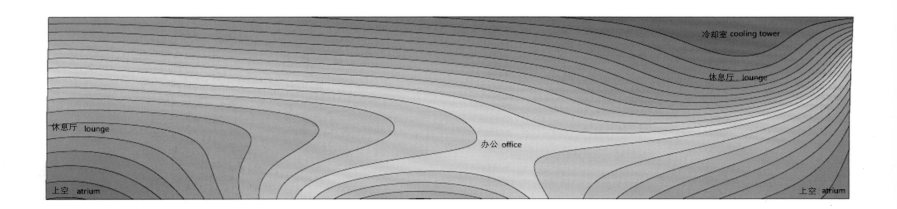

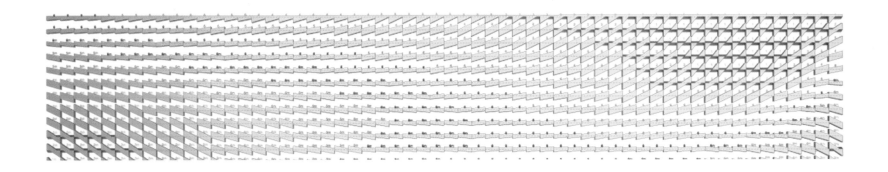

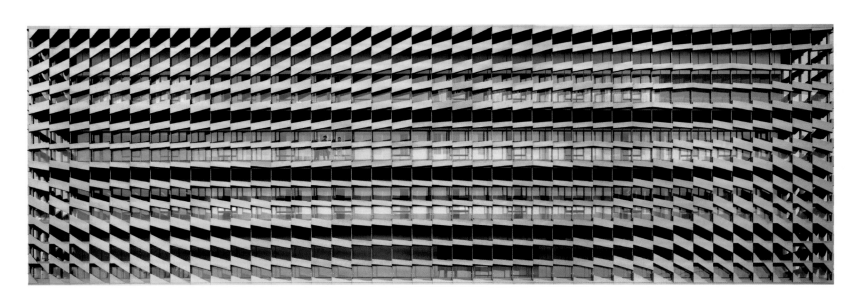

Ice Rink of Liège

列日滑冰场

Location/地点: Liège, Belgium/比利时,列日
Architect/建筑师: L'Escaut-BE Weinand (MA)
Construction area/建筑面积: 7,030m²
Completion date/竣工时间: 2012

Key materials: Façade – aluminium scales
Structure – steel
主要材料: 立面——铝鳞片
结构——钢材

Overview
At the root of the project is a round, fluid and generous shape, as a metaphor of a universe of ice, as its construction progressed: a sea monster, a whale is covered with 200,000 Aluminium scales.

The new ice rink of Liège is embedded into a crevice of the Médiacité, private real estate development project which came to redevelop a former industrial site on the right bank of the Meuse. It inherits several constraints that it clarifies in an autonomous and unitary form, until incorporating the access to the mall's car park in its climax: the whale's head. This dome is the most curved and spectacular area of the façade, where the careful scheme of the scales is revealed. This volume rises gently to allow cars to go. A load bearing element straight out of the 1970's reminds us of the glorious past of Liège, a time when car was queen in town planning as in some architectural projects.

The interior space of the rink focuses on the heart of the matter: functionality, economy and pleasure. In the Médiacité, at the entrance to the ice rink, a white light shower of 1,000 lux indicates the direction of the world of ice. Once past the airlock chamber, vistors dive into an ambient temperature of 16□ all year round.

Detail and Materials
As the opacity of the building is essential to insulate it from the heat, it is its entire body, by its nature, its material and its shape that means the relational dimension it wants to maintain with its environment. Moreover, the composition of its outer shell (on the mass-spring-mass principle) achieves a noise attenuation of 50db and protects the residents of the adjacent street.

Apart from a succession of portholes to the

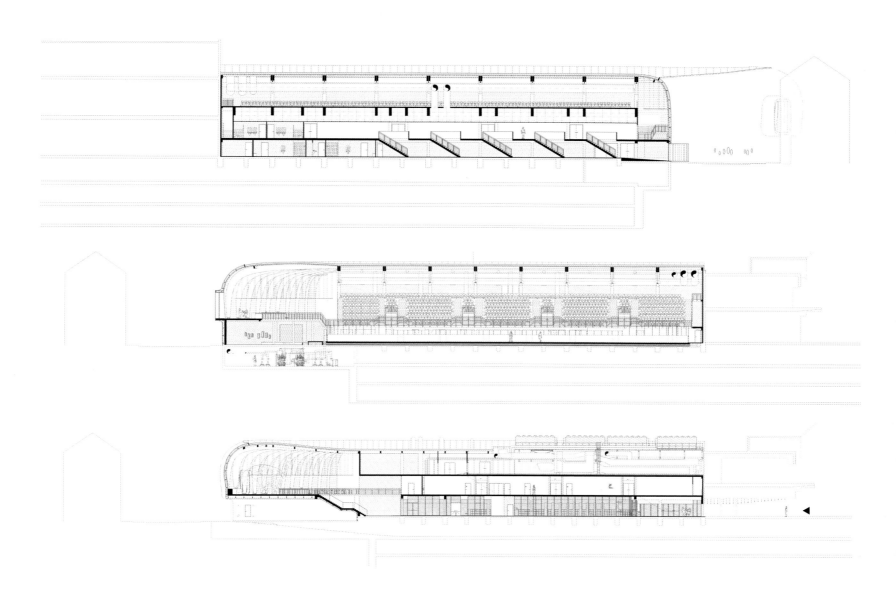

street sidewalk that suggest the activity of the strange object, the only transparent opening in the façade, is this bay as big as an antechamber that realises a frank and larger indoor/outdoor contact.

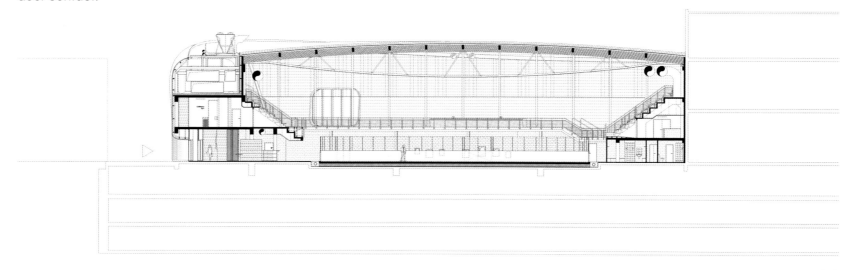

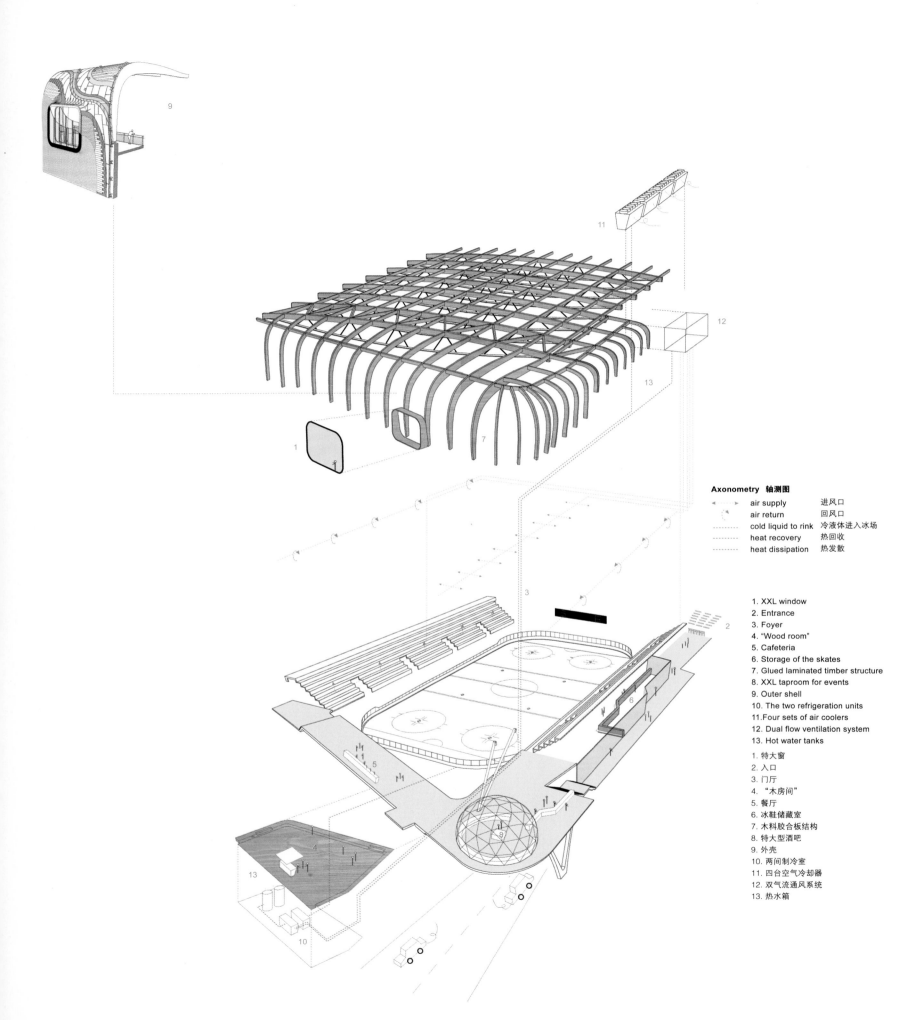

项目概况

项目采用圆滑流畅的庞大造型,象征着冰的世界。远远看去,它像是一只覆盖着200,000块铝制鳞片的巨鲸。

列日滑冰场位于媒体城——一个私有地产开发项目,是默兹河岸边原工业场地的重建。项目呈现为独立的圆滑造型,停车场的入口正好位于巨鲸的头部。这一位置是建筑立面最卷曲、最壮观的一角,呈现了精致的鳞片排列。立面缓缓升起,让车辆进入。源于20世纪70年代的承重零件提醒着人们列日辉煌的过去,当时停车场的设计刚刚进入建筑领域。

滑冰场的内部空间将重点放在功能性、经济性和舒适性三个方面。在滑冰场的入口,光照度为1,000勒克斯的白光提醒着人们进入了冰的世界。通过气闸之后,就进入了全年16℃的室内滑冰场。

细部与材料

建筑的封闭性主要是为了隔绝热量,而它的整个主体结构、材料和形状都表现了与环境相联系的渴望。此外,建筑外壳(采用质点–弹簧–质点设计法则)实现了消减50分贝噪声的效果,保护了周边街区的居民不受噪声干扰。

除了一排朝向街道的舷窗之外,建筑立面上唯一的开口就是建筑的前厅入口,它在室内外之间建立起了紧密的联系。

Façade composition
1. Steel structural decking
2. Stone wool acoustic insulation
3. Stone wool insulation 3cm
4. High density gypsum plasterboards 18mm
5. Class IV vapor barrier
6. Stone wool insulation 12cm
7. Stone wool insulation 6cm
8. High density gypsum plasterboards 2x18mm
9. Sealing PIB membrane
10. Sheathing
11. Aluminium scales

立面组合节点
1. 钢结构甲板
2. 石棉隔音层
3. 石棉隔热层3cm
4. 高密度石膏板18mm
5. 四级隔汽层
6. 石棉隔热层12cm
7. 石棉隔热层6cm
8. 高密度石膏板2x18mm
9. 密封PIB膜
10. 防护层
11. 铝鳞片

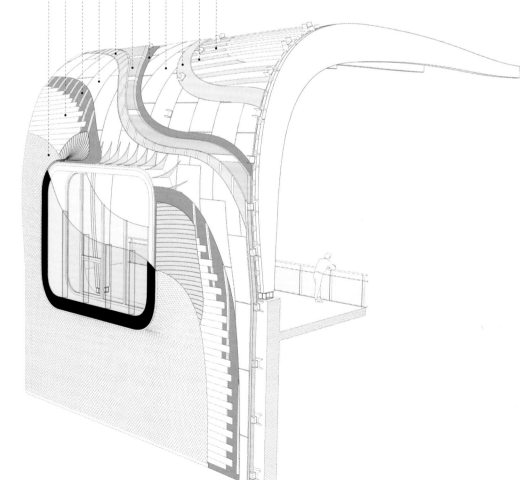

Chapter 3
Basic Information of Copper, Zinc and Titanium

第三章 铜、锌、钛

3.1 Copper and Copper Alloys
Definition

Copper is a chemical element with the symbol Cu (from Latin: cuprum) and density 8.92g/cm³, melting point 1,083℃. It is a ductile metal with very high thermal and electrical conductivity. Pure copper is soft and malleable; a freshly exposed surface has a reddish-orange colour. It is used as a conductor of heat and electricity, a building material, and a constituent of various metal alloys.

Copper alloys are metal alloys that have copper as their principal component. They have high resistance against corrosion. The best known traditional types are bronze, where tin is a significant addition, and brass, using zinc instead. The commonly used in architectural field is brass (alloy of copper with zinc).

3.1 铜与铜合金
定义

铜是一种化学元素，它的化学符号是Cu，密度为8.92g/cm³，熔点为1,083℃。作为一种韧性金属，它具有高导热性和导电性。纯铜是一种坚韧、柔软而富有延展性的金属，其表面呈紫红色，被用作导热体、导电体、建筑材料，以及各种合金的组成成分。

铜合金是以铜为主要成分的金属合金，具有高度的抗腐蚀性。最著名的传统类型是青铜（以锡为重要添加物）和黄铜（以锌为主要添加物）。建筑业最常用的铜是黄铜（锌铜合金）。

Properties

The softness of copper partly explains its high electrical conductivity ($59.6×10^6$ S/m) and thus also high thermal conductivity, which are the second highest among pure metals at room temperature.[1]

Copper does not react with water but it does slowly react with atmospheric oxygen to form a layer of brown-black copper oxide which, unlike the rust which forms when iron is exposed to moist air, protects the underlying copper from more extensive corrosion. A green layer of verdigris (copper carbonate) can often be seen on old copper constructions such as the Statue of Liberty.[2] Copper tarnishes when exposed to sulfides, which react with it to form various copper sulfides.[3]

Like aluminium, copper is 100% recyclable without any loss of quality, regardless of whether it is in a raw state or contained in a manufactured product. In volume, copper is the third most recycled metal after iron and aluminium. It is estimated that 80% of the copper ever mined is still in use today.[4]

Copper has excellent brazing and soldering properties and can be welded.

Application in Architectural Field

Copper and its alloys have been used for thousands of years. In the Roman era, copper was principally mined on Cyprus, hence the origin of the name of the metal as cyprium (metal of Cyprus), later shortened to cuprum. Its compounds are commonly encountered as copper (II) salts, which often impart blue or green colours to minerals such as azurite and turquoise and have been widely used historically as pigments. Architectural structures built with copper corrode to give green verdigris (or patina).

Copper has been used since ancient times as a durable, corrosion resistant, and weatherproof architectural material. [5][6][7][8] Roofs, flashings, rain gutters, downspouts, domes, spires, vaults, and doors have been made from copper for hundreds or thousands of years. Copper's architectural use has been expanded in modern times to include interior and exterior wall cladding, building expansion joints, radio frequency shielding, and antimicrobial indoor products, such as attractive handrails, bathroom fixtures, and counter tops. Some of copper's other important benefits as an architectural material include its low thermal movement, light weight, lightning protection, and its recyclability.

Architectural copper and its alloys can also be "finished" to embark a particular look, feel, and/or colour. Finishes include mechanical surface treatments, chemical colouring, and coatings.[9](See Figure 3.1, Figure 3.2)

3.2 Zinc and Zinc Alloys
Definition

Zinc, in commerce also spelter, is a metallic chemical element; it has the symbol Zn and atomic number 30.

Zinc alloys have been developed for meeting the most challenging specifications in terms of corrosion protection,

属性

铜的高导电性（$59.6×10^6$S/m）和高导热性部分源于它的柔软性。在室温条件下，它是导热性第二强的纯金属。[1]

铜不与水发生反应，但是会与大气中的氧气缓慢发生反应，从而形成一层氧化铜。与铁锈不同，氧化铜层能保护下方的铜不受进一步的腐蚀。老旧的铜结构上还经常出现一层铜绿（碳酸铜），例如自由女神像。[2]铜垢与硫化物反应会形成各种不同的硫化铜。[3]

与铝一样，铜也可以被100%回收，无论是处在未处理的状态还是蕴含在制成品中，都不会损失任何质量。从量上计算，铜是第三大回收金属，仅次于铁和铝。据估计，在有史以来开采的铜中，有80%仍在使用当中。[4]

铜还具有出色的钎焊和低温焊属性，容易焊接。

铜在建筑中的应用

铜和铜合金具有数千年的应用历史。在罗马时代，铜主要开采于塞浦路斯（Cyprus），因此得名cyprium（意为"塞浦路斯的金属"），后来被简写成cuprum。铜化合物常见为铜盐，呈蓝绿色矿物，例如蓝铜矿石（石青）和绿松石，被广泛应用为颜料。由铜建成的建筑结构经腐蚀呈现为铜绿色。

作为一种耐久、抗腐蚀、耐候的建筑材料，铜在古代有着广泛的应用。[5-8]千百年来，建筑的屋顶、防水板、雨水槽、落水管、穹顶、尖顶、拱顶和门都大量采用了铜作为主要材料。铜在现代建筑中的应用被扩展到了室内外墙面包层、建筑伸缩接缝、无线电频率屏蔽和抗菌室内制品（如美观的门把手、卫浴设施、台面等）。作为建筑材料，铜的优点还在于它的低热运动性、轻质量、防雷性和可回收性。

建筑铜和铜合金还可以被赋予独特的外观、质感和色彩。其表面处理方式包括机械表面处理、化学着色和添加涂层。[9]（见图3.1、图3.2）

3.2 锌与锌合金
定义

锌是一种金属化学元素，它的化学符号是Zn，原子序数为30。

锌合金的开发可以满足抗腐蚀、高温和耐磨等方面的

Figure 3.1, Figure 3.2 Buildings enveloped with copper
图3.1、图3.2 建筑由铜覆盖

temperature and wear resistance. Many alloys contain zinc, including brass, an alloy of copper and zinc. Other metals long known to form binary alloys with zinc are aluminium, antimony, bismuth, gold, iron, lead, mercury, silver, tin, magnesium, cobalt, nickel, tellurium and sodium.

Strengths

Exposed to CO_2 and water in the air, zinc produces chemical changes and forms layers of oxy-hydrogen zinc carbonate and zinc carbonate. These dense layers can protect internal structure from further corrosion, thus enhancing self-protection of zinc. Zinc alloys are easy to be formed and welded, environment-friendly and have high recyclability.

Application in Architectural Field

Raw zinc gives a silver white look and will gradually change colour due to oxidation, so it is rarely used in architecture. The most frequently used zinc is pre-oxidised into silver grey. Unlike the oxide layer formed naturally in atmosphere, this oxidised layer can last a long time without colour changes. Moreover, pre-oxidised charcoal grey zinc is more popular among architects due to its unique colour and texture. (See Figure 3.3, Figure 3.4)

要求。许多合金都含锌，例如黄铜（铜锌合金）。其他可与锌组成二元合金的金属包括铝、锑、金、铁、铅、汞、银、锡、镁、钴、镍、碲和钠等。

优点

锌板与空气中的二氧化碳和水接触时，可产生化学变化，形成氢氧碳酸锌层和碳酸锌层。这种致密的氧化层作为一层保护膜防止内部进一步腐蚀，增强了锌板的自我保护特性。锌合金易于成型、焊接，对周围环境不会造成污染，废物回收再利用率高。

锌在建筑中的应用

自然未经处理的原色锌板呈现亮银白色，会在大气中逐渐氧化而变色，在常规建筑中极少使用。建筑中最为常见的锌板多为预氧化处理的银灰色，与在大气中自然形成的氧化层不同，能够持续很长的时间而不变色。除此之外，预制黑灰色锌板由于其独特的色彩及纹理效果更受建筑师钟爱。（见图3.3、图3.4）

Figure 3.3 Zinc panel used on the roof
Figure 3.4 Zinc encloses the building

图3.3 锌板被用于屋顶结构
图3.4 锌板覆盖了建筑

3.3 Titanium and Titanium Alloys

Definition

A metallic element, titanium is recognised for its high strength-to-weight ratio.[10] It is a strong metal with low density that is quite ductile (especially in an oxygen-free environment),[11] lustrous, and metallic-white in colour. The relatively high melting point makes it useful as a refractory metal. It is paramagnetic and has fairly low electrical and thermal conductivity.

Titanium is often alloyed with aluminium (to refine grain size), vanadium, copper (to harden), iron, manganese, molybdenum, and with other metals.

Strengths

With high strength-to-density rate, titanium can meet the requirements of light architecture. Exposed to air, the surface of titanium forms a solid oxidation film with high corrosion resistance. Titanium changes its colours according to different thicknesses of the films. Its expansion factor is $8.2 \times 10^3 mm/(m.℃)$, half to copper or stainless steel and suitable for environment with high temperature variation, such as roofs. Its melting point is as high as 1,660℃. Titanium's heat resistance is much higher than that of aluminium alloy.

Application in Architectural Field

The properties of titanium and its alloys decide they are suitable for roofing structure. Japan is one of the first countries that use titanium and America is the largest consumer of titanium. However, due to the high cost in smelting process, titanium is rarely used except for some buildings with high budget. Museo Guggenheim Bilbao by Frank Owen Gehry used titanium as mian finsh material. The whole walls (except some part of stone base) and roof are covered with titanium sheet with 0.3 to 1.4mm thickness, fairly striking.

Reference / 参考文献

1. Hammond, C. R. (2004). The Elements, in Handbook of Chemistry and Physics 81st edition. CRC press. ISBN 0-8493-0485-7.
2. "Copper.org: Education: Statue of Liberty: Reclothing the First Lady of Metals – Repair Concerns"
3. Rickett, B. I.; Payer, J. H. (1995). "Composition of Copper Tarnish Products Formed in Moist Air with Trace Levels of Pollutant Gas: Hydrogen Sulfide and Sulfur Dioxide/Hydrogen Sulfide". Journal of the Electrochemical Society 142 (11): 3723–3728. doi:10.1149/1.2048404
4. "International Copper Association"
5. Seale, Wayne (2007). The role of copper, brass, and bronze in architecture and design; Metal Architecture, May 2007
6. Copper roofing in detail; Copper in Architecture; Copper Development Association, U.K., www.cda.org.uk/arch
7. Architecture, European Copper Institute;http://eurocopper.org/copper/copper-architecture.html
8. Kronborg completed; Agency for Palaces and Cultural Properties, København,http://www.slke.dk/en/slotteoghaver/slotte/kronborg/kronborgshistorie/kronborgfaerdigbygget.aspx?highlight=copper
9. Finishes – natural weathering; Copper in Architecture Design Handbook, Copper Development Association Inc.,http://www.copper.org/applications/architecture/arch_dhb/fin
10. "Titanium". Columbia Encyclopedia (6th ed.). New York: Columbia University Press. 2000–2006.ISBN 0-7876-5015-3.
11. "Titanium". Encyclopædia Britannica. 2006. Retrieved 2006-12-29. Stwertka, Albert (1998). "Titanium". Guide to the Elements (Revised ed.). Oxford University Press. pp. 81–82. ISBN 0-19-508083-1.

3.3 钛及钛合金

定义

作为一种金属元素，钛以高强度重量比而著称。[10]钛密度高、具有良好的延展性（特别是在无氧环境中）[11]、具金属光泽，呈现为银白色。钛的熔点很高，是一种难熔金属。钛具有顺磁性，导电性和导热性都较差。

钛经常与铝（精炼颗粒大小）、钒、铜（硬化）、铁、锰、钼等元素形成合金。

优点

钛密度小强度高，用于建筑中能够满足建筑轻型化的需要；在自然环境中能够形成坚固的氧化薄膜，具备优异的抗腐蚀性能。氧化膜的薄厚不同，钛可变换出不同的色彩。钛的膨胀系数为$8.2 \times 10^3 mm/(m.℃)$，约为铜或不锈钢的一半，适于温度变化较大的环境，如建筑屋面等；熔点高达1,660℃，钛合金的热强性比铝合金强很多。

钛在建筑中的应用

钛及钛合金的特性决定其非常适用于建筑中的屋面结构。日本是在建筑中最早应用钛材的国家，美国是使用钛材量最大的国家。但由于其在冶炼过程中过高的成本，除少量大投资建筑外，其运用相对较少。弗兰克·盖里设计的位于毕尔巴鄂市的古根海姆博物馆采用钛作为主要饰面材料——整个墙面（除部分石材基座）及屋面采用0.3~0.4毫米厚的钛板覆盖，格外引人注目。

City library in Seinäjoki

塞伊奈约基市图书馆

Location/地点: Seinäjoki, Finland/芬兰，塞伊奈约基
Architect/建筑师: JKMM Architects
Floor area/楼面面积: 4,430m²
Completion date/竣工时间: 2012

Key materials: Façade – copper
Structure – cast-in-place concrete
主要材料：立面——铜
结构——现浇混凝土

Overview

The key point of reference for the new library design was its location in the valuable environment of the civic centre created by Aalto. The aim was to initiate a dialogue between the new and the old part. The new building must respect the protected cultural environment while making a bold statement as a piece of modern architecture – in other words, find the right balance between being conciliatory and challenging. The building's design references the unique characteristics of Aalto's architecture, however strictly avoiding direct quotations or imitation of themes.

Another important challenge for the designers was the changing content of library activities. Dividing the new building into three sculpture-like sections was an important insight that helped to reconcile its large volume with the scale of the civic centre. The building thus relates to its surroundings, and a different statuesque aspect of it is revealed from each direction. On the other hand, the new building discreetly keeps its distance from the outlines and materials of the old civic centre. With liberal generosity, it was placed in the middle of lawns, as though a building in a park.

Detail and Materials

The dark copper of the façades stands out from the whiteness of the surrounding Aalto buildings, the contrast helping to detach the new from the old. A distinctive copper cladding material was designed for the façades that gives the building a unique lively texture.

The building comprises a cast-in-place concrete structure that lent itself to the sculpture-like design. The panoramic, unobstructed views across the interiors were achieved by using challenging long-span beams similar to those used in bridge structures. The visual look is dominated by cast concrete surfaces aiming for an unfussy, rough feel produced by using uneven plank moulds and by leaving the surfaces unfinished after casting. Building maintenance technology was integrated in architectural solutions, meaning that the vaulted ceilings and other internal surfaces could be kept free of technical installations.

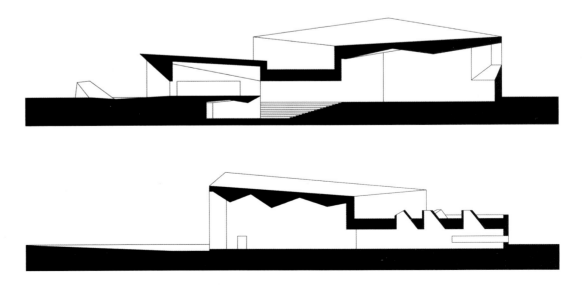

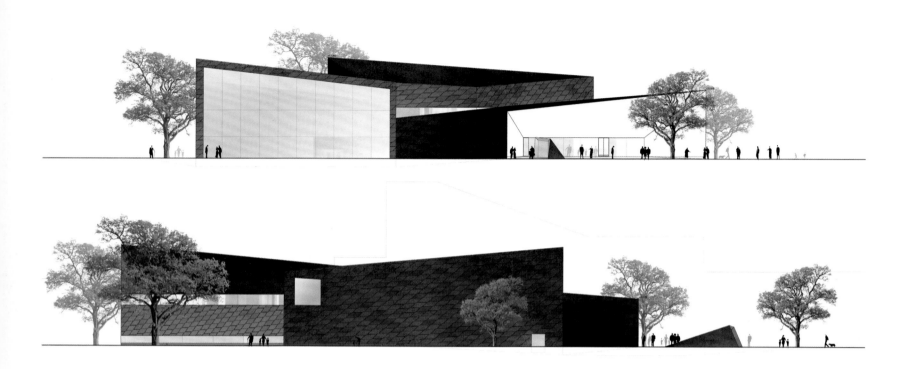

270

项目概况

新图书馆的设计考虑了其珍贵的地理位置——由阿尔托所设计的市政中心。设计的目标是在新旧之间建立起对话。新建筑必须尊重受保护的文化环境,同时又要体现现代建筑的大胆形象。换句话说,它必须在保守和挑战中找到平衡点。建筑的设计参考了阿尔托建筑的特色,同时又避免了直接的引用和模仿。

建筑师所面临的另一个挑战是图书馆中不断变化的活动。新建筑被分成三个部分,使它的规模与市政中心显得更加协调。建筑与周边环境联系起来,但是从各个角度看却各有不同。另一方面,新建筑与旧的市政中心在外形和材料上都保持了一定的距离。建筑场地十分宽敞,因此建筑被设在一片草地中央,就像是在公园中一样。

细部与材料

深铜色的立面从周边一片白色的建筑中脱颖而出,使新旧建筑形成了强烈的对比,特别设计的包铜材料为建筑带来了充满生气的外观。

建筑由现浇的混凝土结构构成,有一种雕塑感。与桥梁结构类似的大跨度横梁的使用保证了室内广阔流畅的视野。未经处理的混凝土表面实现了一种平凡、粗糙的质感。建筑在设计中融入了建筑维护技术,拱形天花板和其他室内界面都能安装技术装置。

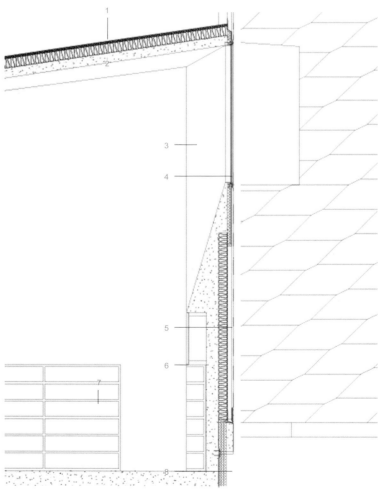

Façade section
1. Pre-oxidised copper roof sheeting
2. Ventilating roof structure
3. Cast-in-place reinforced concrete
4. Low iron oxide triple glass insulated structural glazing
5. Ventilating pre-oxidised copper cladding
6. Birch battened acoustic surface
7. Massive laminated pine bookcase
8. Cast-in-place copper plated wall base

外墙立面剖面
1. 预氧化铜屋顶板
2. 通风屋顶结构
3. 现浇钢筋混凝土
4. 低氧化铁三层绝缘结构玻璃
5. 通风预氧化铜包层
6. 桦木板条隔音表面
7. 大型层压松木书架
8. 现浇镀铜墙座

Sports Hall St. Martin

圣马丁体育馆

Location/地点: Villach, Austria/奥地利，菲拉赫
Architect/建筑师: Dietger Wissounig Architekten
Photos/摄影: Jasmin Schuller, Paul Ott
Built area/建筑面积: 3,170m²
Completion date/竣工时间: 2013

Key materials: Façade – copper sheet
Structure – reinforced concrete, wooden frame
主要材料：立面——铜板
结构——钢筋混凝土、木框架

Façade material producer:
外墙立面材料生产商：
Reinhard Eder Blechbau GmbH, Frankenweg 2, 9100 Völkermarkt

Overview
The old circular sports hall in St. Martiner Street has been replaced by a new hall which is used by the adjoining secondary school as well as various sports clubs and is suitable for international ball game tournaments.

The elongated structure is positioned approximately parallel to the street which establishes a high public presence, protects the outdoor sports ground located behind it and ensures that the existing parking spaces on the northwest side remain available. At the front of the building is the main entrance with a small forecourt. A row of trees and additional parking spaces serve as buffer zone to the busy St. Martiner Street. The three-storey building has been lowered four metres below street level in order to create a direct underground access to the school on the one hand and on the other hand, to give the extensive hall with a total height of 11.8 metres a form which agrees with urban development.

The entire complex has a gross volume of more than 24,000 m³. It provides barrier-free access and is equipped with an orientation system for visually impaired users. Geothermal preheating of supply air, a heat recovery ventilation system and a corresponding insulation ensure high energy efficiency resulting in an annual heating demand of 16.9 kWh/m².

Detail and Materials
The buildings appearance is characterised by a copper façade made of folded perforated metal plates covering the hall like a semi-transparent veil. The copper sheets are staggered by one folded element at each storey which structures the front horizontally. The façade is interrupted by glass fronts on the upper floor at the northeast side and on the ground floor at the southeast side. The former ensures an even and glare-free incidence of daylight in addition to the numerous skylights, while the latter provides a view into the sports hall from the schoolyard.

A maintenance walkway also covered with copper sheet protectively overhangs the main entrance and the glass front.

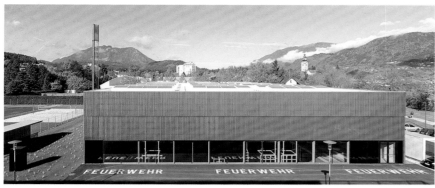

项目概况

新建的体育馆取代了圣马丁纳街上的圆形体育馆,供隔壁的中学和各种体育俱乐部使用,能够举办国际级球类运动锦标赛。

细长的结构基本与街道平行,形成了高大的公众形象,既保护了后方的露天体育场,又能保证西北面的停车场可以继续使用。在建筑前方是主入口和小小的前院。一排树木和若干个停车位形成了缓冲带,将建筑与繁忙的圣马丁纳街隔开。三层高的建筑比街道低4米,这样,一方面,形成了一个直接与学校相连的地下通道;另一方面,这为体育馆提供了高达11.8米的高挑空间,同时又符合了城市开发的要求。

整个体育馆的总空间体积大于24,000平方米,提供无障碍出入通道,并配有简单明晰的导向系统。送风地热预热、热回收通风系统和相应的绝缘设施保证了建筑具有极高的能源效率,年度热需求量仅为16.9 kWh/m^2。

细部与材料

建筑外观以折叠穿孔铜板立面为特色,像是在体育馆外蒙上了一层半透明的薄纱。铜板随着楼层水平排列,错落有致。东北侧立面的上方楼层和东南侧立面的一楼均设有玻璃窗。前者与天窗共同保证了室内的自然采光,后者则让学校操场的视野与体育馆内部联系起来。

养护通道上方同样设有铜板保护,下方是建筑的主入口和玻璃立面。

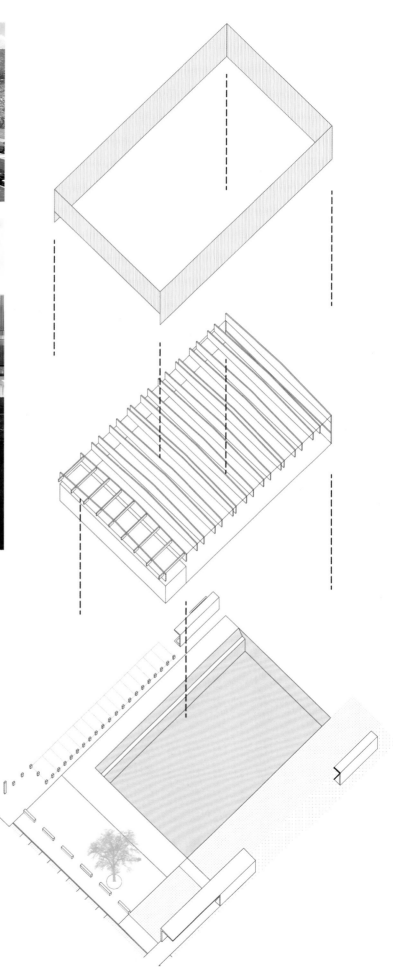

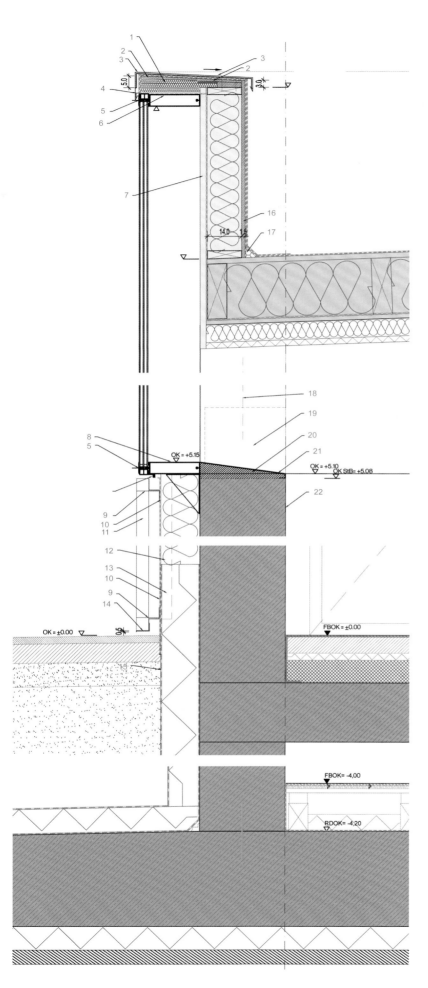

Façade detail
1. Insulation
2. Multi-layer panel
3. Cap flashing copper
4. Fastening element suction resistant
5. Pressure plate profile A6/C34
6. Transom
   Pressure relief profile
   Cream white
7. Multi-layer panel
   Water resistant
   Latex coating cream white
8. Transom
   Cream white
9. Insect screen
10. Façade sealing membrane, black, unprinted
11. Copper sheets, folded, perforated
12. L profile on thermally separated substructure
13. Recess in insulation for L profile
14. Dip edge, serving as edge protection for copper façade
15. Dimpled sheeting
16. OSB panel
17. Insulation wedge
18. Partial windbracing
19. Concealed beam tie on welding base
20. Wooden substructure space between filled with insulation material
21. Aluminium, cream white
22. Exposed concrete sprayed cream white

立面节点
1. 隔热层
2. 多层板
3. 铜防水顶盖
4. 防吸固定元件
5. 压板型材 A6/C34
6. 横梁
   降压型材
   乳白色
7. 多层板
   防水层
   乳胶涂料，乳白色
8. 横梁
   乳白色
9. 纱窗
10. 立面密封膜，黑色，无印花
11. 折叠穿孔铜板
12. L形隔热下层结构
13. L形隔热结构凹槽
14. 下沉边缘，作为铜立面的边缘保护
15. 凹纹板
16. 定向刨花板
17. 隔热楔
18. 局部抗风支撑
19. 焊接底座上的隐梁
20. 隔热材料填充之间的木下层结构
21. 铝，乳白色
22. 露石混凝土，喷涂成乳白色

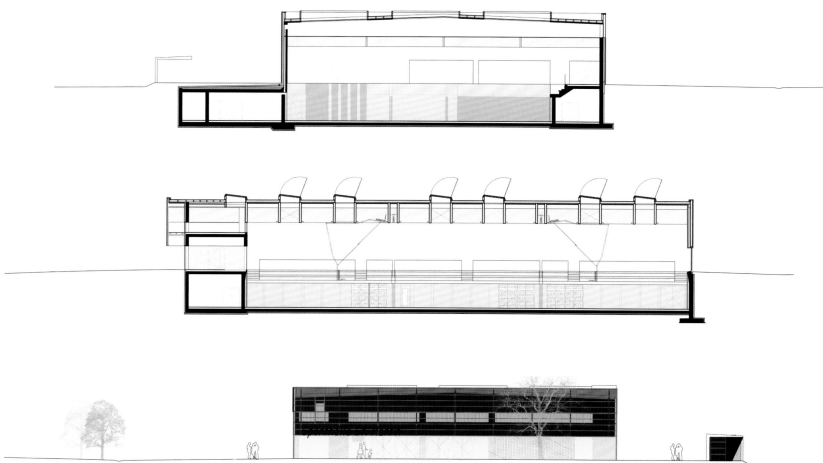

Tower Euravenir

欧拉维尼尔大厦

Location/地点: Lilles, France/法国，里尔
Architect/建筑师: Benoit Jallon et Umberto Napolitano/ LAN Architecture
Photos/摄影: Julien Lanoo
Built area/建筑面积: 3,486m²
Completion date/竣工时间: 2014

Key materials: Façade – copper, glass
主要材料： 立面——铜、玻璃

Overview

This architecture has created a new urban space that combines private and public, vertical and horizontal. The base of the project provides inhabitants and office workers a public space that fosters social interaction; it functions on a human scale. Due to the prohibition from building out to the edge of the parcel, a kind of portico provides a sense of porosity as well as protection from inclement weather. It is a lively outdoor space where people who live and work in the building can mingle with passers-by and shop customers.

This office project has a very flexible program; form dictates use, and not vice-versa. Each level is organised around a central core that holds all the servant spaces and vertical circulations. The office floor areas were conceived to allow for a flexible, rational layout and to encourage the division of the floor spaces into two equivalent surfaces. Moreover, the tower's geometric faceting at once frees up the views and opens up the entire intersection to the wooded background of the cemetery to the north.

In order to complete this process of interrelation, the façades were designed to become a series of windows that provide a 360-degree panorama of the city, framing views of the city's newer parts, its green spaces, and the downtown.

Detail and Materials

The building's envelope was designed as a way to visually reinvent the city. The façades are characterised by different designs in response to their orientation, their usage, and their thermal requirements. In this way, areas that are predominantly glass, some parts of which have a double skin, are juxtaposed with different forms

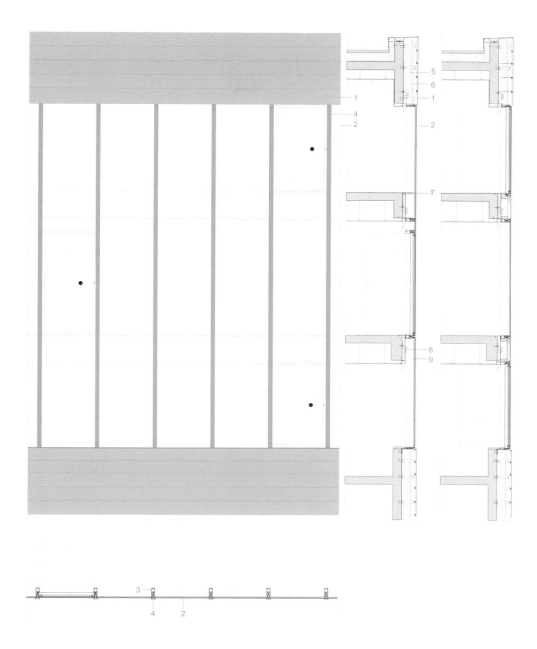

Façade detail
1. Spines thick copper: 15/10th
2. Double glazed insulating laminated
3. Amount Aluminium façade
4. Skin folded sheet copper pre-oxidised 15/10th
5. Mineral insulated 200mm + rain barrier
6. 20/10th white sheet steel
7. Galvanised steel grating
8. Steel fasteners integrated amounts
9. Support framework for horizontal siding installation

立面节点
1. 厚铜骨架：15/10th
2. 双层绝缘玻璃
3. 铝立面
4. 预氧化褶皱铜板15/10th
5. 石棉隔热层200mm+防雨层
6. 20/10th白钢板
7. 镀锌钢格栅
8. 钢连接件
9. 水平侧线的支撑框架

of copper cladding that are more or less porous.

The design of the façades and the building spaces is governed by the principal lattice pattern, 1.35 metres in height, which runs around the entire top of the building. It is marked by a U-shaped metal component to which the various elements that make up the envelope are attached.

This vertical motif is interrupted by three different bands that emphasise the building's horizontal composition and form a sort of crown at the level of the acroterium. A secondary motif formed by the bands and the trumeaus cuts out the façades.

The copper is used as a kind of fixed siding along the opaque or semi-glassed stretches of the façade. It is also present in the form of perforated panels that helps precisely regulate the amount of light penetrating the building, depending on the orientation.

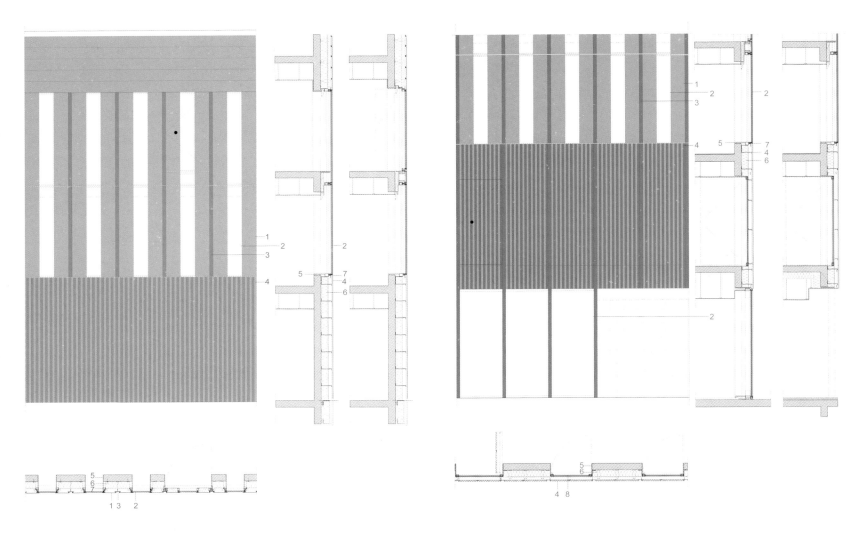

Façade detail
1. Smooth nested pre-oxidised copper box: 20/10
2. Double glazing interlayer insulation railings, acoustic and solar control
3. Hollow Aluminium connector plate bending (15/10 Black)
4. Interlocking board slot copper: 20/10
5. Wood shelf
6. Mineral insulated 200mm + rain barrier
7. Chassis, Aluminium finishing snare

立面节点
1. 光滑嵌套式预氧化铜箱：20/10
2. 双层绝缘玻璃围栏，隔音控热
3. 中空弯曲铝连接板（15/10黑色）
4. 铜连锁板槽：20/10
5. 木架
6. 矿物棉隔热层200mm+防雨层
7. 铝饰面底架

Façade detail
1. Smooth nested pre-oxidised copper box: 20/10
2. Double glazing interlayer insulation railings, acoustic and solar control
3. Hollow Aluminium connector plate bending (15/10 Black)
4. Cladding in niche with perforation rate of 40% thick: 20/10th
5. Wood shelf
6. Mineral insulated 200mm + rain barrier
7. Chassis, Aluminium finishing snare
8. Chassis, aluminium thermolaque

立面节点
1. 光滑嵌套式预氧化铜箱：20/10
2. 双层绝缘玻璃围栏，隔音控热
3. 中空弯曲铝连接板（15/10黑色）
4. 立面包层，穿孔率40%，厚20/10th
5. 木架
6. 矿物棉隔热层200mm+防雨层
7. 铝饰面底架
8. 隔热铝底架

项目概况

建筑打造了全新的城市空间,融合了公与私、水平与垂直。项目的底座为居民和办公人员提供了社交活动的公共空间,十分人性化。由于市政规划不允许建筑超出地块边缘,建筑师设计了一个门廊,既让建筑显得通透,又能遮风挡雨。在建筑中生活和工作的人们可以在这个活跃的露天空间与行人和商店的顾客实现互动。

这个办公项目十分灵活。建筑的形式决定了它的用途;但是用途并不影响形式。每层空间都围绕着一个中央核展开,中央空间内设置着所有服务区和垂直交通流线。办公楼层采用灵活、合理的布局,每层空间可以分成两等分。此外,建筑的几何切面打开了视野,让整个交叉空间都朝向北面墓园的树林。

为了进一步完善设计,建筑立面上设置了大量开窗,提供了360度城市全景,囊括了城市新区、绿地以及市中心的景色。

细部与材料

建筑的外壳在视觉上对城市进行了重新改造。对应不同的朝向、功能区以及热需求,立面拥有不同的设计特征。这样一来,建筑表皮由大面积的玻璃和铜皮交错而成,实现了多变的造型。

立面和建筑空间的设计以格子图案为基础。1.35米高的方格贯穿了整座建筑。组成立面的各种不同的元件通过格子上的U形金属组件安装起来。

三条不同的水平带打乱了这个垂直图案,突出了建筑的水平构成,在山墙底座形成了一个顶冠。水平带和间柱所形成的图案将立面切割成不同的部分。

铜被应用在立面不透明或半透明的固定侧端上,同时也以穿孔铜板的形式为建筑的某些朝向提供遮阳保护。

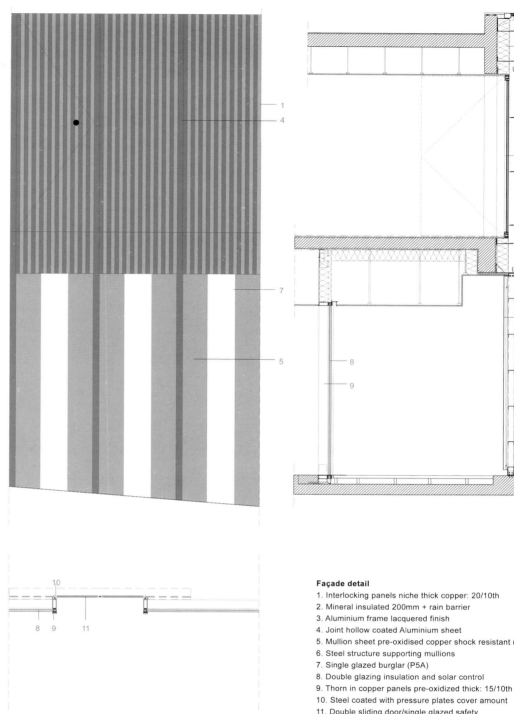

Façade detail
1. Interlocking panels niche thick copper: 20/10th
2. Mineral insulated 200mm + rain barrier
3. Aluminium frame lacquered finish
4. Joint hollow coated Aluminium sheet
5. Mullion sheet pre-oxidised copper shock resistant (Q4)
6. Steel structure supporting mullions
7. Single glazed burglar (P5A)
8. Double glazing insulation and solar control
9. Thorn in copper panels pre-oxidized thick: 15/10th
10. Steel coated with pressure plates cover amount
11. Double sliding door/single glazed safety

立面节点
1. 连锁厚铜板:20/10th
2. 矿物棉隔热层200mm+防雨层
3. 涂漆铝框
4. 涂层中空铝板
5. 预氧化防震铜板竖框(Q4)
6. 钢结构支撑竖框
7. 单层防盗玻璃(P5A)
8. 双层绝缘遮阳玻璃
9. 预氧化铜板,厚15/10th
10. 压板涂层钢
11. 双拉门/单层安全玻璃

Hiiu
希尤住宅

Location/地点: Tallinn, Estonia/爱沙尼亚，塔林
Architect/建筑师: Alver Architects (Andres Alver, Tarmo Laht, Indrek Rünkla, Ulla Mets, Sven Koppel)
Photos/摄影: Alver Architects, Kaido Haagen
Site area/占地面积: 18,000m²
Completion date/竣工时间: 2013

Key materials: Façade – copper
主要材料：立面——铜

Façade material producer:
外墙立面材料生产商：
WICKEDER WESTFALENSTAHL GmbH, Germany.

Overview

"Hiiu" is an ongoing residential development by OÜ Tardamel located in Tallinn, Estonia. The complex includes 8,500 m² of residential units – 23 apartments (phase 1 - built) and 10 single-family houses (currently under construction), all integrated into the surrounding landscape of the Tallinn's picturesque historic garden neighborhood of Nõmme. The project was defined by its site and the local zoning regulations which restricted the amount of developable space to a maximum of 600 m2 per unit with a very low total FAR. Thus, the overall concept of the project became to be "Buildings in the park".

Detail and Materials

The façade solution for the apartment houses originates from the architects' will to integrate the complex into the landscape. The apartment houses are clad in copper, which will gradually get oxidized and turn green, blending into the park.

There were a number of factors that defined the solution for the façade of the Hiiu residential complex.

First and foremost, were the picturesque surroundings in which the complex was to be placed. The architects wanted to blend the complex into the environment, but to do it in a modern way. Thus, instead of painting the buildings green, or specifying "default" natural materials like wood, it was decided to go with copper cladding. Given its chemical properties, copper will turn green over time, letting the building age elegantly while slowly blending into the environment.

Using copper was also dictated by the will of the architects to merge the roofs and walls of the buildings. In Nõmme, the picturesque historic area of Tallinn, the zoning regulations restrict the maximum bulk of buildings to two floors with a half-floor attic. While typically this results in a de-

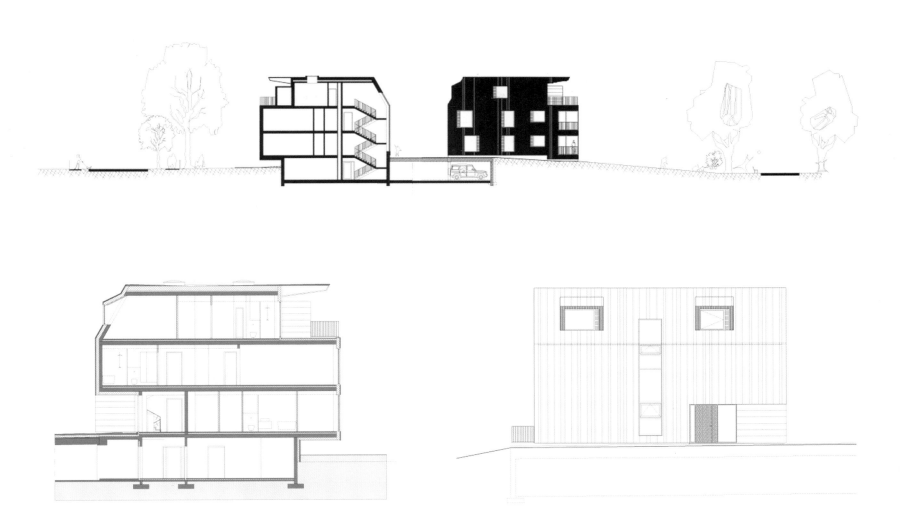

fault gable-roof typology, Alver Architects used the restriction as a creative form-generating tool, while using copper to visually merge the walls and the roofs of the buildings. The result is a development which while conforming to the regulations, still boasts a unique character, and generates extra value for the developer.

The stripy texture of the façade of the apartment houses was a solution to the tight budget constraints. As it is cheaper to procure thin stripes of copper rather than large uniform panels or sheets, the façade was composed of thin copper ribbons of different length and width. Once the copper façade was laid out, the windows were placed into it in such a way that they matched the module of the façade, while following the inner logic and insulation requirements of the apartments inside."

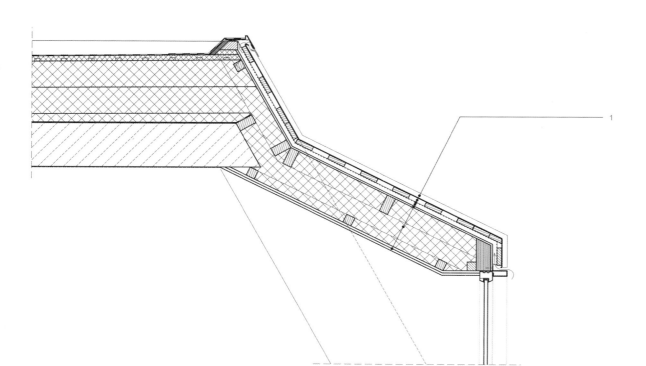

项目概况
希尤住宅位于爱沙尼亚首都塔林。整个项目包含8,500平方米的住宅单元，其中有23套公寓（一期工程，已建成）和10套独栋住宅（在建中），所有住宅都由风景如画的古典花园景观围绕。受场地和地区规划规定所限，每个单元的可开发空间被控制在600平方米以内，建筑高度也不能过高。因此，整个项目变成了"公园中的小楼"。

细部与材料
公寓楼的立面设计灵感来源于建筑师将建筑与景观结合起来的想法。公寓采用铜包层，铜经过氧化变成绿色，逐渐与公园景观结合起来。希尤住宅的立面设计有以下几个设计特色：

首先是住宅所在地风景优美的环境。建筑师希望以现代的方式让住宅融入环境。因此，他们没有将建筑涂成绿色或是利用木材等天然材料进行装饰，而是选择了铜包层。铜的化学属性让它会随着时间的推移而变绿，让建筑可以优雅地老化，慢慢融入环境。

铜的使用还让建筑的屋顶与墙壁融为一体。在建筑所在的历史区域，建筑规划十分严格，要求建筑至多高两层，带有半层阁楼。这种规定通常会衍生出山形屋顶，而建筑师却根据这一限制创造性地将墙壁与屋顶结合起来。最终，设计既遵循了规定，又显得别具一格，为开发商提供了附加的建筑价值。

公寓楼里面的条纹纹理解决了紧张的预算问题。细铜条的价格比大块平整铜板的价格要低，因此整个立面都由不同宽度和长度的铜条组成。窗户的安装在铜立面布置完毕之后，保证了它们与立面的造型一致，同时也满足了内部公寓的空间和保温要求。

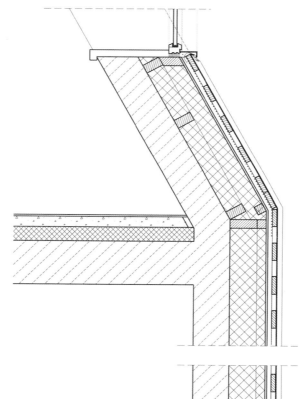

Section detail
1. Roof construction
   - 2x layer of drywall (26mm)
   - Thermal insulation: stone wool (250mm)
   - Timber substructure
   - Wind barrier (13mm)
   - Air cavity (25mm)
   - Diffusion membrane
   - Façade substructure of treated wood (25mm)
   - Copper façade cladding

2. External wall construction
   - Structural R/C
   - Thermal insulation stone wool (200mm)
   - Wind barrier (13mm)
   - Air cavity (25mm)
   - Façade substructure of treated wood (25mm)
   - Copper façade cladding

剖面节点
1. 屋顶结构
   双层干式墙（26mm）
   隔热层：石棉（250mm）
   木下层结构
   挡风板（13mm）
   空气层（25mm）
   扩散膜
   防腐木材立面下层结构（25mm）
   铜立面包层

2. 外墙结构
   结构钢筋混凝土
   隔热层：石棉（200mm）
   挡风板（13mm）
   空气层（25mm）
   防腐木材立面下层结构（25mm）
   铜立面包层

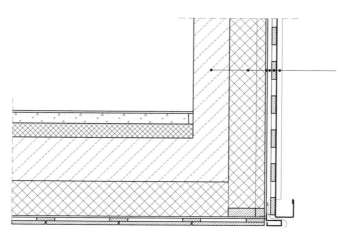

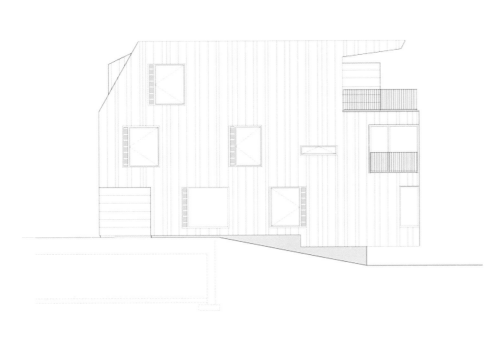

Louisiana State Museum and Sports Hall of Fame
路易斯安那州博物馆和体育名人堂

Location/地点: Natchitoches, USA/美国,纳契托什
Architect/建筑师: Trahan Architects
Built area/建筑面积: 2,601m²

Key materials: Façade – copper, glass
Structure – cast stone
主要材料: 立面——铜、玻璃
结构——铸石

Exterior Rainscreen Manufacturer/
外墙立面雨幕系统生产商:
A2MG (www.a2mg.com)

Overview
The Louisiana State Museum and Sports Hall of Fame in historic Natchitoches, Louisiana merges two contrasting collections formerly housed in a university coliseum and a nineteenth century courthouse, elevating the visitor experience for both. Set in the oldest settlement in the Louisiana Purchase on the banks of the Cane River Lake, the design mediates the dialogue between sports and history, past and future, container and contained.

The exploration focuses on three questions. How does the design explore the client brief to exhibit sports and history simultaneously? How does it respond to the historic building fabric? How does it make a connection to context?

The resolution is to interpret athletics as a component of cultural history rather than as independent themes. While sports and regional history may appeal to different audiences, the exhibits and configuration explore interconnections between the two. The spaces flow visually and physically together, configured to accommodate state-of-the-art exhibits, education and support functions. Visitors however can experience both narratives either separately or simultaneously.

Detail and Materials
The "simple" exterior, clad with pleated copper panels, alluding to the shutters and clapboards of nearby plantations, contrasts with and complements the curvaceous interior within. The louvered skin controls light, views and ventilation, animates the façade, and employs surface articulation previously achieved by architectural ornamentation. The flowing interior emerges at the entrance, enticing visitors to leave the walking tour and into the evocative exhibit spaces within.

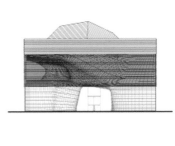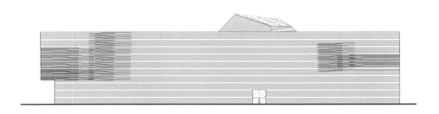
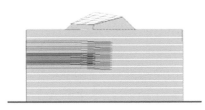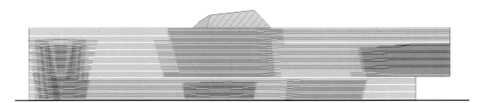

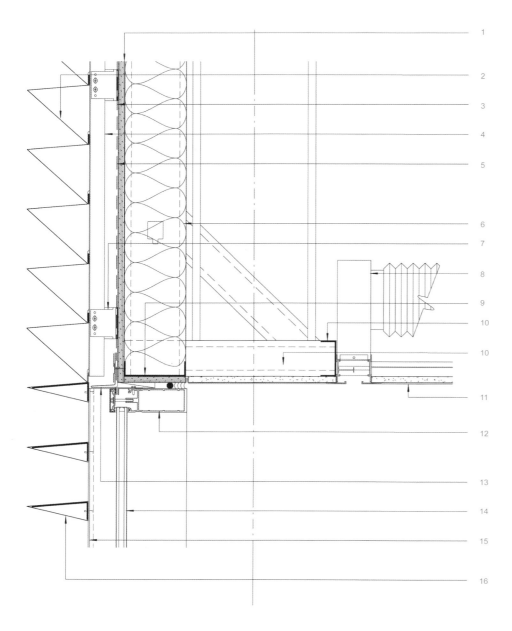

Façade detail

1. 5/8" exterior fire retardant-treated plywood sheathing
2. Offset flat seam pleated copper panel
3. Liquid applied weather barrier
4. Copper panel system vertical "I" profile, per manufacturer
5. Liquid applied weather barrier
6. Full thick batt insulation
7. Copper panel system clip, spacing per manufacturer
8. Continuous linear diffuser, re: mechanical
9. 6" metal stud track
10. 3 5/8" metal studs @ 16" o.c.
11. 5/8" paperless gypsum board at linear diffuser, 4" min. From edge
12. Aluminium curtain wall system with brushed stainless steel cladding, re: schedule
13. Head flashing, finish to match copper panels
14. 1" clear insulated glass unit with low-e coating
15. Formed 2 5/8" x 2 7/16" stainless steel cover to match mullion
16. Copper louver

立面节点

1. 5/8"外层防火夹层包板
2. 叠加平缝褶皱铜板
3. 液面耐候层
4. 铜板系统，L型材
5. 液面耐候层
6. 全厚隔热层
7. 铜板系统夹
8. 连续条形散流器
9. 6"金属立杆轨道
10. 3 5/8"金属立杆@ 16" o.c.
11. 5/8"无纸石膏板，与条形散流器相连，最小厚度4"
12. 铝幕墙系统，带有拉丝不锈钢包层
13. 顶部防水板，饰面与铜板相配
14. 1"低辐射透明绝缘玻璃
15. 成型2 5/8" x 2 7/16"不锈钢盖板，与竖窗框相配
16. 铜百叶

项目概况

之前，路易斯安那州博物馆和体育名人堂分别位于一座建于19世纪的政府大楼和一所大学体育馆内。该项目将二者合并起来，提升了游客的体验。项目位于凯恩河畔，在体育与历史、过去与未来、包容和被包容之间建立起了对话。

设计主要面临三个挑战：如何满足客户同时展示体育和历史的要求？如何应对项目所处的历史环境？如何使建筑与环境联系起来？

最终，设计将体育运动诠释为文化历史的一部分，而不是独立出来。虽然体育和地区历史所吸引的观众略有不同，展览试图寻找到二者的交叉点。空间结构流畅自然，配置了最先进的展览、教育和辅助功能。游客可以选择只参观一项，或者同时参观两项。

细部与材料

建筑"简单"的外观包裹在褶皱的铜板中，与附近种植园中的百叶窗和护墙板有着异曲同工之妙，并且与柔和曲折的室内空间形成了对比。百叶表皮能够控制光线、视野和通风，令建筑立面充满活力，同时也为建筑添加了美观的装饰。流畅的室内空间于入口处吸引着游客进入内部精彩的展览。

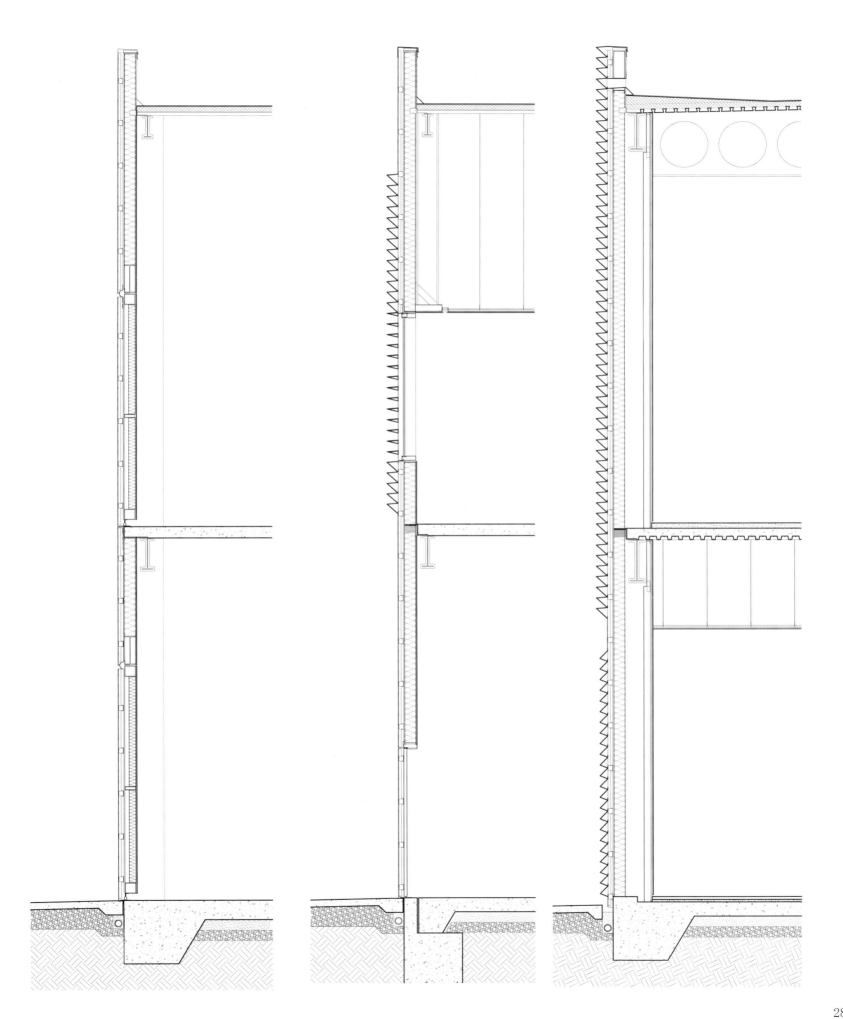

YapıKredi Banking Academy

雅皮科里迪金融学院

Location/地点: Gebze, Turkey/土耳其，盖布泽
Architect/建筑师: Mehmet Kütükçüoğlu, Ertuğ Uçar
Photos/摄影: Cemal Emden, www.cemalemden.com
Built area/建筑面积: 9,500m²

Key materials: Façade – copper panel
主要材料：立面——铜板

Overview

The Academy Building by TEGET Architects is an addition to the already existing banking Centre designed by British Architect, John McAslan. As an addition, the new building is not simply in continuity with the conglomerate of the existing 10 cubes; but rather an abstract statement of its own using the main complex as a backdrop. It introduces a focal point to an otherwise non-hierarchical organisation via a pair of scaleless copper beams extending out to the landscape. The program is the academy, the gate for the bank, where new comers will be educated as well as the 'in-service' employees. A composition of classrooms, social areas and meeting rooms. It is handled in a straight forward manner as the pair of copper bars are allocated for education whereas the central void for socializing.

Detail and Materials

The two program beams, stemming from the complex, stretches out on the plane of a missing cube, cannot be stopped and cantelevers as topography drops. They are covered by stretched copper panels as the outer layer of the façade, stripping it off from any hint of scale. Due to the stretching process, each panel has a double-fold direction from left to right. Making use of it, the panels are checkered with rotations to create the pixelation effect. The double façade controls excessive light in the classrooms. Overall, it is opaque from outside during the day, ambiguous in twilight, transparent in the evening.

The central space is treated as an autonomous strip clearly distinguished from outside. Here, the copper peels off, and is replaced by translucent movable fabrics in institutional colors. The interior façade begins earlier and stops before the copper bars. The roof and the façade is sealed with transparent air cushions, ETFE. The void is further accentuated, pushing the feeling of vertigo, by a series of suspended social spaces and their alternative circulation made of steel and glass, filtering light to the street level. The ground, out of concrete, facilitates another level of social life, with volcanoes providing light to the conference room below, water, giant stairs leading to the canteen which turns out to be an informal odeon, and a wood deck as a lounge.

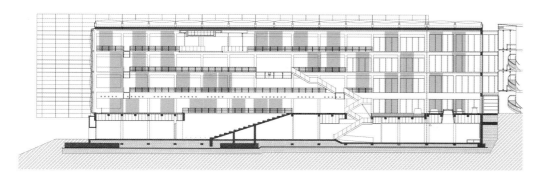

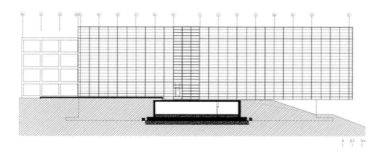

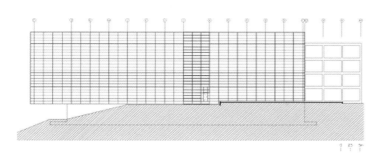

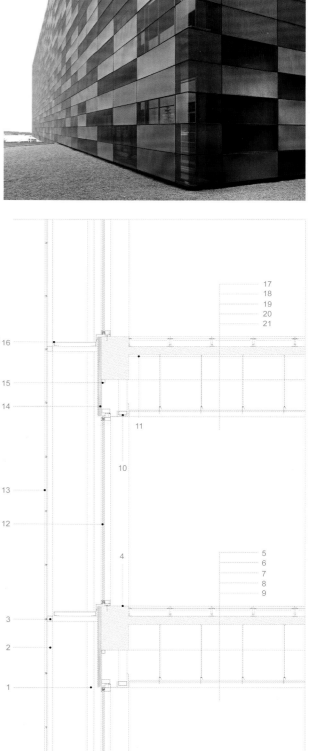

项目概况

这座由TEGET建筑事务所打造的学院建筑是现存的金融中心（由英国建筑师约翰·麦卡斯兰设计）的附属建筑。作为附属建筑，新建的学院大楼不仅在造型上与原有的金融中心相一致，还利用主楼作为背景，呈现出独特的风格。它通过一对延伸到外部景观中的铜结构引入了一个焦点，从而划分出了空间结构。

学院是银行的门户，为新进职员和在职员工提供培训教育，包括教室、社交空间和会议。建筑采用直线布局，由一对平行的铜结构组成，二者之间中空的空间被设为社交空间。

细部与材料

两个平行的体块从建筑中一直延伸到外围的景观之中。它们表面覆盖着拉伸的铜板，弱化了建筑的宏大感。拉伸过程让每块铜板都从左到右实现了双向折叠。铜板形成的方格交错排列，形成了像素的效果。双层立面控制了教室的采光。白天，从外面看它是不透明的；黄昏则显得模糊不清；夜晚则呈现出透明感。

中央空间被处理成独立地带，与外面界限分明，褪去了铜板表皮，以半透名彩色纤维板替代。内部立面的长度要短于铜结构。屋顶和立面密封在透明的四氟乙烯气垫中。一系列悬空的社交空间和由钢材、玻璃所组成的交通流线让中央空间呈现出混乱而统一的感觉。

YKB system section
1. Drainage system 125 mm
2. Steel profile 80/80
3. Steel profile 80/80
4. Polished concrete floor
5. Megalit flooring
6. Void
7. Reinforced concrete slab
8. Suspended ceiling system
9. Linear aluminum ceiling
10. Plasterboard
11. Acoustical pad
12. Façade module: Solar Low-E Glass Aluminium framing
13. Copper panel 1 mm
14. OSB panel
15. Steel profile
16. Catwalk: perforated galvanized steel sheet
17. Megalit flooring
18. Void
19. Reinforced concrete slab
20. Suspended ceiling system
21. Linear Aluminium ceiling

系统结构节点
1. 排水系统125cm
2. 钢材80/80
3. 钢材80/80
4. 抛光混凝土地面
5. Megalit地面
6. 中空空间
7. 钢筋混凝土板
8. 吊顶系统
9. 线形铝吊顶
10. 石膏板
11. 隔音垫
12. 立面模块：低辐射玻璃 铝框架
13. 铜板1mm
14. 定向刨花板
15. 钢材
16. 天桥：穿孔镀锌钢板
17. Megalit地面
18. 中空空间
19. 钢筋混凝土板
20. 吊顶系统
21. 线形铝吊顶

Bu Yeon Dang

浮烟堂

Location/地点: Gyeounggi-do, Korea/韩国，京畿道
Architect/建筑师: HyoMan Kim/IROJE KHM Architects
Photos/摄影: JongOh Kim
Site area/占地面积: 330m²

Key materials: Façade – copper
主要材料：立面——铜

Overview
This site that is located in city has the surrounding context of country and that is an inclined and an irregular form of land. In order to maximize the efficiency of land use, the irregular form of site was converted to the form of architecture. As a result, the shape of site became the shape of architecture.

The house is a complex place with living and working. Because of setting the boundary of house and office, the architects separated this building into upper layer and underground layer which is able to separate the gate of house and office along the level of inclined access road.

Skip floor system that ease the psychological burden of vertical mobility maintain the dramatic promenade with producing the visual and spatial continuity.

Floating bamboo garden
Floating wooden boxes built-in small bamboo garden on the open space in each level, are dynamic and accidental program of space and this scenery of inner space is the major impression of this complex house.

Dramatic sequence of vertical story of exterior space
By landscaping all of the rooftop of this house, the dramatic exterior space that is composed of various type of the space, from underground level to rooftop level, are combined with dynamic indoor space and surrounding scenery of nature, and produced the rich story of space.

Detail and Materials
With the intent to harmony with the surrounding context of nature, flexible configuration of internal space, architecturizing the curved line of site, the vocabulary of architectural form made the non-architectural shaped architecture. As if "architectural nature", it is covered with curved copper skin.

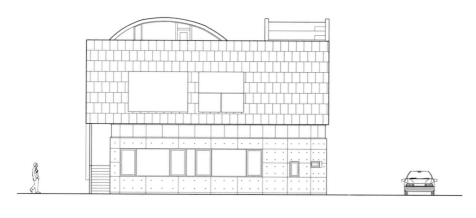

细部与材料

为了实现与周边的自然环境融为一体、内部空间的灵活配置以及将场地的曲线特征转化为建筑，建筑采用了与众不同的建筑形式。作为一个"建筑自然体"，它的外立面全部采用弯曲的铜皮覆盖。

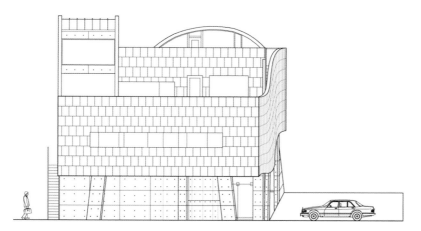

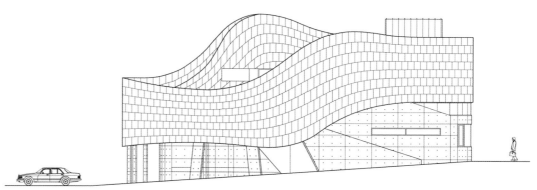

项目概况

项目场地位于城市附近，周边一派乡村氛围，呈倾斜的不规则形状。为了最大限度地利用土地，场地的不规则形状被转化到了建筑上。因此，场地的形状就变成了建筑的形状。

别墅综合了居住与办公空间。为了区分住宅与办公的功能，建筑师将建筑分为地上和地下两层，住宅与办公的入口沿着倾斜的入口通道分隔开来。

跳层系统减轻了垂直移动的心理负担，以夸张的长廊营造出视觉和空间上的连续感。

浮竹花园

嵌在小型竹子花园中的悬浮木箱为空间带来了动感和意外感，是别墅空间内最有趣的室内景象。

外部空间的动感排列

别墅的屋顶进行了景观设计，富有动感的外部空间包含各种各样的空间类型。从地下到屋顶，外部景观与活跃的室内空间与周围的自然景观相互结合，形成了丰富的空间感。

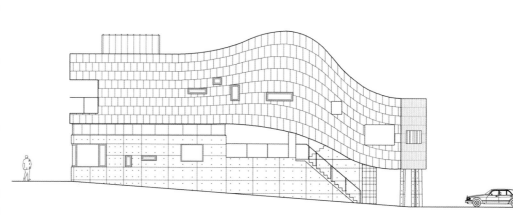

Section
1. Living room
2. Stand living area
3. Home bar
4. Family room
5. Bamboo garden
6. Roof garden
7. Child's bed room
8. Entrance court
9. Entrance
10. Pond
11. Guest bedroom
12. Parking
13. Plaza
14. Reception hall
15. Office

剖面图
1. 起居室
2. 站立式起居区
3. 家庭酒吧
4. 家庭娱乐房
5. 竹子花园
6. 屋顶花园
7. 儿童房
8. 入口庭院
9. 入口
10. 池塘
11. 客房
12. 停车场
13. 广场
14. 接待厅
15. 办公室

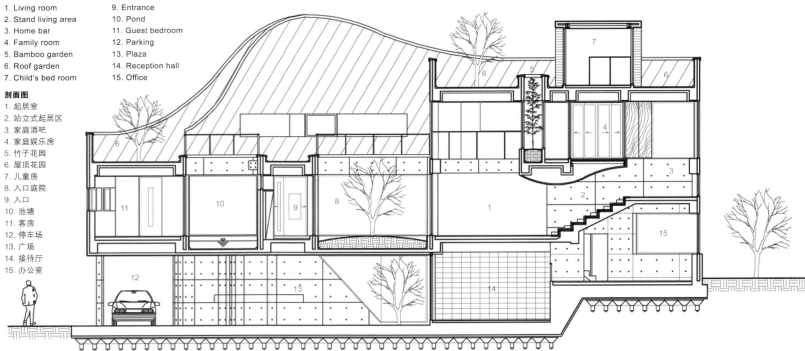

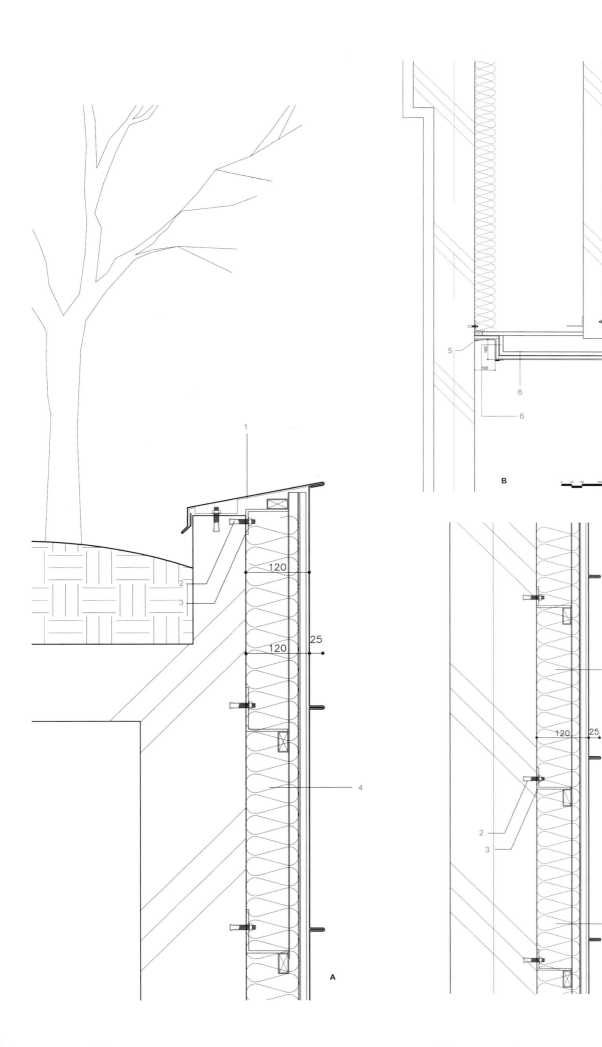

Detail A: sectional detail of roof/
Detail B,C: sectional detail of wall
1. THK0.5 copper sheet flashing
   1.0T galvanised zinc stiffner
2. Set anchor
3. THK4.5 bracket
4. THK0.5 copper sheet standing-seam@430
   H=10mm deltamenbrain
   2T asphalt sheet of water proof
   THK12 water proof plywood
   THK120 insulation
   40x20x1.6T coloured pipe@610
   40x20x1.6T coloured pipe@1200
5. THK0.5 copper sheet folder flashing
6. THK0.5 copper sheet flat-screen
   2T asphalt sheet of water proof
   THK12 water proof plywood
   40x20x1.6T coloured pipe@906

节点A：屋顶剖面节点/节点B、C：墙面剖面节点
1. 0.5厚铜板
   1.0厚镀锌加劲杆
2. 固定锚点
3. 4.5厚支架
4. 0.5厚铜板立缝@7430
   H=10mm薄膜
   12厚防水夹层板
   120厚隔热层
   40x20x1.6T彩色管@610
   40x20x1.6T彩色管@1200
5. 0.5厚折叠铜板
6. 0.5厚平铜板
   2厚防水沥青层
   12厚防水胶合板
   40x20x1.6T彩色管@906

Théâtre 95

95剧场

Location/地点: Cergy-Pontoise, France/法国，蓬图瓦兹
Architect/建筑师: GPAA GaellePéneauArchitectesAssociés
Photos/摄影: 11h45
Gross floor area/总楼面面积: 3,000m²
Site area/占地面积: 10,000m²
Completion date/竣工时间: 2012

Key materials: Façade – copper
主要材料：立面——铜

Overview
The challenges of the project
The contractor's definition of the project is articulated around a close link between reshaping the Théâtre 95, outlining an urban morphology, and finding free spaces in the urban island. The theatre's extension is framed by a dialectic vision of interactions between town and theatre: the theatre stretches beyond its limits while the town must find in it a porous space, a space of journeys, echoes, dreams, encounters and everyday life, devoted to contemporary modes of expression and deliberately placed at the very heart of the town.

A peculiar face-off
The building's pleated roof is the first component which strikes the visitor's eye: this is the outline which the extension has borrowed to link old and new. The connection consists in a "semi public" hall-atrium, which follows the "FilD'Ariane" - a public footpath which winds its way without interruption from the South-East to the North-West of the town, and is thus "integrated" into the building.

The existing pleated outline of the roof is continued in the hall-atrium volume, where it transforms into juxtaposed strips which create shafts of light entering the hall. The pleats are also echoed in the new auditorium, facing South, creating a new rhythm which emphasizes the choice of erecting the new structure out of line with Cergy's traditional orthogonal grid. The pleated outline has become the "crown" which is found in the volume of the new auditorium.

The new volume rises in an almost baroque posture, as if in confrontation with what is already there: the existing building conserves its identity, the atrium linking it to the new, and setting up a "face-off" relationship between two visions which mix, stand in opposition and join together in a boldly chaotic statement.

Detail and Materials

The new extension will house a "flexible" 400 seat auditorium: the volume, which includes stage and technical areas, is blind, and covered with golden scales which bring light to a fairly colourless urban environment.

The choice of materials is based on a certain number of considerations which include questions of maintenance, longevity, environment, energy efficiency, but also appearance and style. Copper offers a response to all these questions. A performance space is also a strong presence at the heart of a town: it is a playful space, designed for culture and leisure which, like a magic lantern, must shine and draw all eyes towards it. Thus the envelope is made up of smooth, diamond-shaped scales, made from a copper aluminium alloy lending them a golden shade which will fade very little over time, and also contributing to lighting the hall-atrium.

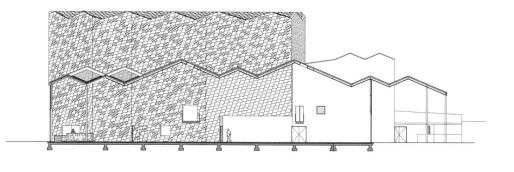

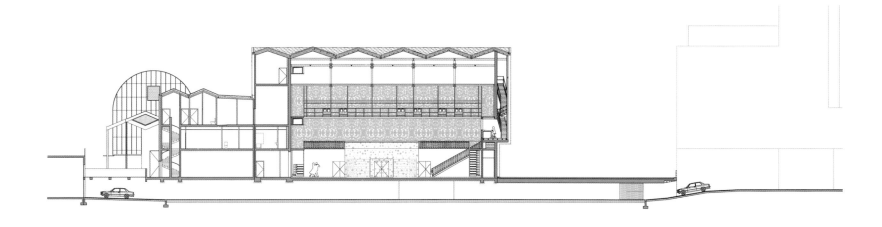

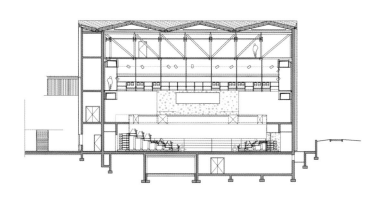

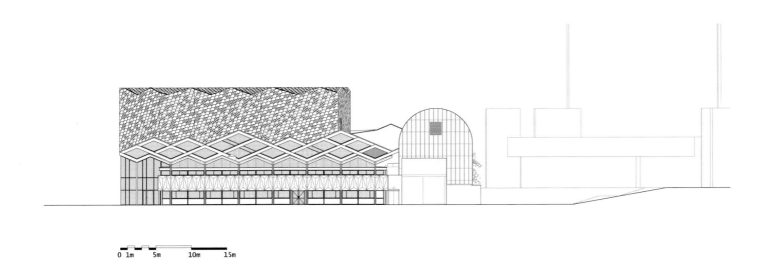

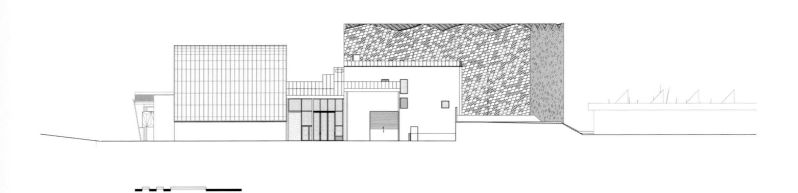

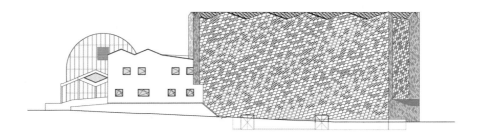

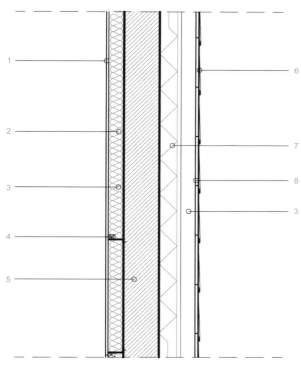

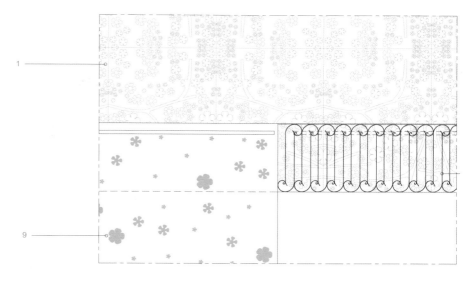

Detail
1. Wood panels etched plywood 18mm + acoustic fleece
2. Acoustic insulation
3. Blade ventilation air
4. Batten fixing 6x4mm mounted on steel cornier
5. Sailing concrete
6. Cladding copper gold poses ecalille
7. Insulation
8. Sheathing
9. Storage chests bleachers

节点
1. 木板蚀刻夹层板18mm+隔音起绒布
2. 隔音层
3. 叶片通风口
4. 板条固定6x4mm,安装在钢角上
5. 海工混凝土
6. 金色铜包层
7. 隔热层
8. 包板
9. 露天看台储藏箱

项目概况

项目面临的挑战

项目要求重塑95剧场的形象,刻画出独特的城市形态并且在城市岛屿中找到自由的空间。剧院的扩建工程在剧场与市政部门的引导下进行:一方面,剧场扩展了自己的空间;另一方面,城市从中获得了多方面的功能:旅行、聚会、寻梦、日常生活。位于市中心的剧院必须体现现代的生活方式。

独特的应对方案

建筑的褶皱式屋顶令人耳目一新:它以独特的轮廓连接了新与旧。半开放的中庭沿着贯穿城市东南—西北方向的阿里安人行道而建,将人行道与建筑融为一体。

屋顶的褶皱造型在中庭中得到了延续,并且让阳光进入了大厅。新的礼堂朝南而建,重复了褶皱的形式,在中规中矩的传统城市网格中创造出全新的旋律。褶皱式轮廓成了新礼堂的"王冠"。

新建的结构接近于巴洛克风格,与原有的建筑形成了对比:建筑颠覆了形象,通过中庭与新形式连接起来,将新旧混合起来,形成了大胆而混沌的表达。

细部与材料

扩建结构内将设置一个可容纳400人的礼堂(其中包括舞台和技术空间),整体结构由金色鳞片覆盖,为平淡无色的城市环境带来了光亮和色彩。

材料的选择考虑了维修养护、使用寿命、环境、节能、外观、风格等多方面问题。铜是最好的解决方案。作为一个演出空间,它在市中心呈现出了强大的存在感:它像是一台为文化和休闲而生的幻灯机,闪闪发光,吸引着所有人的眼球。建筑外壳由光滑的菱形鳞片构成,铜铝合金赋予了它们金色的色彩。随着时间的流逝,色彩会逐渐褪色。这种材料还有利于点亮中庭空间。

Platform of Arts and Creativity

艺术与创意平台

Location/地点: Guimarães, Portugal/葡萄牙，吉马良斯
Architect/建筑师: Pitágoras Architects
Completion date/竣工时间: 2012

Key materials: Façade – brass, glass
主要材料: 立面——铜、玻璃

Overview

The program provided a clear concept and defined the objectives intended to achieve with this infrastructure, listing a series of skills and spaces that constitute the functional program for both the new and the existing buildings, as well as the adjacent plots of land. For this purpose three major program areas were defined: Art Centre, Creative Labs, Workshops to Support Emerging Creativity.

Finally, the intent to recover the existing building on the eastern side, trying to promote the installation of additional commercial activities that could enhance the creation of a space with a broad scope in regards to multidisciplinary cultural activities.

The whole structure, according to the program would complement the existing equipment in the city, as well as those which are under development within the European Capital of Culture. When interpreting the program, we aimed to allow for the possibility of each one of its components to function independently and simultaneously, creating accesses to each of the various services and support areas, as well as to the outdoor square and garden.

The architects opted for a methodology of intervention that involves the rehabilitation of the existing building to the east, keeping the materials and textures, but redoing the entire inside at level 0. For the building at north, and for reasons previously mentioned, the façade towards the Avenue, which characterises the building, is renovated, but its interior and façade facing the square were object of and almost complete demolition and redesign. Although it is intended to maintain the scale and the existing formal relations, we propose a new solution for the building that promotes a strong relationship with the square and emphasizes the relationship of this structure with the outer space.

Detail and Materials

The new building takes a radically different language, by contrast with its surroundings,

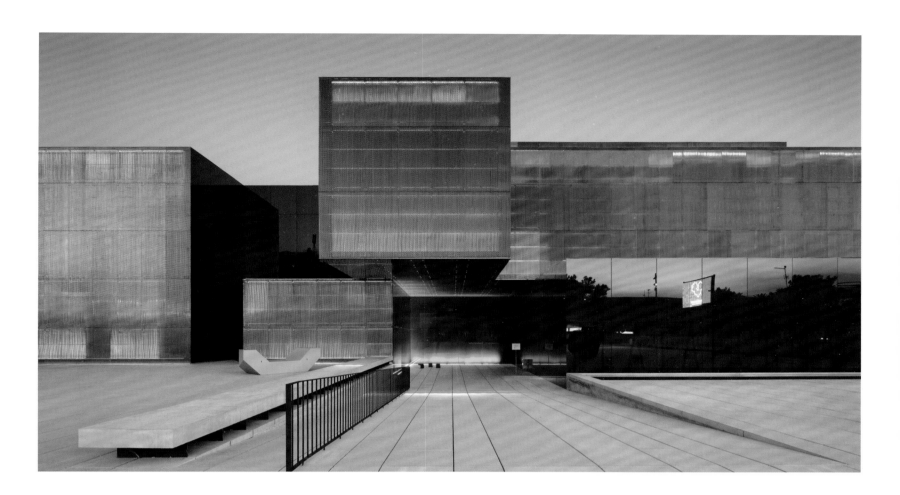

both from the standpoint of their language and image, discrete, repetitive, as well as by the succession of volumes, with full and empty, marked by the juxtaposition of contrasting surfaces. The coatings, a grid of metal profiles in brass and glass surfaces chromatised on ventilated façades, accentuates a range of textures that is intended display, more dense and opaque in the majority of faces in the case of the metal structure, and transparent when it covertly comes to glass surfaces that intentionally conceal the few openings that the building comprises. This series of volumes and dissonant elements, which result from decomposition of the initial volume, was originated by the need to create a variety of different spaces.

in the exhibition area, creating a tension evident in the volume of the building and the relationship with the space of the square, making it the main feature of its design.

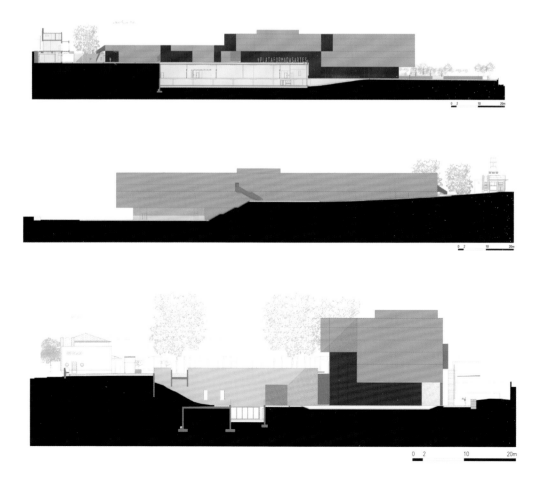

项目概况

项目规划清晰地列出了设计主题和设计目标，为新旧建筑以及周边地块提供了一系列技术与空间规划。项目主要分为三个功能区：艺术中心、创意实验室和新兴创意辅助工作室。整个结构将完善城市结构的已有设施和在建设施，为这座欧洲文化之都带来新的活力。

建筑师选择将原有建筑向东进行重建，保留原有的建筑材料和材质并且对室内进行重建。对于建筑的北面，建筑师进行了翻新，但是几乎全部拆除并重新设计了室内和朝向广场的建筑立面。在保留原有的规模和形式关系的前提下，建筑师提出了一个新的解决方案，在建筑和广场之间建立了更紧密的联系，突出了建筑结构与外围空间的关系。

细部与材料

新建筑采用了完全不同的建筑语言，与周边的环境形成了鲜明对比。建筑采用了反复、离散的结构方式，实现了空间序列的连续性，虚实结合。建筑包层由铜和玻璃表面构成，通风透气。建筑立面在金属部分显得密集而厚重，在玻璃部分则显得通透而明亮。空间的序列和离散的元素将建筑分解成独立的个体，满足了展览区各种各样的空间需求。这种设计既在建筑内部形成了紧凑的空间关系，又在建筑和广场之间建立起了紧密的联系，是设计的点睛之笔。

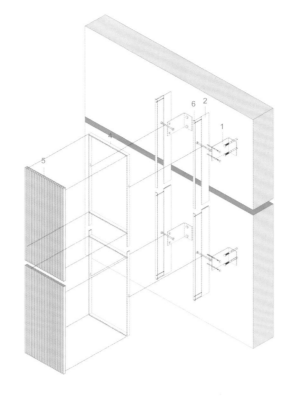
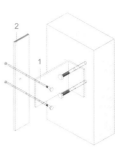

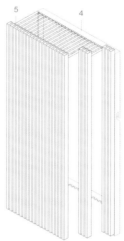
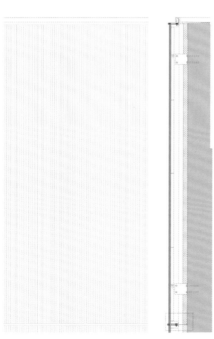
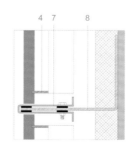
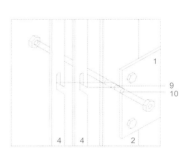

Façade detail
1. Sheet-iron plate 5mm thick, punctual, for metalizing and painting
2. Iron bar 70x6mm for metalizing and painting
3. Brass bar 25x4mm
4. Brass toothed bar 4mm thick, fits the tubular brass, welded to the vertical bars
5. Façade cladding panels brass type "CC Wolf, ref. LA 040 "
6. Brass bar 20x2mm
7. Sheet-metal brass, 2mm thick
8. Sheet-iron plate 5mm thick, punctual, for metalizing and painting
9. Neoprene pad
10. Threaded rod of iron 8mm in diameter, welded to the iron rod, for metalizing and painting

立面节点
1. 薄铁板5mm，电镀涂漆
2. 铁条70x6mm，电镀涂漆
3. 铜条25x4mm
4. 铜齿条4mm，安装在铜管上，焊接在竖条上
5. 立面覆面镶板，铜CC Wolf, ref. LA 040型
6. 铜条20x2mm
7. 薄铜板2mm
8. 薄铁板5mm，电镀涂漆
9. 橡胶支座
10. 铁螺纹杆，直径8mm，焊接在铁杆上，电镀涂漆

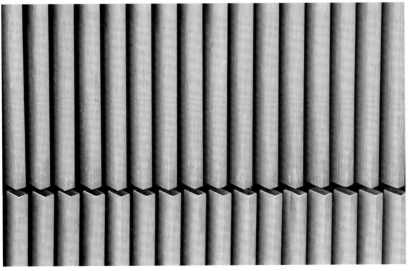

Tales Pavilion
朝阳态思故事厅

Location/地点: Beijing, China/中国，北京
Architect/建筑师: Luca Nichetto
Photos/摄影: Jonathan Leijonhufvud
Completion date/竣工时间: 2013

Key materials: Façade – brass
主要材料: 立面——铜

Overview

The goal of the Tales Pavilion is to establish itself as a design stage of international significance. The pavilion will become a dynamic and vibrant new complex for excellence in design. Tales' philosophy of "story-telling design," in unison with its acute attention in selecting collaborators and brands, perfectly suited Nichetto's innovative approach to design and vice versa. In the words of Tales Co-founder and Creative Director, Terence Yeung, "It all came down to a matter of trust and good feeling," how it all came to be was "like a natural connection."

Standing in great contrast with the "organic feel" of the building, a bronze framework of large straight-cut windows gives onlookers a glimpse of a warm and cozy interior. Visitors will stroll through a non-linear concrete promenade that leads them to the front door, while enjoying the façade's perfect balance created between man-made modernity and natural randomness.

Detail and Materials

The Tales Pavilion came into being through a creative process inclusive of both design and enterprise. Nichetto developed the façade of the pavilion by converting 1,200 pieces of brass tubes into "grass leaves" that camouflage the entire structure. The "brass leaves" oxidize and change color naturally with the passing of the seasons, merging with the natural scenery of the Beijing Lido Garden and giving a sense of life to the pavilion. The architectural design is Nichetto's depiction of the young and avant-garde Tales, which, much like grass, is free, natural, and full of life with a grand desire of growing.

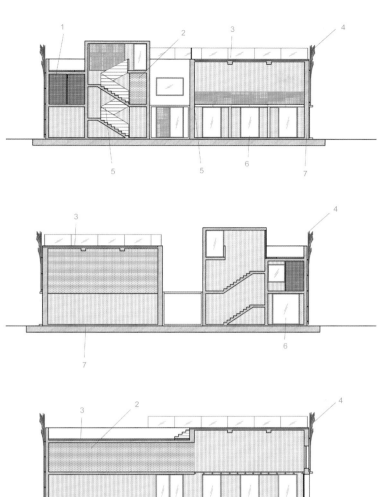

Section detail
1. Concrete tiles
2. Plaster bricks
3. Wood
4. Brass leaves oxidize
5. Elm wood recycled
6. Glass
7. Concrete

剖面节点
1. 混凝土砖
2. 石膏砖
3. 木材
4. 氧化铜条
5. 回收榆木
6. 玻璃
7. 混凝土

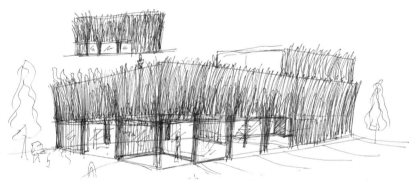

项目概况

态思故事厅的设计目标是打造自身在国际设计舞台上的影响力。故事厅是卓越设计的展示空间。态思执着于"有故事的设计",在选择合作者和品牌的过程中贯穿这一哲学,这与建筑师卢卡·尼切托的创新设计概念不谋而合。用态思创始人兼创意总监杨智杰的话来说,"这一切都源于直觉和信任",这就是一种"设计的默契"。

与建筑的"有机形态"形成鲜明对比的是,线条凌厉、极具现代感的古铜色巨大窗框,温暖而舒适的室内环境隐约可见。漫步通过曲折的小径,一边欣赏现代感与自然随性完美结合的建筑外观,一边来到故事厅的大门。

细部与材料

态思故事厅的设计筹划是一个极具创造性的过程,其风格兼顾了设计和商用。尼切托用超过1,200根铜管变作覆满整座展厅的"草",随着时间与季节的变化,这些"铜草"会自然地变幻颜色。依托丽都花园别致的自然氛围,"铜草"静静地散发着一股蓬勃的生命气息。这一建筑设计便是尼切托描绘的那个年轻而前卫的态思,正如风中劲草一样,自由、自然、充满朝气而渴望成长。

Socio-Cultural Centre in Mulhouse
牟罗兹社会文化中心

Location/地点: Mulhouse, France/法国，牟罗兹
Architect/建筑师: Paul Le Quernec
Executor architects/执行建筑师: Fabrice Wianni & Banjamin Ringeisen
Photos/摄影: ©11h45
Floor area/楼面面积: 1,250m²
Overall cost/总成本: 2,444,000 euros/2,444,000 欧元
Completion date/竣工时间: 2013
Key materials: Façade – zinc
主要材料: 立面——锌

Overview

As the architectural brief covered all the plot's area, the architects did not really have a choice for the siting or the orientation of the building. So they focused on the functions of the cultural Centre even before studying its functioning.

A socio-cultural Centre embodies the spirit of initiative and solidarity. This building plays a convening key role in the same way as a church, the only difference being that it would not necessarily gather the believers but "the people who believe in". This reflection is fundamental to attach the greatest value to the content as well as the container.

After placing all the required spaces, we have noticed that the entire plot was filled. So they had the idea to "go out of the grid". Two reasons justify this intention. The first one is to avoid that the building turns into a "parallelepiped box", like one of these austere housing projects. The second one is to express the dynamism of a cultural and social Centre as well as its users.

The ground floor is composed of two parts that can function together or separately. These parts are rigorously aligned with each other as well as the plot's limits. As they rise, these parts twist on the first floor, escape from the plot's limits in order to move towards to the square as a flower twists to catch the sunlight. As a result, the building is deformed as if a positive force evoking the presence of an energy that needs to be expressed drove it…

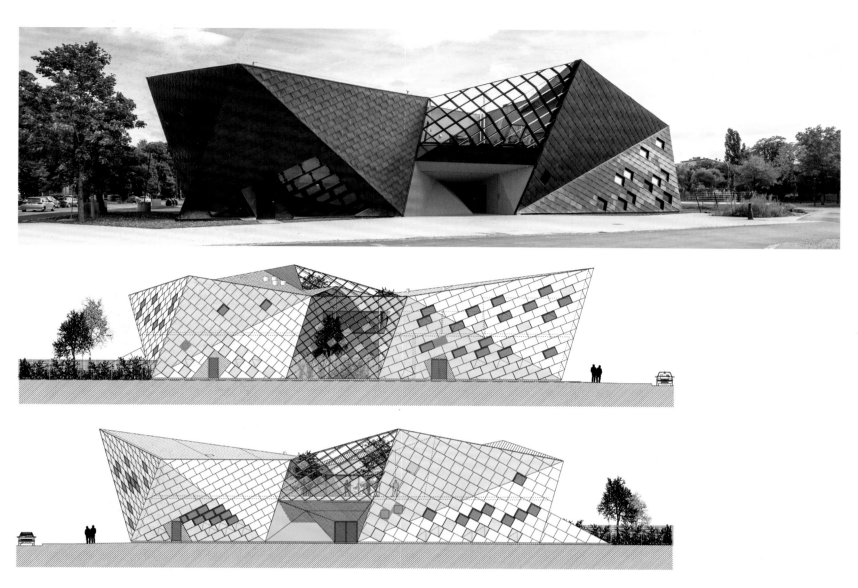

Detail and Materials

The architects have imagined creating a vegetal moat that would protect the lower parts of the building. In order to strengthen the building's shell that they initially wanted to cover with a classic coating, at the request of client, the architects have also set up a system consisting of zinc scales (pre-weathered in black in the absence of being able to finance inox scales) on wooden sheathing. Moreover, all glazed elements have been designed as modules of 1,5 or 2m² in order to avoid having much excessively large glass surfaces. All these arrangements put together, have generated a reptilian-like building that imposes some kind of natural and spontaneous respect.

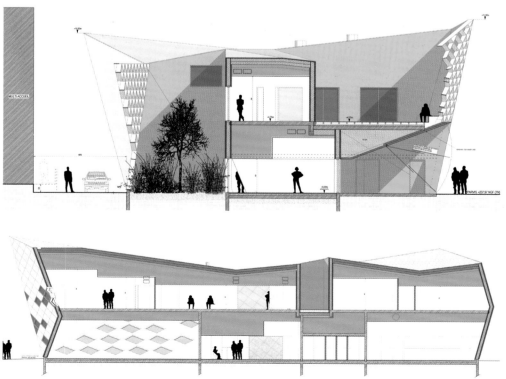

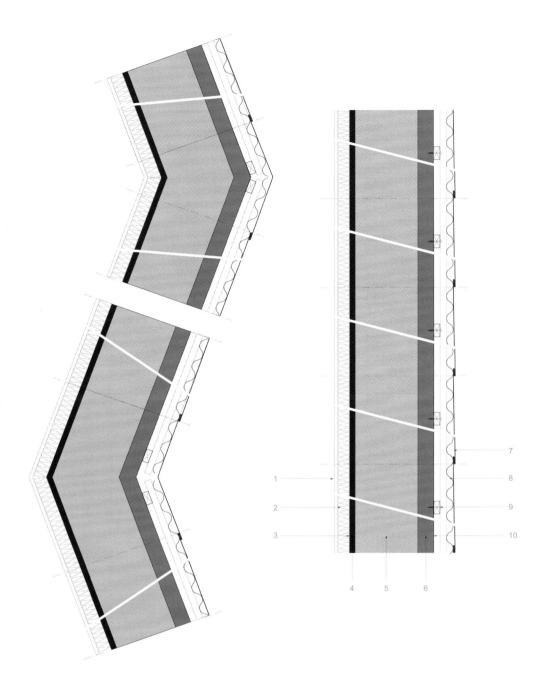

Composition of wood framed façade walls

1. Plaster board
2. Acoustic insulation: semi-rigid mineral wool (45mm) blocked between metallic structure
3. Damp-proof membrane and vapour barrier
4. Lateral bracing panel (attached to the main structure)
5. Primary structure:
   glue-laminated framework (240x60mm, interaxial distance: 60cm) fixed to a metallic structure on the upper and lower parts that are attached to a concrete base or a wood joist
   secondary wooden structure (240x750mm):
   fixed between each wooden post. The whole system is embedded in cellulose wadding (24cm)
6. Wood fibre-based insulation panels, attached to the primary structure (60cm)
   Panels with score edge
   Perfect joint and sealing between panels with cover strips
7. Rain screen membrane stretched on wooden laths
8. Fongcide and insecticide – threated wooden spacers (25x50mm, interaxial distance: 500mm) attached to the main structure by stainless steel screws through insulation panels
9. Metal sheet to support metallic coating
10. Pre-weathered zinc sheets superimposed in fish scales

木框立面墙的构成

1. 石膏板
2. 隔音层：半刚性矿物棉（45mm），夹在金属结构之间
3. 防潮膜和隔汽层
4. 横向支撑板（安装在主结构上）
5. 一级结构：
   胶合夹层框架（240x60mm，轴间距：60cm），上方固定在金属结构上，下部固定于混凝土底座或木龙骨上
   二级木结构（240x750mm）：
   固定在木杆之间。整个系统都嵌入纤维板填塞物（24cm）
6. 木纤维基隔热板，安装在一级结构上（60cm）
   面板带有刻痕边缘
   面板以覆盖条进行完美结合和密封
7. 防雨膜，拉伸在木板条上方
8. 经防虫防霉处理的木隔离条（25x50mm，轴间距：500mm），通过不锈钢螺丝穿过隔热板固定在主结构上
9. 支撑金属包层的金属板
10. 预氧化锌板，叠加成鱼鳞状

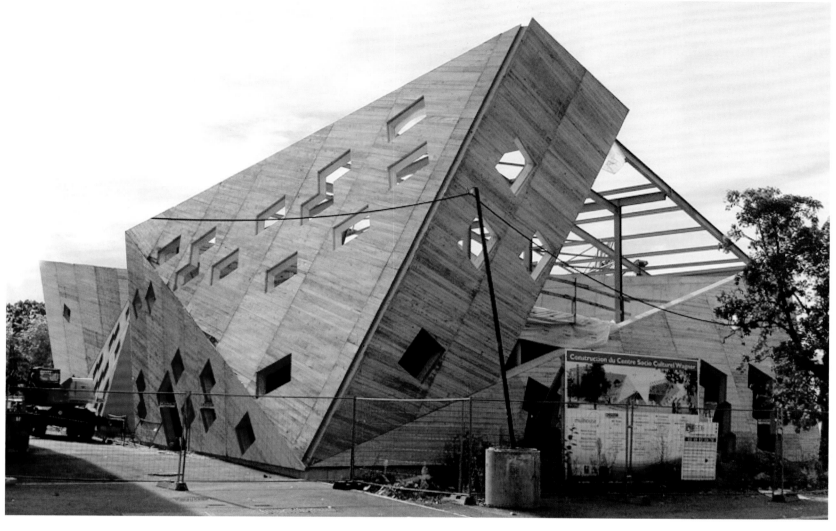

项目概况

由于建筑规划已经覆盖了整个场地，建筑师无需对建筑的选址和朝向进行设计。因此，他们将重点放在了文化中心的功能设置上。

社会文化中心象征着创新和团结的精神。这座建筑的意义与教堂类似，唯一不同之处在于它所召集的不是教徒，而是信仰文化的人们。这一想法是建筑设计的基础，也是它的核心价值所在。

在设置了所有必要的空间后，建筑师发现整个场地都已经被填满，因此他们产生了"走出网格"的想法。这种想法有两个好处：一是能避免建筑变成与住宅相似的"平行六面体的盒式结构"，二是能体现社会文化中心及其使用者的活力。

一楼由两部分组成，可以分开运作也可以合并，在场地的限制下紧紧相连。二楼的空间脱离了场地的限制，向花儿一样朝着阳光绽放开来。建筑实现了变形，将内在的活力呈现了出来。

细部与材料

建筑师曾试图打造一个植物带来保护建筑的底层。为了强化建筑的外壳，他们最初想根据客户的要求利用古典的包层来对其进行覆盖。建筑师在木包板上打造了一个锌鳞片（预氧化成黑色，与不锈钢鳞片相配）系统。此外，所有玻璃组件都被设计成了1.5~2平方米的模块，避免了面积过大的玻璃表面。这些设置结合起来，共同组成了一座带有鳞片的建筑，体现了设计对自然随性的敬意。

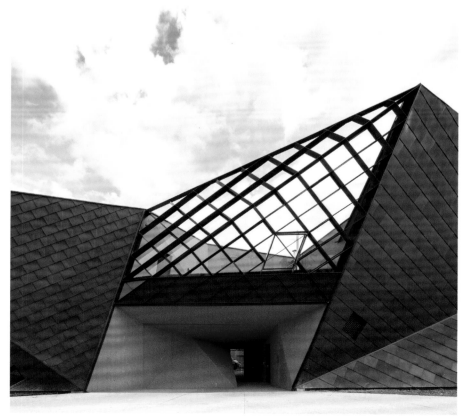

Municipal Market
市政市场

Location/地点: Barcelona, Spain/西班牙，巴塞罗那
Architect/建筑师: Comas-Pont aquritectes slp
Photos/摄影: Jordi Comas, Pere Tordera
Built area/建筑面积: 2,187m²
Completion date/竣工时间: 2011

Key materials: Façade – zinc, wood slats
Structure – metal
主要材料：立面——锌、木板条
结构——金属

Overview
This building is a key item to rehabilitate the neighbourhood. The urbanisation respects streets rank and emphasises Pintor Guardia Street like a central neighbourhood axis, it is a pedestrian street now and it connects equipments and public spaces.

Base floor is almost transparent and emphasise visual connection between interior and exterior, this is reinforced with continuity on the interior pavement to the exterior below porches. The cover strips are projected on the pavement with granite black lines which continue at exterior, therefore connectivity is emphasised and make easy to define back park zone, tree areas, the hard urbanisations surface and all exterior furniture situation.

Detail and Materials
The new fresh products market is a Mediterranean building: the sales area is a large open space with the light as a protagonist, a controlled light filtered through a wooden slats system of the façades and inner skin panels and slats of wood shavings.

A continued zinc skin unifies roof and façades and it folds generating different highs that allow light entry and the all building ventilation. This skin folds itself creating porches to the main entrances. Fragmented cover dialogs in height with environment's buildings and hide its big scale.

A dry semi-industrialised construct system was used to fulfil short timing (8months) and minimize the waste: metallic structure, sandwich wood panels in big format, wood slats, zinc roof and façades and interior OSB panels. All materials have natural finishes (wood and zinc). Passive systems (renewable energy heating and cooling systems (geothermal energy, heat recover machines…), roof water collect tank and the possibility of dismantle and recycle all building materials) let to catch with an excellent energetic building work.

项目概况

建筑是复兴周边社区的重要项目之一。城市化进程以画家守卫街为中轴,将其划为步行街,起到了连接重要设施和公共空间的作用。

建筑底层基本是透明的,强调室内外的视觉连接。门廊下连接室内外的走道进一步突出了二者的连续性。屋顶的陶文投射到走道的黑色花岗岩分割线上,延续了室内外的协调性。这种设计也突出了后方的公园区、树木、市政地面铺设和所有室外休闲设施。

细部与结构

新建的生鲜市场为典型的地中海建筑:大型开放式售货区以光线为特色,光线透过外立面的木条和内墙的壁板及刨花板渗透进来。

连续的锌板将屋顶和外立面统一成一个整体,锌板的折叠起伏让阳光进入室内,并保证了建筑的通风。这层建筑外壳通过折叠形成了主入口的门廊。分散起伏的屋顶与周边的建筑高低呼应,也掩藏了建筑的巨大规模。

设计师采用半工业化建造系统来实现短期施工,同时减少了建筑废料的产生。建造系统包括金属结构、大型夹心模板、木板条、锌屋顶和内外墙的刨花板。所有材料都有天然装饰(木纹和金属纹)。被动系统(可再生能源加热和制冷系统、地热能源、热恢复机等)、屋顶雨水收集系统以及对建筑材料的回收都使得项目成为了一座优秀的节能建筑。

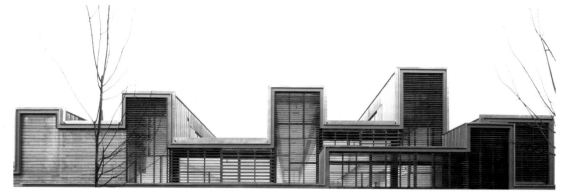

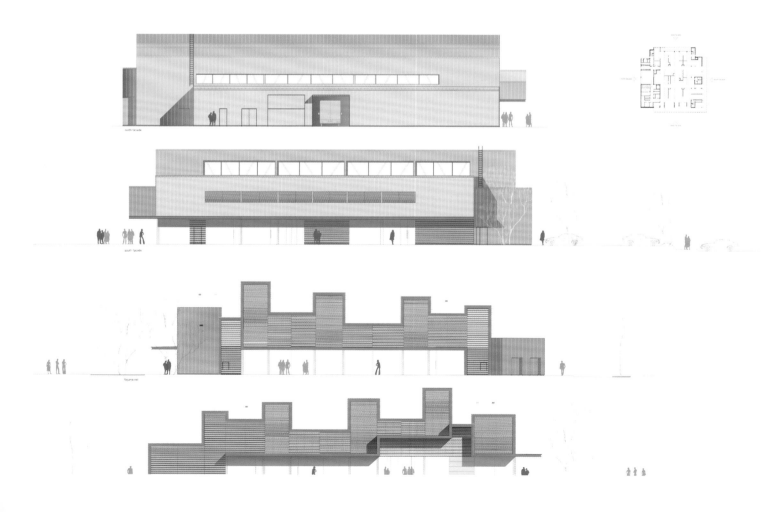

Roof (CO)

01 - Sandwich panel with thermal insulation
02 - Double waterproofing sheet
03 - High density pholtethylene sheet
04 - Zinc sheet 0.8 mm thick
05 - Water resistant panel 22mm
06 - Thermal insulation, rigid mineral wool
07 - Zinc sheet stitch
08 - Stainless steel drain
09 - Subbase drain
10 - Zinc drain
11 - Stainless steel ring and cable

Cladding (RE)

01 - Subbasement
02 - Waterproof textile sheet
03 - Reinforced concrete girder 20cm thick
04 - Layer of mortar
05 - Terrazzo pavement 40x40cm
06 - Mat pavement
07 - Lacquered iron sheet ceiling
08 - Lacquered iron sheet
09 - Galvanised steel profile
10 - OSB panel 15mm thick

屋顶（CO）

01．隔热夹层板
02．双层 防水板
03．高密度 聚乙烯板
04．0.8mm锌板
05．22mm防水板
06．隔热矿物棉
07．锌片夹缝
08．不锈钢排水槽
09．底基层排水槽
10．锌排水槽
11．不锈钢环和缆绳

包层（RE）

01．底基层
02．防水纺织薄板
03．20cm厚钢筋混凝土梁
04．砂浆层
05．水磨石铺面40x40cm
06．衬垫铺面
07．喷漆铁皮天花板
08．涂漆铁板
09．镀锌钢
10．15mm厚OSB板

Le Carré en Seine
塞纳河畔卡雷住宅

Location/地点: Seine River, France/法国，塞纳河
Architect/建筑师: PietriArchitectes
Photos/摄影: Thierry Favatier, Vincent Fillon
Area/面积: 15,000m²
Completion date/竣工时间: 2013
Cost/成本: 21 million€/2,100万欧元

Key materials: Façade – zinc, concrete
Structure – concrete
主要材料：立面——锌、混凝土
结构——混凝土

Façade material producer:
外墙立面材料生产商：
QUARTZ-ZINC® chez VMZinc

Overview
The project is a part of a renewal of some central quarters in Stockholm close to the railway yard. The client asked for a modern, interesting office building in this central and well exposed site just beside the railway station. At the same time the building must adjust to the character and scale of the surroundings. Prefabricated concrete elements were required; because of the nearness of the railroad a fast erection of the building was necessary and high security required of the completed building. The elements were delivered ready made from factory to minimise the assembling process in the risky environment.

Detail and Materials
To avoid the impression of piled up façade elements they are placed as a stretcher bond. In addition to this the windows shift placing from one floor to the other. The structural concept was to make the elements to act as boards instead of the traditional pillar-beam structure. This makes it possible for the forces to spread sideways. No consistently stacked element is needed. The matrixes are reduced to only three types.

The element's surface is divided with fake joints to resemble a traditional stone wall. Randomly selected parts are treated with acid to vary the expression. The inner load bearing wooden framed window is double glazed. The size of this frame is maximised regarding structural strength and indoor climate. In the outer window frame a third larger glass is placed. This solution solves some of the fundamental problems with the site, both esthetical and practical. The outer glass gives some noise and risk reduction, and adds a wider lookout. The large glass area also provides a lighter expression to the façade and brings out new nuances to the concrete.

   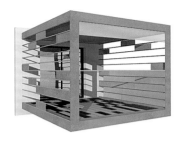

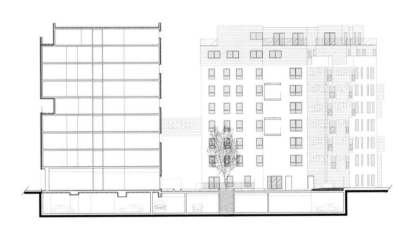 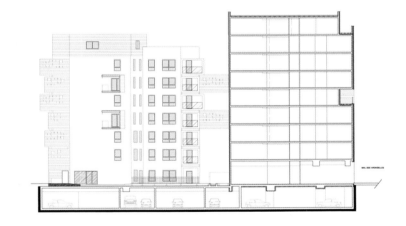

项目概况
项目坐落在塞纳河畔，正对圣日耳曼岛，符合可持续开发指标的严格要求。三座建筑通过朝向塞纳河的天井实现了通透感。

细部与材料
住宅的建筑结构简单，其特色在于对外层空间的处理。建筑师在立面之外利用附加的"锌盒"来实现对视野和日光的应用。

双层混凝土墙的使用提供了具有吸引力的装饰和绝缘，在此类高密度项目中尚属首创。为了保证城市规划的统一感，住宅酒店也采用了同种建筑材料。

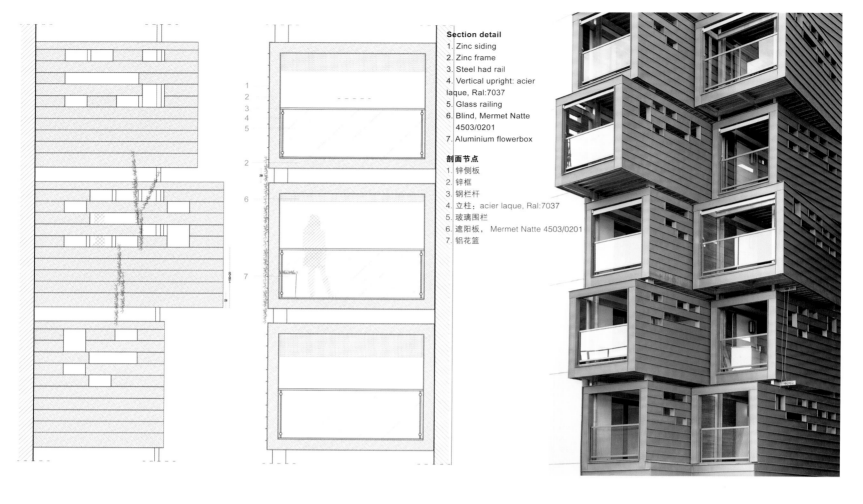

Section detail
1. Zinc siding
2. Zinc frame
3. Steel had rail
4. Vertical upright: acier laque, Ral:7037
5. Glass railing
6. Blind, Mermet Natte 4503/0201
7. Aluminium flowerbox

剖面节点
1. 锌侧板
2. 锌框
3. 钢栏杆
4. 立柱：acier laque, Ral:7037
5. 玻璃围栏
6. 遮阳板，Mermet Natte 4503/0201
7. 铝花篮

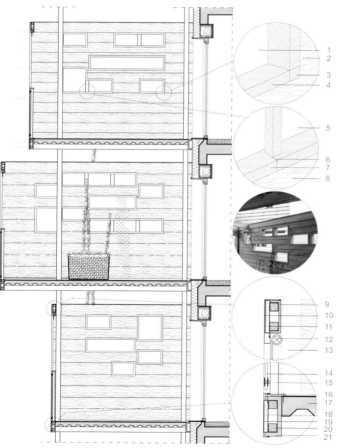

Façade detail
1. Exterior
2. Wood siding
3. Added welded piece
4. Zinc sheet folded
5. Exterior
6. Zinc cut and welded
7. Zinc sheet folded
8. Wood cladding
9. Secondary steel structure
10. Zinc siding
11. Sheathing
12. Blind
13. Primary steel structure
14. Glass railing
15. Vertical post of railing
16. Installation pad
17. Collaborative sheet floor system
18. Sheathing
19. Zinc siding
20. Strip
21. Secondary steel structure

立面节点
1. 外墙
2. 木侧板
3. 附加焊接件
4. 折叠锌板
5. 外墙
6. 切割焊接锌材
7. 折叠锌板
8. 木包层
9. 次级钢结构
10. 锌侧板
11. 外层包板
12. 遮阳板
13. 主要钢结构
14. 玻璃围栏
15. 栏杆立柱
16. 隔热衬垫
17. 协同钢楼面系统
18. 外层包板
19. 锌侧板
20. 密封条
21. 次级钢结构

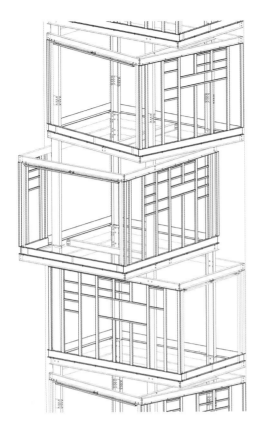

Cultural Civic Centre
市民文化中心

Location/地点: Palenica, Spain/西班牙，帕尔尼卡
Architect/建筑师: EXIT ARCHITECTS
Photos/摄影: FG + SG FOTOGRAFÍA DE ARQUITECTURA, EXIT ARCHITECTS
Gross floor area/总楼面面积: 5,077m²
Completion date/竣工时间: 2011

Key materials: Façade – zinc, aluminium, glass, brick
主要材料：立面——锌、铝、玻璃、砖

Overview
The proposal intends to convert the former prison into a meeting place, recovering some of the old spaces, and creating at the same time new structures that make possible the new planned activities. It is a project that respects the existing building, which is given a contemporary, lighter appearance, and where the natural light will play a key role.

The entire building is organised around a great hall that connects the 4 pavilions of the former prison. It is a diaphanous space based only on a few mild cylindrical courtyards of glass that illuminate and provide the backbone of the stay. Due to its central location in relation to the pavilions, this space acts as a nerve Centre and distributor of users, across the Pavilion access and reception, directed towards the rest of the areas of the Centre.

Detail and Materials
The use of metallic materials in all intervention, as the zinc in façades and roofs, glass and u-glass in the lower bodies and skylights and the Aluminium lattices as light filters also contributes to this.

项目概况

项目计划将一座监狱改造成聚会场所，使其重获新生，同时打造新的建筑结构来进行计划活动。项目既尊重了原有建筑，又赋予了它现代、轻快的外观，使自然采光成为了设计的重要组成部分。

整座建筑围绕着大厅展开，大厅连接了前监狱的四个区域。大厅空间十分通透，仅有几个圆柱形玻璃庭院提供采光和支撑。出于其中央的地理位置和与其他各区的联系，这一空间起到了控制中心和疏散人流的作用。人们通过大厅进入市政中心的其他区域。

细部与材料

建筑屋顶使用了锌板，底层结构和天窗使用了玻璃和U形玻璃，而窗户遮阳则选择了铝格栅。这些材料的运用都帮助建筑与城市连接起来。

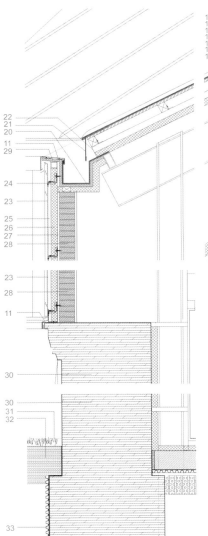

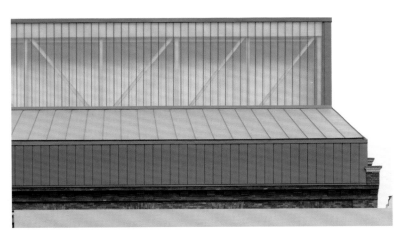

节点

1. 聚氯乙烯板
2. 土工布过滤材料
3. 石棉隔热层，带有隔汽层
4. 镀锌折叠钢板
5. 镀锌涂漆折叠钢板，用于立面顶端
6. 镀锌板，用于屋顶顶盖
7. U形玻璃265/60/7mm
8. 镀锌钢支架
9. 外围铝框
10. 镀锌钢方管
11. 钢板滴水檐
12. 折叠钢板
13. 钢屋架
14. 锌板，双锁结构
15. 聚乙烯板
16. 防水刨花板
17. 木板条
18. 镀锌钢
19. 聚苯乙烯刚性泡沫绝缘
20. 锌板槽
21. 镀锌钢屋檐
22. 镀锌钢板，用于屋顶通风
23. 锌板立面包层
24. 镀锌钢
25. 空洞
26. 喷涂聚氨酯泡沫绝缘
27. 砂浆石膏
28. 多孔砖
29. 中密度纤维板
30. 原有砖墙，待修复
31. 绿地种植
32. 种植土
33. 聚苯乙烯刚性泡沫绝缘
34. 石膏天花板
35. 隔音石膏天花板

Detail

1. PVC sheet
2. Geotextile filter
3. Rock-wool insulation with vapour barrier
4. Galvanised steel plate folded
5. Galvanised and painted steel sheet folded for façade crown
6. Galvanised steel sheet folded for roof crown
7. U-profiled glass 265/60/7mm
8. Galvanised steel bearers
9. Perimeter aluminium frame
10. Galvanised steel square tube
11. Steel sheet drip edge
12. Steel sheet folded
13. Steel roof truss
14. Zinc sheeting, double lock
15. Polythene sheet
16. Waterproof particle board
17. Timber battens
18. Galvanised steel profiles
19. Polystyrene rigid-foam insulation
20. Zinc sheet gutter
21. Galvanised steel sheet roof flap
22. Galvanised steel sheet for roof ventilation
23. Zinc-sheet façade cladding
24. Galvanised steel profile
25. Cavity
26. Spray polyurethane foam insulation
27. Mortar plaster
28. Perforated brickwork
29. Medium density fibre board
30. Existing brick wall to be rehabilitated
31. Green lawn planting
32. Vegetal soil
33. Polystyrene rigid-foam insulation
34. Plasterboard ceiling
35. Acoustic plasterboard ceiling

Index 索引

a- lab
www.a-lab.no

ACM Architects
www.acm-architects.com

Alver Architects
www.ata.ee

Andre Kikoski Architect
www.akarch.com

Andrej Kalamar/Studio Kalamar
www.kalamar.si

Ángela García de Paredes. Ignacio Pedrosa
www.paredespedrosa.com

Angelo Lunati, Luca Varesi
www.onsitestudio.it

Arkitema Architects
http://arkitema.dk

Atelier du Pont
www.atelierdupont.fr

Bang Architectes
http://bangarchitectes.fr

BATLLE I ROIG ARQUITECTES
www.batlleiroig.com

BCQ arquitectura Barcelona
http://bcq.es

BE Weinand (MA)
www.weinand.be

BP Architectures – Mandataire
http://jb-a.fr

Brooks + Scarpa
www.brooksscarpa.com

Caramel architekten ZT gmbh
www.caramel.at

Colboc Franzen & Associates
www.cfa-arch.com

COLL-BARREU ARQUITECTOS
www.coll-barreu-arquitectos.com

Comas-Pont aquritectes slp
www.comas-pont.com

David Sebastian + Gerard Puig Arquitectes
www.sebastianpuig.com

De Zwarte Hond
www.dezwartehond.nl

Dico si Tiganas Design Office
www.dicositiganas.ro

Dietger Wissounig Architekten
www.wissounig.com

Durisch + Nolli Architetti Sagl, Lugano
www.durischnolli.ch

ELÍAS RIZO ARCHITECTS
www.eliasrizo.com

Elkus Manfredi Architects
www.elkus-manfredi.com

Endo Shuhei
www.paramodern.com

Enric Batlle, Joan Roig architects
www.batlleiroig.com

EXIT ARCHITECTS
www.exit-architects.com

Feilden Clegg Bradley Studios
http://fcbstudios.com

FLINT, architect
http://flintarch.com

FOJAB arkitekter
www.fojab.se

GPAA-GaellePéneauArchitectesAssociés
www.gpaa.fr

Guillermo Hevia H., Gh+A Arquitectos
www.guillermohevia.cl

HHD_FUN
www.hhdfun.com

HyoMan Kim/IROJE KHM Architects
www.irojekhm.com

Ignacio Quemada Arquitectos Slp
www.ignacioquemada.com

Jaume Bach, Eugeni Bach (Bach Arquitectes)
www.bacharquitectes.com

JHK Architecten
www.jhk.nl
JKMM Architects
www.jkmm.fi
John McAslan + Partners
www.mcaslan.co.uk
Jose Escobedo/ JAS ARQUITECTURA
www.jasarquitectura.com
KLab Architecture
www.klab.gr
lab Modus
www.labmodus.net
LAN Architecture
www.lan-paris.com
L'Escaut Architectures
www.escaut.org
Luca Nichetto
www.lucanichetto.com
Lyons
www.lyonsarch.com.au
MEDIOMUNDO arquitectos
www.mediomundo.es
Mehmet Kütükçüoğlu, Ertuğ Uçar
www.teget.com
Mei architecten en stedenbouwers
www.mei-arch.eu
MITarh
www.mitarh.rs
MVA / Marin Mikelić, Tomislav Vreš
www.mva.hr
NAM ARQUITECTURA
www.namarquitectura.com
Nicolas Guillot architect
www.nicolasguillot.com
Park Associati
www.parkassociati.com
Paul Le Quernec
www.paul-le-quernec.fr

Pepe Gascón/ Pepe Gascón Arquitectura
www.pepegascon.com
PETER HOGG + TOBY REED ARCHITECTS
www.phtr.com.au
Pietri Architectes
www.pietriarchitectes.com
Pitágoras Architects
www.pitagoras.pt
Alessandro Ripellino Arkitekter
http://a-ripellino.se/
Rudiš-Rudiš architekti s.r.o
www.rudis-rudis.cz
Salvador Mata Pérez (Mata y Asociados)
http://matayasociados.com
Savioz Fabrizzi Architectes
www.sf-ar.ch
SO – IL
http://so-il.org
Takashi Yamaguchi (TAKASHI YAMAGUCHI & ASSOCIATES)
www.yamaguchi-a.jp
Trahan Architects
www.trahanarchitects.com
Tsushima Design Studio
www.tdstudio.jp
Urbanus
www.urbanus.com.cn
Vereniging van Studiebureaus 'ABSCIS - PROVOOST – TECHNUM'
http://abscis-architecten.be
Woods Bagot – Bruno Mendes
www.woodsbagot.com
Zbigniew Wroński, Szczepan Wroński, Wojciech Conder/WXCA www.wxca.pl
Zechner & Zechner ZT GmbH, Vienna
www.zechner.com

图书在版编目（CIP）数据

建筑材料与细部结构. 金属 /（西）佩雷斯编；常文心译. —
沈阳：辽宁科学技术出版社，2016.3（2016.12 重印）
　ISBN 978-7-5381-9217-9

Ⅰ.①建… Ⅱ.①佩… ②常… Ⅲ.①建筑材料－金属材料
Ⅳ.① TU5

中国版本图书馆 CIP 数据核字（2015）第 272623 号

出版发行：辽宁科学技术出版社
　　　　　（地址：沈阳市和平区十一纬路 25 号 邮编：110003）
印　刷　者：上海利丰雅高印刷有限公司
经　销　者：各地新华书店
幅面尺寸：245mm×290mm
印　　张：20
字　　数：200 千字
出版时间：2016 年 3 月第 1 版
印刷时间：2016 年 12 月第 2 次印刷
责任编辑：鄢　格
封面设计：何　萍
版式设计：何　萍
责任校对：周　文

书　　号：ISBN 978-7-5381-9217-9
定　　价：368.00 元

联系电话：024-23280367
邮购热线：024-23284502
E-mail: 1207014086@qq.com
http://www.lnkj.com.cn